Civil Engineering Solved Problems

Eighth Edition

Michael R. Lindeburg, PE

The Power to Pass®
www.ppi2pass.com

Professional Publications, Inc. • Belmont, California

Benefit by Registering This Book with PPI

- Get book updates and corrections.
- Hear the latest exam news.
- Obtain exclusive exam tips and strategies.
- Receive special discounts.

Register your book at **ppi2pass.com/register**.

Report Errors and View Corrections for This Book

PPI is grateful to every reader who notifies us of a possible error. Your feedback allows us to improve the quality and accuracy of our products. You can report errata and view corrections at **ppi2pass.com/errata**.

CIVIL ENGINEERING SOLVED PROBLEMS

Eighth Edition

Current printing of this edition: 1

Printing History

date	edition number	printing number	update
Aug 2012	6	1	New edition. Code updates. Copyright update.
Jun 2014	7	1	New edition. Code updates. Copyright update. New content.
Nov 2015	8	1	New edition. Code updates. Copyright update. New content.

Printed in the United States of America.

PPI
1250 Fifth Avenue
Belmont, CA 94002
(650) 593-9119
ppi2pass.com

ISBN: 978-1-59126-512-2

Library of Congress Control Number: 2015950982

F E D C B A

Table of Contents

TOPIC V: Transportation

TOPIC VI: Systems, Management, and Professional

Preface

When I wrote the first edition of *Civil Engineering Solved Problems*, I had been teaching civil engineering PE courses for many years. At the time, I was accustomed to a classroom setting, where the method of learning tends to be deductive: I taught the important concepts to my students, and then I presented examples to reinforce those concepts. However, I consistently found that students needed more than a presentation of important concepts and reinforcing examples. What they needed were problems to solve. And so I wrote this book.

Whereas my classroom was an exercise in deductive learning, this book is an exercise in inductive learning. You are presented first with problems to solve, and then, from the provided solutions, you can check your answers and learn the concepts that are required to derive the correct answer. Seeing these concepts in action will help you to come to conclusions on your own and learn basic concepts in a different manner than is usually presented in a classroom. In a sense, you are seeing all of the details before being awakened to the bigger picture. Through this book, that is what I strive to do: awaken you to the bigger picture.

The problems in this book have always been more challenging and esoteric than those you will encounter on the exam. This is good practice. On the morning breadth section of the exam, you will not just be able to work the easy problems. And, no matter which afternoon section you choose, you will need exposure to some of the tough breadth subjects.

Though the problems in this eighth edition are not multiple-choice, I do not see this as a significant issue. The subject matter remains appropriate for the civil PE exam, and code-dependent problems are up to date with NCEES codes and specifications. That the problems are more challenging than those you will find on the exam makes this book great practice for the exam.

I invite you to visit PPI's website at **ppi2pass.com/cefaq** to keep up with the current exam format and codes, as well as view or post errata at **ppi2pass.com/errata**.

Michael R. Lindeburg, PE

Acknowledgments

Ever since I wrote *101 Solved Mechanical Engineering Problems* (a similar book for mechanical engineers), I have wanted to bring out a book of representative civil exam problems. The writing, editing, and production work on the first edition of this book spanned five years. Now, this book is in its eighth edition, primarily due to the dedication and assistance of many people.

In response to the NCEES codes and standard changes, I enlisted the substantial assistance of Akash Rao, PE, LEED AP, who updated the structural portion of this book; and Ralph Arcena, who checked this book's calculations.

In PPI's product development and implementation department, I would like to acknowledge Sarah Hubbard, director of product development and implementation; Cathy Schrott, production services manager; Jenny Lindeburg King, associate editor-in-chief; Nicole Evans, associate project manager; Sierra Cirimelli-Low, Tyler Hayes, Tracy Katz, Ellen Nordman, Heather Turbeville, and Ian A. Walker, copy editors; Kate Hayes, production associate; and Tom Bergstrom, technical illustrator, for helping give the eighth edition of this book tangible form.

This book has been used by hundreds of engineers who, while preparing for their civil PE exam, sent me their comments and suggestions. Their comments were invaluable, and their suggestions have been incorporated into the book. Their names are too numerous to list here, but their contributions have made the book much easier for you to use.

Thanks to you all!

Michael R. Lindeburg, PE

How to Use This Book

This book is a collection of problems categorized by exam subject. Just as a fish is born knowing how to swim, you probably feel that you already know how to use this book. Actually, there are several ways to use it, and some of them are even pretty good. Only one way, however, will maximize the return on your investment in time.

At one end of the effort spectrum are people who will work through the book from beginning to end, studying every detail, duplicating every calculation, and making sure that they know the underlying subject. At the other end of the effort spectrum are people who will simply take the book into the exam with them hoping to find a similar problem whose solution they can copy.

Some will study using this book alone. Others will use this book in conjunction with the *Civil Engineering Reference Manual.*

Some will go through this book from beginning to end, using it as a secondary review. Others will go in and out of it, alternating their time between this book and others.

Admittedly, when I wrote the book, I had a vision of how it would be used. I was pretty certain that most people would study a chapter from the *Civil Engineering Reference Manual*, solve all the homework problems in the companion book *Practice Problems for the Civil Engineering PE Exam*, and then turn to this book for additional exposure to exam-level problems. Then they would return to their *Reference Manual* to study the next chapter.

This book was never intended to cover every type of exam problem or to be an all-in-one review. Though I tried to include extremely relevant problems, I did not write this book to be a diagnostic tool. You should not solve these problems and then design your review around what you did not know. If you do not do well on a particular problem, I would not want you to spend the next three months preparing for that type of problem.

The tried-and-true method of exam preparation is a systematic, thorough, and complete approach based on long-term exam trends rather than transient and odd-ball fads. That is how the *Civil Engineering Reference Manual* is intended to be used.

The value of a collection of problems such as this does not lie in its ability to guide your preparation. Rather, the value is in giving you an opportunity to consolidate all of your knowledge and to practice your test-taking skills.

The three most important skills are (1) learning which problems to answer and which to guess at or skip, (2) organizing your references and other resources, and (3) managing your time. That means you have to attempt the problems when you are ready to practice those skills: after you have studied the underlying subject matter.

How to Use This Book

Codes and Standards

The information that was used to write and update this book was based on the exam specifications at the time of publication. However, as with engineering practice itself, the PE exam is not always based on the most current codes or cutting-edge technology. Similarly, codes, standards, and regulations adopted by state and local agencies often lag issuance by several years. It is likely that the codes that are most current, the codes that you use in practice, and the codes that are the basis of your exam will all be different.

PPI lists on its website the dates and editions of the codes, standards, and regulations on which NCEES has announced the PE exams are based. It is your responsibility to find out which codes are relevant to your exam.

CONSTRUCTION DESIGN STANDARDS

ACI 318: *Building Code Requirements for Structural Concrete*, 2011. American Concrete Institute, Farmington Hills, MI.

ACI 347: *Guide to Formwork for Concrete*, 2004. American Concrete Institute, Farmington Hills, MI. (In ACI SP-4, Seventh ed. App.)

ACI SP-4: *Formwork for Concrete*, Seventh ed., 2005. American Concrete Institute, Farmington Hills, MI.

AISC: *Steel Construction Manual*, Fourteenth ed., 2011. American Institute of Steel Construction, Inc., Chicago, IL.

ASCE 37: *Design Loads on Structures During Construction*, 2002. American Society of Civil Engineers, Reston, VA.

CMWB: *Standard Practice for Bracing Masonry Walls Under Construction*, 2012. Council for Masonry Wall Bracing, Mason Contractors Association of America, Lombard, IL.

MUTCD-Pt 6: *Manual on Uniform Traffic Control Devices*—Part 6, Temporary Traffic Control, 2009. U.S. Department of Transportation, Federal Highway Administration, Washington, DC.

NDS: *National Design Specification for Wood Construction ASD/LRFD*, 2012 ed. American Wood Council, Washington, DC.

OSHA 1926: *Occupational Safety and Health Regulations for the Construction Industry* (U.S. Federal version). U.S. Department of Labor, Washington, DC.

GEOTECHNICAL DESIGN STANDARDS

ASCE/SEI7: *Minimum Design Loads for Buildings and Other Structures*, 2010. American Society of Civil Engineers, Reston, VA.

OSHA 1926: *Occupational Safety and Health Regulations for the Construction Industry* (U.S. Federal version). U.S. Department of Labor, Washington, DC.

STRUCTURAL DESIGN STANDARDS

AASHTO *LRFD*: *AASHTO LRFD Bridge Design Specifications*, Sixth ed., 2012. American Association of State Highway and Transportation Officials, Washington, DC.

ACI 318[1]: *Building Code Requirements for Structural Concrete*, 2011. American Concrete Institute, Farmington Hills, MI.

ACI 530/530.1[2]: *Building Code Requirements and Specification for Masonry Structures* (and companion commentaries), 2011. The Masonry Society, Boulder, CO; American Concrete Institute, Detroit, MI; and Structural Engineering Institute of the American Society of Civil Engineers, Reston, VA.

AISC: *Steel Construction Manual*, Fourteenth ed., 2011. American Institute of Steel Construction, Inc., Chicago, IL.

ASCE/SEI7: *Minimum Design Loads for Buildings and Other Structures*, 2010. American Society of Civil Engineers, Reston, VA.

AWS D1.1/D1.1M[3]: *Structural Welding Code—Steel*, Twenty-second ed., 2010. American Welding Society, Miami, FL.

AWS D1.2/D1.2M[4]: *Structural Welding Code—Aluminum*, Sixth ed., 2014. American Welding Society, Miami, FL.

AWS D1.4/D1.4M[5]: *Structural Welding Code—Reinforcing Steel*, Seventh ed., 2011. American Welding Society, Miami, FL.

---

[1]ACI 318 App. C does not apply to the Civil PE structural depth exam.

[2]Only the Allowable Stress Design (ASD) method may be used on the structural depth exam, except that ACI 530 Sec. 3.3.5 (strength design) may be used for walls with out-of-plane loads.

[3]AWS D1.1, AWS D1.2, and AWS D1.4 are listed in the Codes, Standards, and Documents subsection of NCEES's Civil PE structural depth exam specifications.

[4]See Ftn. 3.

[5]See Ftn. 3.

IBC: *2012 International Building Code* (without supplements). International Code Council, Inc., Falls Church, VA.

NDS[6]: *National Design Specification for Wood Construction ASD/LRFD*, 2012 ed., and *National Design Specification Supplement, Design Values for Wood Construction*, 2012 ed. American Wood Council, Washington, DC.

OSHA 1910[7]: *Occupational Safety and Health Standards* (U.S. Federal version). U.S. Department of Labor, Washington, DC.

OSHA 1926: *Occupational Safety and Health Regulations for the Construction Industry* (U.S. Federal version) Subpart E, Personal Protective and Life Saving Equipment, 1926.95–1926.107; Subpart M, Fall Protection, 1926.500–1926.503, App. A–E; Subpart Q, Concrete and Masonry Construction, 1926.700–1926.706, with App. A; and Subpart R, Steel Erection, 1926.750–1926.761, with App. A–H. U.S. Department of Labor, Washington, DC.

PCI: *PCI Design Handbook: Precast and Prestressed Concrete*, Seventh ed., 2010. Precast/Prestressed Concrete Institute, Chicago, IL.

TRANSPORTATION DESIGN STANDARDS

AASHTO *GDPS*: *AASHTO Guide for Design of Pavement Structures* (GDPS-4-M), 1993, and 1998 supplement. American Association of State Highway and Transportation Officials, Washington, DC.

AASHTO *Green Book*: *A Policy on Geometric Design of Highways and Streets*, Sixth ed., 2011. American Association of State Highway and Transportation Officials, Washington, DC.

AASHTO: *Guide for the Planning, Design, and Operation of Pedestrian Facilities*, First ed., 2004. American Association of State Highway and Transportation Officials, Washington, DC.

HSM: *Highway Safety Manual*, First ed., 2010. American Association of State Highway and Transportation Officials, Washington, DC.

AASHTO *MEPDG*: *Mechanistic-Empirical Pavement Design Guide: A Manual of Practice*, Interim ed., 2008. American Association of State Highway and Transportation Officials, Washington, DC.

AASHTO: *Roadside Design Guide*, Fourth ed., 2011. American Association of State Highway and Transportation Officials, Washington, DC.

AI: *The Asphalt Handbook* (MS-4), Seventh ed., 2007. Asphalt Institute, Lexington, KY.

FHWA: *Hydraulic Design of Highway Culverts*, Hydraulic Design Series no. 5, Publication no. FHWA-HIF-12-026, Third ed., 2012. U.S. Department of Transportation, Federal Highway Administration, Washington, DC.

HCM: *Highway Capacity Manual*, 2010 ed. Transportation Research Board, National Research Council, Washington, DC.

MUTCD: *Manual on Uniform Traffic Control Devices*, 2009 (including Revisions 1 and 2, May 2012). U.S. Department of Transportation, Federal Highway Administration, Washington, DC.

PCA: *Design and Control of Concrete Mixtures*, Fifteenth ed., 2011. Portland Cement Association, Skokie, IL.

---

[6]Only the ASD method may be used for wood design on the structural depth exam.

[7]Part 1910 is listed in the Codes, Standards, and Documents subsection of NCEES's Civil PE structural depth exam specifications.

Nomenclature

a	acceleration	ft/sec^2
a	depth	in
A	annual amount	\$
A	area	ft^2
A	watershed area	ac
A_g	area of gross cross section	in^2
A_q	pile tip cross-sectional area	in^2
A_s	area of laterally tied longitudinal steel in a column or pilaster	in^2
b	dimension	in
b_a	tensile force on an anchor bolt	lbf
b_o	punching shear area perimeter	in
b_v	shear force on an anchor bolt	lbf
b'	adjusted width	in
B	benefit	\$
B	tensile capacity	kips/bolt
B	width	ft
BOD	biochemical oxygen demand	mg/L
c	capacity	–
c	coefficient	–
c	cohesion	lbf/ft^2
c	column parameter	–
c	depth to neutral axis	in
c	wave speed	ft/sec
c_p	specific heat	Btu/lbm-°F
C	adjustment factor	–
C	average runoff coefficient	–
C	Chezy coefficient	–
C	cost	\$
C	cycle length	sec
C	long chord distance	ft
C	rational coefficient	–
C	roughness coefficient	–
C_1	correction for embedment	–
C_2	correction for creep	–
C_b	bending coefficient	–
C_c	compression index	–
C_D	drag coefficient	–
C_p	column stability factor	–
C_u	uniformity coefficient	–
C_w	correction coefficient	–
d	control delay	sec/veh
d	depth	ft
d	diameter	in
d	dimension	in
d	distance	ft
d	sight distance	ft
d_1	uniform control delay	sec/veh
d_2	incremental delay	sec/veh
d_3	initial queue delay	sec/veh
d_b	nominal diameter of reinforcement	in

D	degree of curve	deg
D	density	veh/mi
D	depreciation	\$
D	depth	ft
D	diameter	ft
D	distance	ft
D	drag	lbf
D_o	oxygen deficit after initial mixing	mg/L
D_p	phase duration	sec
D_t	oxygen deficit at time t	mg/L
DO	dissolved oxygen	mg/L
e	eccentricity	in
e	extension of effective green	sec
e	inflation	decimal
e	superelevation rate	ft/ft
e	void ratio	–
E	average imperviousness	%
E	efficiency of waste utilization	%
E	external distance	ft
E	modulus of elasticity	lbf/in^2
E	passenger car equivalent	–
E'	adjusted modulus of elasticity	lbf/in^2
f	adjustment factor	various
f	bearing stress	lbf/in^2
f	Darcy friction factor	–
f	effective friction angle	deg
f	side friction factor	–
f	temperature factor	–
f_a	allowable bending stress	lbf/in^2
f_b	actual bending stress	lbf/in^2
f_c	actual compressive stress	lbf/in^2
f'_c	specified compressive strength of concrete	lbf/in^2
$f'_{c,i}$	initial concrete compressive strength	lbf/in^2
f'_m	specified compressive strength of masonry	lbf/in^2
f_{pe}	compressive stress at service limit state after loss	lbf/in^2
f_{pi}	initial steel prestress	lbf/in^2
f_{ps}	stress in prestressing tendons	lbf/in^2
f_{pf}	final steel prestress	lbf/in^2
f_{pu}	specified ultimate tensile strength of prestressing steel	lbf/in^2
f_{py}	yield tensile strength of prestressing steel	lbf/in^2
f_s	tensile steel stress	lbf/in^2
f_{se}	effective stress in prestressed reinforcement after all prestress losses	lbf/in^2
f_v	actual or required shear stress	lbf/in^2

f_y	specified yield strength of nonprestressed steel	lbf/in^2
F	force	lbf
F	future worth	$\$$
F_a	compressive stress due to axial load	lbf/in^2
F_b	reference design value for bending	lbf/in^2
F_b'	allowable bending stress	lbf/in^2
F_b^*	tabulated bending design value multiplied by all applicable adjustment factors except C_{fu}, C_v, and C_L	lbf/in^2
F_c	reference design value for compression	lbf/in^2
F_c'	allowable compression design value	lbf/in^2
F_c^*	reference compression design value multiplied by all applicable adjustment factors except C_P	lbf/in^2
$F_{c\perp}$	reference design value for compression perpendicular to grain	lbf/in^2
$F_{c\perp}'$	allowable compression perpendicular to grain	lbf/in^2
F_{cE}	critical buckling design value	lbf/in^2
F_{Exx}	minimum specified strength	lbf/in^2
F_{nt}	allowable tensile bolt stress	lbf/in^2
F_{nt}'	nominal tensile stress for combined tension and stress	lbf/in^2
F_{nv}	allowable shear bolt stress	lbf/in^2
F_s	allowable reinforcement tensile stress	lbf/in^2
F_v	reference design value for shear	lbf/in^2
F_v'	allowable shear design value	lbf/in^2
F_y	yield strength	lbf/in^2
FS	factor of safety	–
g	effective green time	sec
g	grade	–
g	gravitational acceleration, 32.2	ft/sec^2
g_c	gravitational constant, 32.2	$lbm\text{-}ft/lbf\text{-}sec^2$
g/C	green ratio	–
G	end condition coefficient	–
G	grade	decimal
G	mean velocity gradient	$1/sec$
G	specific gravity	–
G	uniform gradient amount	$\$$
h	couple separation distance	in
h	head	ft
h	height	ft
H	height	ft
H	horizontal force	lbf
H	relative humidity	%
i	effective interest rate	decimal
i'	effective rate	decimal
$i\%$	effective interest rate	%
I	index	–
I	interior angle	deg
I	moment of inertia	in^4
I	rainfall intensity	in/hr
I	upstream filtering/metering adjustment factor	–
I_z	strain influence factor	–

j	lever-arm factor, ratio of distance between centroid of flexural compressive forces and centroid of tensile forces to effective depth	–
j	number of trips	–
J	polar moment of inertia of punching shear surface	ft^4
k	adjustment factor	–
k	coefficient of permeability	ft/sec
k	earth pressure coefficient	–
k	first order reaction constant	–
k	incremental delay adjustment factor	–
k	ratio of the distance between compression face of wall and neutral axis to the effective depth	–
k	stiffness	ft/lbf
k	substrate utilization coefficient	–
k_1	Terzaghi's subgrade modulus for a 1 ft^2 plate	$tons/ft^3$
k_d	kinetic coefficient	–
k_d	microbial delay coefficient	–
K	effective length factor	–
K	minor loss coefficient	–
K_D	deaeration coefficient	$1/day$
K_e	buckling length coefficient	–
K_p	hydraulic conductivity	$gal/ft\text{-}day$
K_p	permeability	ft/day
K_R	reaeration coefficient	$1/day$
l	length	ft
l	lost time	sec
l_p	post spacing	ft
L	length	various
L	loading	$lbm/ac\text{-}day$
L	support force at the left	kips
L_b	embedment length	ft
L_f	load factor	–
m	horizontal-to-vertical side slope	–
m	mass	lbm
$\dot{m}$	mass flow rate	lbm/sec
M	maintenance costs	$\$$
M	middle ordinate	ft
M	moment	various
M'	adjusted moment capacity	$ft\text{-}lbf$
n	number	–
n	Manning's roughness coefficient	–
n	modular ratio	–
n_h	coefficient of modulus variation	–
N	amount of nitrogen	mg/L
N	minimum bearing length	in
N	number of events	–
N_L	number of lanes	–
N_o	stability number	–
N_q	bearing capacity factor	–
p	population	–
p	pressure	lbf/ft^2
p	probability	–
p	tributary length per bolt pair	in
p_1	inlet pressure	lbf/in^2

p_2	outlet pressure	lbf/in^2
P	amount of phosphorus	mg/L
P	axial load	lbf
P	power	hp
P	present worth	$
P	wetted perimeter	ft
P_a	compressive force due to axial load	lbf
$P_{n,max}$	column capacity	kips
P_u	design load	kips
P_x	net mass of cell tissue produced	lbm/day
PHF	peak hour factor	–
q	heat required	Btu/hr
q'	effective vertical stress at pile tip	lbf/ft^2
Q	flow rate	ft^3/sec or MGD
Q	peak discharge	ft^3/sec
Q_n	downdrag force	lbf
Q_p	tip capacity of the pile	lbf
r	radius	ft
r	radius of gyration	in
r_1	radius of influence of the well	ft
r_2	radius of excavation	ft
R	BOD reaction constant	1/day
R	curve radius	ft
R	gas constant	ft-lbf/ lbm-°R
R	hydraulic radius	ft
R	radius of curvature of inside face of laminations	ft
R	rate	ft/hr
R	reaction	lbf
R	recirculation ratio	–
R	stiffness factor	–
R	support force at the right	lbf
R	total runoff	ac-ft or in
R_d	endogenous coefficient	1/day
Re	Reynolds number	–
s	distance	ft
s	saturation flow rate	vph
s	spacing	ft
s_o	base saturation flow rate	vph
S	degree of saturation	–
S	section modulus	in^3
S	settlement	in
S	slope	–
S	speed	mph
S	strength	lbf/ft
S	surface area	ft^2
S_0	influent soluble BOD_5	mg/L
S_0	ultimate BOD	lbm/gal
S_y	yield strength	lbf/in^2
SG	specific gravity	–
SLR	surface loading rate	$gal/day-ft^2$
t	tax rate	%
t	thickness	in
t	time	various
T	analysis period duration	hr
T	stiffness factor	–
T	tangent length	ft
T	temperature	°F
T	tensile force	lbf
T_{50}	time from beginning of runoff to 50% cumulative runoff	hr
T_{75}	time from beginning of runoff to 75% cumulative runoff	hr
T_d	time from beginning of runoff to end of channel discharge	hr
T_p	time from beginning of runoff to peak discharge	hr
U	overall coefficient of heat transfer	Btu/ft^2-°F-hr
U	specific substrate utilization rate	1/day
v	velocity	ft/sec
v	shear stress	lbf/in^2
v^*	overflow rate	$gal/day-ft^2$
v/c	volume-capacity ratio	–
V	shear	lbf
V	volume	various
VLR	volumetric loading rate	$lbm/day-ft^3$
w	limit	–
w	load per unit length	kips/ft
w	water content	decimal
w	width	ft
W	weight	lbf
W	width	ft
x	price per item	$
x	tangent distance	ft
x	species parameter for volume factor	–
x_r	return sludge volatile suspended solids	mg/L
X	biomass concentration	mg/L
X	lane group v/c ratio or degree of saturation	–
y	aquifer thickness after drawdown	ft
y	distance	in
y	kinetic coefficient	lbm VSS/ lbm BOD_5
y	tangent offset	ft
y_1	original thickness of saturated layer	ft
y_2	thickness of aquifer at edge of excavation	ft
Y	original aquifer phreatic zone thickness	ft
Y	yield coefficient	lbm/lbm
z	elevation	ft
Z	plastic section modulus	in^3

Symbols

α	angle	deg
α	strength enhancement factor	–
β	slope angle	deg
β_1	equivalent stress block ratio	–
β_c	ratio of long side to short side of a column	–
δ	Hardy Cross correction	–
δ	soil-pile friction angle	deg
Δ	angle	deg
Δ	deflection	deg
ϵ	specific roughness	ft
ϵ	strain	–

η	efficiency	%
γ	angle	deg
γ	specific weight	lbf/ft^3
γ_s	specific weight of particles	lbf/ft^3
γ_v	fraction of moment transferred by shear	–
λ	long-term deflection factor	–
μ_{max}	maximum specific growth rate	day^{-1}
ω	$\rho f_y / f'_c$	–
Ω	safety factor	–
ϕ	angle	deg
ϕ	soil friction angle	deg
ϕ	strength reduction factor	–
ρ	density	lbm/ft^3
ρ	reinforcement ratio	–
ρ_t	ratio of steel area to gross area of masonry	–
σ	stress	lbf/in^2
$\overline{\sigma}_v$	effective vertical stress	–
θ	angle	deg
θ	hydraulic retention time	days
θ	temperature coefficient	–
θ_c	mean cell residence time	days
v	kinematic viscosity	ft^2/sec
ξ	time-dependent factor	–

Subscripts

0	initial
1	start-up
2	clearance
5	5-day
a	acceleration, added, allowable, area type, or axial
as	ambient summer air
ave	average
aw	ambient winter air
A	added or aluminum
b	bearing, bending, biological, bolt, bottom, bulk, or masonry breakout
ba	absorbed asphalt
bb	bus blocking
B	user benefits
BOD	biochemical oxygen demand
c	available, closure, concrete, critical, concentration, or masonry crushing
calc	calculated
cr	cracked or critical
C	curvature or combustibles
d	dead, dry, mean detention, or microbial decay
D	dead, degree of curve, load duration, or longitudinal drag
e	effective or effluent
eff	effective or effluent
EW	east-to-west
f	concrete strength effect, final, flange, floor, or friction

F	ferrous materials or size
g	gap, grade, or gross
G	glass
h	horizontal
hs	shrinkage from humidity
HV	heavy vehicle
i	lane group i or initial
I	influent or investment costs
l	live or liquid
L	lanes, left, length, live, or lumber sheathing
Lpb	left, pedestrians and bicycles
LT	left-turn
LU	lane utilization
m	masonry
max	maximum
mb	bulk
min	minimum
mm	maximum
M	maintenance costs or wet use
n	net, nominal, or nozzle
nc	unconfined compressive
NS	north-to-south
o	normal
obs	observed
opt	optimal
p	column stability, parking, peak, plastic, plasticity, or polar
pry	pryout
r	required or root
req	required
R	right
Rpb	right, pedestrians and bicycles
RT	right-turn
s	raw sludge, settling, spiral, steel, studs, or volume-surface ratio effects
sat	saturated
sb	bulk
se	effective
sh	shrinkage
SN	south-to-north
t	tensile, top, or total
td	time-development
T	temperature, theoretical, or truck and bus
u	ultimate or unfactored loads
v	vertical or shear
V	volume
VSS	volatile suspended solids
w	lane width, sludge wasting, stem, walls, water, or weld
ww	winter wastewater
ws	summer wastewater
WE	west-to-east
x	strong axis
y	weak axis

Topic I: Water Resources

Chapter

1 Fluid Mechanics

PROBLEM 1

As a condition of obtaining a business license, a factory was required by the planning commission to install a private water supply to supplement what it draws from the municipal supply. It did so, using new PVC pipe exclusively. Its reservoir (filled from a local well) is open to the atmosphere. No water is added to or taken from the system at any point between the pipe entrance at point A and the factory at point C. The water temperature is 50°F.

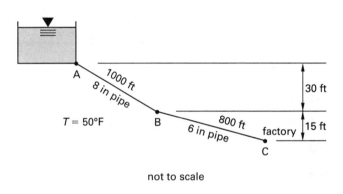

not to scale

1.1. If the water discharges freely at point C, how high must the water level in the reservoir be to maintain a static pressure of 40 psig at the reservoir outlet (point A)?

For the remaining questions, assume the reservoir is 70 ft deep at point A.

1.2. What is the maximum flow rate if the static pressure at point C must not drop below 40 psig?

1.3. To make the factory self-sufficient, what size pipe should be installed directly from point A to point C (parallel to the existing line) to provide 1900 gal/min at a static pressure of 40 psig?

PROBLEM 2

An elevated tank feeds a simple pipe system as shown. All pipes are 5–10 yr old cast iron. There is a fire hydrant at point C. The minimum allowable pressure at point C is 22 psig for firefighting requirements. All minor losses are insignificant. Neglect friction losses from the tank to point A.

2.1. What is the maximum static head at point C?

2.2. What is the maximum flow out of the hydrant?

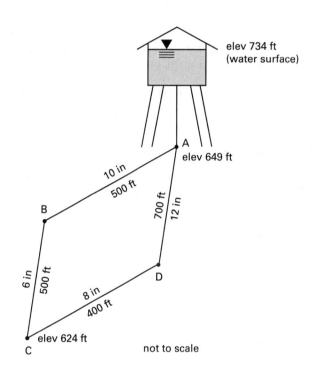

not to scale

PROBLEM 3

The water pressure at point D in a new PVC pipe (Hazen-Williams coefficient, $C = 150$) water distribution system is too low.

section	length	diameter	average flow removed
AB	16,000 ft	6 in	140 gal/min at B
BC	14,000 ft	6 in	130 gal/min at C
CD	8000 ft	4 in	70 gal/min at D

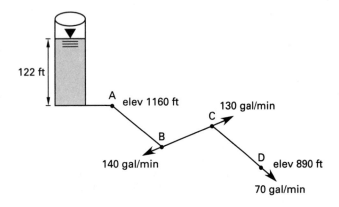

3.1. Based on the average flow removed, what is the water pressure at point D?

3.2. If the unacceptably low pressure is experienced at the average flows given, what might be the causes?

3.3. Suppose the excessively low pressure is experienced only during the hours of daily peak demand. What is the water pressure at point D?

3.4. How would you eliminate the low-pressure problem?

PROBLEM 4

Pipes from three reservoirs meet at point D at elevation 1200 ft as shown. All pipes are new PVC.

leg	length	diameter	C
A	5000 ft	12 in	150
B	4000 ft	8 in	150
C	10,000 ft	10 in	150

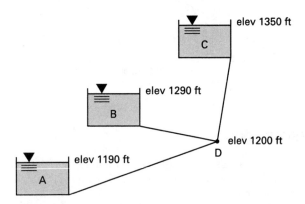

4.1. What is the flow direction (away from or into point D) in each leg?

4.2. What is the head loss in each pipe between the reservoirs and point D?

PROBLEM 5

The main water pipeline makes a complete loop around a small community. The pipe is rigid cast iron. There is no elevation change around the loop. Minor losses can be neglected.

section	length	diameter	C
AB	800 ft	8 in	140
BC	1000 ft	8 in	140
AE	1400 ft	6 in	120
ED	2000 ft	10 in	160
DC	1200 ft	6 in	120

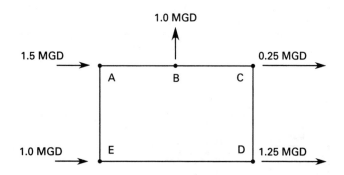

5.1. What is the flow quantity between A and B?

5.2. What is the flow direction between A and B?

5.3. What is the pressure drop between A and B?

5.4. What is the flow quantity between B and C?

5.5. What is the flow quantity between A and E?

5.6. What is the flow quantity between E and D?

5.7. What is the flow quantity between C and D?

5.8. What is the pressure drop between C and D?

5.9. What is the flow direction between C and D?

5.10. What is the minimum time to close a valve at point D without experiencing water hammer at point E?

PROBLEM 6

A pump supplies 50°F water to a residential community through a parallel level (equal-elevation) network as shown. All pipe is 5 yr old cast iron. Minor losses can be neglected. The static pressure drop between the pump and the outlet must not exceed 10 psig.

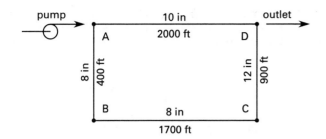

not to scale

6.1. What is the maximum flow the pump can operate at without the network exceeding the pressure-drop limitation?

6.2. What is the flow velocity in leg AD?

6.3. What is the pressure drop in leg AD?

6.4. What is the pressure drop in leg ABCD?

6.5. What is the pressure head (in feet) at the pump discharge?

PROBLEM 7

A 240 ft long ductile iron pipe runs between the bottom drains of two sludge processing basins. There are two 90° elbows in the line, and all other minor losses should be disregarded. The surface elevations of the two basins differ by 12 ft. There is no pump, and flow is by gravity from the higher basin to the lower basin. The sludge temperature varies from 45°F (in the winter) to 80°F (in the summer). Depending on the time of day, the required transfer flow rate varies from a low of 3 MGD to a high of 6 MGD.

7.1. What standard diameter should be specified for the pipe to connect the sludge basins?

7.2. Explain the factors to consider when choosing a diameter.

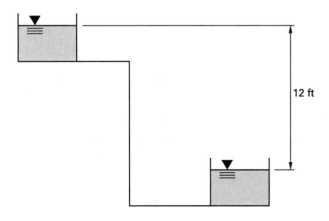

SOLUTION 1

1.1. With the energy-datum zero level at point C, the energy equation between this point and the reservoir level is

$$h + 30 \text{ ft} + 15 \text{ ft} = \frac{v_{BC}^2}{2g} + \frac{p_C}{\gamma} + h_f$$

h is the water depth in the reservoir, and $p_C = 0$, assuming that the system discharges to the atmosphere. The local losses have been neglected. The friction losses are

$$h_f = f_{AB} \left(\frac{L_{AB}}{D_{AB}} \right) \left(\frac{v_{AB}^2}{2g} \right) + f_{BC} \left(\frac{L_{BC}}{D_{BC}} \right) \left(\frac{v_{BC}^2}{2g} \right)$$

The specific roughness for PVC is $\epsilon = 5 \times 10^{-6}$ ft, so the relative roughnesses are

$$\frac{\epsilon}{D_{AB}} = \frac{5 \times 10^{-6} \text{ ft}}{\dfrac{8 \text{ in}}{12 \dfrac{\text{in}}{\text{ft}}}} = 7.5 \times 10^{-6}$$

$$\frac{\epsilon}{D_{BC}} = \frac{5 \times 10^{-6} \text{ ft}}{\dfrac{6 \text{ in}}{12 \dfrac{\text{in}}{\text{ft}}}} = 1.0 \times 10^{-5}$$

From the Moody diagram, the pipes are smooth for most of the Reynolds number range. As a first approximation, choose

$$f_{AB} = f_{BC} = 0.013$$

The friction losses are

$$h_f = (0.013) \left(\frac{1000 \text{ ft}}{\dfrac{8 \text{ in}}{12 \dfrac{\text{in}}{\text{ft}}}} \right) \left(\frac{v_{AB}^2}{(2)\left(32.2 \dfrac{\text{ft}}{\text{sec}^2} \right)} \right)$$

$$+ (0.013) \left(\frac{800 \text{ ft}}{\dfrac{6 \text{ in}}{12 \dfrac{\text{in}}{\text{ft}}}} \right) \left(\frac{v_{BC}^2}{(2)\left(32.2 \dfrac{\text{ft}}{\text{sec}^2} \right)} \right)$$

$$= 0.303 v_{AB}^2 + 0.323 v_{BC}^2$$

The continuity equation relates the velocities v_{AB} and v_{BC}.

$$\left(\frac{\pi D_{AB}^2}{4}\right)v_{AB} = \left(\frac{\pi D_{BC}^2}{4}\right)v_{BC}$$

$$v_{AB} = \left(\frac{D_{BC}}{D_{AB}}\right)^2 v_{BC}$$

$$= \left(\frac{6 \text{ in}}{8 \text{ in}}\right)^2 v_{BC}$$

$$= 0.563 v_{BC}$$

Substituting this into the friction loss equation,

$$h_f = (0.303)(0.563 v_{BC})^2 + 0.323 v_{BC}^2$$

$$= 0.419 v_{BC}^2$$

The energy equation becomes

$$h + 45 \text{ ft} = \frac{v_{BC}^2}{(2)\left(32.2 \frac{\text{ft}}{\text{sec}^2}\right)} + 0.419 v_{BC}^2$$

$$h = 0.435 v_{BC}^2 - 45 \text{ ft} \qquad \text{[Eq. 1]}$$

A second equation containing h and v_{BC} is obtained from the required 40 lbf/in² pressure at point A. The energy equation between the reservoir level and A is

$$h + 45 \text{ ft} = 45 \text{ ft} + \frac{v_{AB}^2}{2g} + \frac{p_A}{\gamma}$$

Using the continuity equation and the known values produces

$$h = \frac{(0.563 v_{BC})^2}{(2)\left(32.2 \frac{\text{ft}}{\text{sec}^2}\right)} + \frac{\left(40 \frac{\text{lbf}}{\text{in}^2}\right)\left(12 \frac{\text{in}}{\text{ft}}\right)^2}{62.4 \frac{\text{lbf}}{\text{ft}^3}}$$

$$= 0.0049 v_{BC}^2 + 92.3 \text{ ft} \qquad \text{[Eq. 2]}$$

Solving Eq. 1 and Eq. 2 simultaneously yields

$$h = 93.9 \text{ ft}$$

$$v_{BC} = 17.9 \text{ ft/sec}$$

From the continuity equation,

$$v_{AB} = 0.563 v_{BC} = (0.563)\left(17.9 \frac{\text{ft}}{\text{sec}}\right) = 10.1 \text{ ft/sec}$$

In order to check the friction parameters, the Reynolds numbers are needed. The water viscosity at 50°F is $\nu = 1.41 \times 10^{-5}$ ft²/sec.

$$Re_{AB} = \frac{v_{AB}D_{AB}}{\nu} = \frac{\left(10.1 \frac{\text{ft}}{\text{sec}}\right)\left(\frac{8 \text{ in}}{12 \frac{\text{in}}{\text{ft}}}\right)}{1.41 \times 10^{-5} \frac{\text{ft}^2}{\text{sec}}}$$

$$= 4.8 \times 10^5$$

$$Re_{BC} = \frac{v_{BC}D_{BC}}{\nu} = \frac{\left(17.9 \frac{\text{ft}}{\text{sec}}\right)\left(\frac{6 \text{ in}}{12 \frac{\text{in}}{\text{ft}}}\right)}{1.41 \times 10^{-5} \frac{\text{ft}^2}{\text{sec}}}$$

$$= 6.3 \times 10^5$$

From the Moody diagram, the friction factors chosen are correct. The minor losses can now be determined.

$$\text{entrance loss} \approx 0.5\left(\frac{v_{AB}^2}{2g}\right)$$

$$= (0.5)\left(\frac{\left(10.1 \frac{\text{ft}}{\text{sec}}\right)^2}{(2)\left(32.2 \frac{\text{ft}}{\text{sec}^2}\right)}\right)$$

$$= 0.8 \text{ ft}$$

$$K_{\text{contraction}} = \left(\frac{1}{2}\right)\left(1 - \left(\frac{D_1}{D_2}\right)^2\right)$$

$$= \left(\frac{1}{2}\right)\left(1 - \left(\frac{6 \text{ in}}{8 \text{ in}}\right)^2\right)$$

$$= 0.22$$

$$\text{contraction loss} \approx 0.22\left(\frac{v_{BC}^2}{2g}\right)$$

$$= (0.22)\left(\frac{\left(17.9 \frac{\text{ft}}{\text{sec}}\right)^2}{(2)\left(32.2 \frac{\text{ft}}{\text{sec}^2}\right)}\right)$$

$$= 1.1 \text{ ft}$$

(Contraction head loss coefficients vary widely.)

The losses are small compared to the friction losses but should be added to the water depth in the reservoir.

$$h = 93.9 \text{ ft} + 0.8 \text{ ft} + 1.1 \text{ ft}$$

$$= \boxed{95.8 \text{ ft}}$$

1.2. The energy equation between the reservoir and point C is

$$70 \text{ ft} + 45 \text{ ft} = \frac{v_{BC}^2}{2g} + \frac{p_C}{\gamma} + f_{AB}\left(\frac{L_{AB}}{D_{AB}}\right)\left(\frac{v_{AB}^2}{2g}\right)$$
$$+ f_{BC}\left(\frac{L_{BC}}{D_{BC}}\right)\left(\frac{v_{BC}^2}{2g}\right)$$

From Sol. 1.1,

$$v_{AB} = 0.563 v_{BC}$$

Assuming $f_{AB} = 0.016$ and $f_{BC} = 0.015$ yields

$$\frac{v_{BC}^2}{(2)\left(32.2 \frac{\text{ft}}{\text{sec}^2}\right)} + \frac{\left(40 \frac{\text{lbf}}{\text{in}^2}\right)\left(12 \frac{\text{in}}{\text{ft}}\right)^2}{62.4 \frac{\text{lbf}}{\text{ft}^3}}$$

$$+ (0.016)\left(\frac{1000 \text{ ft}}{\frac{8 \text{ in}}{12 \frac{\text{in}}{\text{ft}}}}\right)\left(\frac{(0.563 v_{BC})^2}{(2)\left(32.2 \frac{\text{ft}}{\text{sec}^2}\right)}\right)$$

$$+ (0.015)\left(\frac{800 \text{ ft}}{\frac{6 \text{ in}}{12 \frac{\text{in}}{\text{ft}}}}\right)\left(\frac{v_{BC}^2}{(2)\left(32.2 \frac{\text{ft}}{\text{sec}^2}\right)}\right) = 115 \text{ ft}$$

Solving for the velocity,

$$v_{BC} = 6.7 \text{ ft/sec}$$

From the continuity equation,

$$v_{AB} = (0.563)\left(6.7 \frac{\text{ft}}{\text{sec}}\right) = 3.8 \text{ ft/sec}$$

In order to check the friction factors, calculate the Reynolds numbers.

$$Re_{AB} = \frac{v_{AB} D_{AB}}{\nu} = \frac{\left(3.8 \frac{\text{ft}}{\text{sec}}\right)\left(\frac{8 \text{ in}}{12 \frac{\text{in}}{\text{ft}}}\right)}{1.41 \times 10^{-5} \frac{\text{ft}^2}{\text{sec}}}$$
$$= 1.8 \times 10^5$$

$$Re_{BC} = \frac{v_{BC} D_{BC}}{\nu} = \frac{\left(6.7 \frac{\text{ft}}{\text{sec}}\right)\left(\frac{6 \text{ in}}{12 \frac{\text{in}}{\text{ft}}}\right)}{1.41 \times 10^{-5} \frac{\text{ft}^2}{\text{sec}}}$$
$$= 2.4 \times 10^5$$

Referring to the Moody diagram, the friction factor is correct. The flow through the system is

$$Q_{ABC} = \left(\frac{\pi D_{BC}^2}{4}\right) v_{BC}$$

$$= \left(\frac{\pi\left(\frac{6 \text{ in}}{12 \frac{\text{in}}{\text{ft}}}\right)^2}{4}\right)\left(6.7 \frac{\text{ft}}{\text{sec}}\right)$$

$$\times \left(7.4805 \frac{\text{gal}}{\text{ft}^3}\right)\left(60 \frac{\text{sec}}{\text{min}}\right)$$

$$= \boxed{590 \text{ gal/min}}$$

1.3. Pipeline AC must carry the flow not provided by pipeline ABC.

$$1900 \frac{\text{gal}}{\text{min}} - 590 \frac{\text{gal}}{\text{min}} = \frac{1310 \frac{\text{gal}}{\text{min}}}{\left(7.4805 \frac{\text{gal}}{\text{ft}^3}\right)\left(60 \frac{\text{sec}}{\text{min}}\right)}$$

$$= 2.9 \text{ ft}^3/\text{sec}$$

Try $D_{AC} = 10$ in. Using the specific roughness in Sol. 1.1, the relative roughness is

$$\frac{\epsilon}{D_{AC}} = \frac{5 \times 10^{-6} \text{ ft}}{\frac{10 \text{ in}}{12 \frac{\text{in}}{\text{ft}}}} = 6 \times 10^{-6}$$

From the Moody diagram, choose an initial value for the friction factor $f_{AC} = 0.014$. The energy equation along pipeline AC is

$$70 \text{ ft} + 45 \text{ ft} = \frac{v_{AC}^2}{2g} + \frac{p_C}{\gamma} + f_{AC}\left(\frac{L_{AC}}{D_{AC}}\right)\left(\frac{v_{AC}^2}{2g}\right)$$

$$115 \text{ ft} = \frac{v_{AC}^2}{(2)\left(32.2 \frac{\text{ft}}{\text{sec}^2}\right)} + \frac{\left(40 \frac{\text{lbf}}{\text{in}^2}\right)\left(12 \frac{\text{in}}{\text{ft}}\right)^2}{62.4 \frac{\text{lbf}}{\text{ft}^3}}$$

$$+ (0.014)\left(\frac{1800 \text{ ft}}{\frac{10 \text{ in}}{12 \frac{\text{in}}{\text{ft}}}}\right)\left(\frac{v_{AC}^2}{(2)\left(32.2 \frac{\text{ft}}{\text{sec}^2}\right)}\right)$$

Solving for the velocity yields

$$v_{AC} = 6.8 \text{ ft/sec}$$

The diameter can be calculated from the continuity equation.

$$Q = \left(\frac{\pi D_{AC}^2}{4}\right) v_{AC}$$

$$2.9 \ \frac{\text{ft}^3}{\text{sec}} = \left(\frac{\pi D_{AC}^2}{4}\right)\left(6.8 \ \frac{\text{ft}}{\text{sec}}\right)$$

Solving for D_{AC} yields

$$D_{AC} = (0.74 \ \text{ft})\left(12 \ \frac{\text{in}}{\text{ft}}\right) = 8.8 \ \text{in}$$

This is close to the assumed value. Select the next largest available pipe size: 10 in. The final step is to calculate the Reynolds number to check the friction factor.

$$\text{Re}_{AC} = \frac{v_{AC} D_{AC}}{\nu}$$

$$= \frac{\left(6.8 \ \frac{\text{ft}}{\text{sec}}\right)\left(\frac{10 \ \text{in}}{12 \ \frac{\text{in}}{\text{ft}}}\right)}{1.41 \times 10^{-5} \ \frac{\text{ft}^2}{\text{sec}}}$$

$$= 4.0 \times 10^5$$

The friction factor from the Moody diagram is close to the initial choice. Therefore, the new PVC pipe should have a standard diameter of

$$D_{AC} = \boxed{10 \ \text{in}}$$

SOLUTION 2

2.1. The maximum static head at C occurs in the absence of flow through the system. In this case, the friction losses are zero, and the energy equation between the tank and point C is

$$734 \ \text{ft} = z_C + \frac{p_C}{\gamma}$$

$$= 624 \ \text{ft} + \frac{p_C}{\gamma}$$

Solving for the static head yields

$$\frac{p_C}{\gamma} = \boxed{110 \ \text{ft}}$$

2.2. The maximum flow takes place when C is at its minimum allowable pressure of 22 psig.

For path ABC, the continuity equation is

$$\left(\frac{\pi D_{AB}^2}{4}\right) v_{AB} = \left(\frac{\pi D_{BC}^2}{4}\right) v_{BC}$$

$$v_{AB} = \left(\frac{D_{BC}}{D_{AB}}\right)^2 v_{BC}$$

$$= \left(\frac{6 \ \text{in}}{10 \ \text{in}}\right)^2 v_{BC}$$

$$= 0.36 v_{BC}$$

The energy equation is

$$734 \ \text{ft} = 624 \ \text{ft} + \frac{p_C}{\gamma} + \frac{v_{BC}^2}{2g}$$

$$+ f_{AB}\left(\frac{L_{AB}}{D_{AB}}\right)\left(\frac{v_{AB}^2}{2g}\right)$$

$$+ f_{BC}\left(\frac{L_{BC}}{D_{BC}}\right)\left(\frac{v_{BC}^2}{2g}\right)$$

Introducing the continuity equation and simplifying,

$$110 \ \text{ft} = \frac{p_C}{\gamma} + \frac{v_{BC}^2}{2g}\left(1 + (0.36)^2\left(\frac{f_{AB}L_{AB}}{D_{AB}}\right)\right.$$

$$\left. + \frac{f_{BC}L_{BC}}{D_{BC}}\right)$$

For a 5–10 yr old cast-iron pipe, the specific roughness is $\epsilon = 0.001$ ft. The relative roughnesses are

$$\frac{\epsilon}{D_{AB}} = \frac{0.001 \ \text{ft}}{\dfrac{10 \ \text{in}}{12 \ \frac{\text{in}}{\text{ft}}}} = 0.0012$$

$$\frac{\epsilon}{D_{BC}} = \frac{0.001 \ \text{ft}}{\dfrac{6 \ \text{in}}{12 \ \frac{\text{in}}{\text{ft}}}} = 0.002$$

From the Moody diagram, initial estimates for the friction factors are

$$f_{AB} = 0.021$$

$$f_{BC} = 0.024$$

Substituting all known values in the energy equation yields

$$110 \text{ ft} = \frac{\left(22 \frac{\text{lbf}}{\text{in}^2}\right)\left(12 \frac{\text{in}}{\text{ft}}\right)^2}{62.4 \frac{\text{lbf}}{\text{ft}^3}} + \left(\frac{v_{BC}^2}{(2)\left(32.2 \frac{\text{ft}}{\text{sec}^2}\right)}\right)$$

$$\times \left(1 + (0.36)^2 (0.021)\left(\frac{500 \text{ ft}}{\frac{10 \text{ in}}{12 \frac{\text{in}}{\text{ft}}}}\right)\right.$$

$$\left. + (0.024)\left(\frac{500 \text{ ft}}{\frac{6 \text{ in}}{12 \frac{\text{in}}{\text{ft}}}}\right)\right)$$

Solving for the velocity v_{BC} yields

$$v_{BC} = 12.0 \text{ ft/sec}$$

From the continuity equation,

$$v_{AB} = 0.36 v_{BC} = (0.36)\left(12.0 \frac{\text{ft}}{\text{sec}}\right) = 4.3 \text{ ft/sec}$$

The friction factors are checked by calculating the Reynolds numbers.

$$\text{Re}_{AB} = \frac{v_{AB} D_{AB}}{\nu} = \frac{\left(4.3 \frac{\text{ft}}{\text{sec}}\right)\left(\frac{10 \text{ in}}{12 \frac{\text{in}}{\text{ft}}}\right)}{1.41 \times 10^{-5} \frac{\text{ft}^2}{\text{sec}}}$$

$$= 2.5 \times 10^5$$

$$\text{Re}_{BC} = \frac{v_{BC} D_{BC}}{\nu} = \frac{\left(12.0 \frac{\text{ft}}{\text{sec}}\right)\left(\frac{6 \text{ in}}{12 \frac{\text{in}}{\text{ft}}}\right)}{1.41 \times 10^{-5} \frac{\text{ft}^2}{\text{sec}}}$$

$$= 4.3 \times 10^5$$

The Moody diagram indicates that the friction factors remain unchanged with these values. The flow through branch ABC is

$$Q_{ABC} = \left(\frac{\pi D_{AB}^2}{4}\right) v_{AB}$$

$$= \left(\frac{\pi \left(\frac{10 \text{ in}}{12 \frac{\text{in}}{\text{ft}}}\right)^2}{4}\right)\left(4.3 \frac{\text{ft}}{\text{sec}}\right)$$

$$\times \left(7.4805 \frac{\text{gal}}{\text{ft}^3}\right)\left(60 \frac{\text{sec}}{\text{min}}\right)$$

$$= 1053 \text{ gal/min}$$

The continuity equation for branch ADC yields

$$v_{AD} = \left(\frac{D_{DC}}{D_{AD}}\right)^2 v_{DC}$$

$$= \left(\frac{8 \text{ in}}{12 \text{ in}}\right)^2 v_{DC}$$

$$= 0.44 v_{DC}$$

The energy equation is

$$734 \text{ ft} = 624 \text{ ft} + \frac{p_C}{\gamma} + \frac{v_{DC}^2}{2g} + f_{AD}\left(\frac{L_{AD}}{D_{AD}}\right)\left(\frac{v_{AD}^2}{2g}\right)$$

$$+ f_{DC}\left(\frac{L_{DC}}{D_{DC}}\right)\left(\frac{v_{DC}^2}{2g}\right)$$

After introducing the continuity equation and collecting terms, this becomes

$$110 \text{ ft} = \frac{p_C}{\gamma} + \frac{v_{DC}^2}{2g}\left(1 + (0.44)^2\left(\frac{f_{AD} L_{AD}}{D_{AD}}\right)\right.$$

$$\left. + \frac{f_{DC} L_{DC}}{D_{DC}}\right)$$

The relative roughnesses are

$$\frac{\epsilon}{D_{AD}} = \frac{0.001 \text{ ft}}{\frac{12 \text{ in}}{12 \frac{\text{in}}{\text{ft}}}} = 0.001$$

$$\frac{\epsilon}{D_{DC}} = \frac{0.001 \text{ ft}}{\frac{8 \text{ in}}{12 \frac{\text{in}}{\text{ft}}}} = 0.0015$$

From the Moody diagram, estimate initial values for the friction factors.

$$f_{AD} = 0.020$$

$$f_{DC} = 0.022$$

Replacing all known values in the energy equation produces

$$110 \text{ ft} = \frac{\left(22 \ \frac{\text{lbf}}{\text{in}^2}\right)\left(12 \ \frac{\text{in}}{\text{ft}}\right)^2}{62.4 \ \frac{\text{lbf}}{\text{ft}^3}} + \left(\frac{v_{DC}^2}{(2)\left(32.2 \ \frac{\text{ft}}{\text{sec}^2}\right)}\right)$$

$$\times \left(1 + (0.44)^2 (0.020)\left(\frac{700 \text{ ft}}{\frac{12 \text{ in}}{12 \ \frac{\text{in}}{\text{ft}}}}\right)\right.$$

$$\left. + (0.022)\left(\frac{400 \text{ ft}}{\frac{8 \text{ in}}{12 \ \frac{\text{in}}{\text{ft}}}}\right)\right)$$

Solving for the velocity,

$$v_{DC} = 15.0 \text{ ft/sec}$$

From the continuity equation,

$$v_{AD} = 0.44 v_{DC} = (0.44)\left(15.0 \ \frac{\text{ft}}{\text{sec}}\right) = 6.6 \text{ ft/sec}$$

The Reynolds numbers for this branch are

$$\text{Re}_{AD} = \frac{v_{AD} D_{AD}}{\nu} = \frac{\left(6.6 \ \frac{\text{ft}}{\text{sec}}\right)\left(\frac{12 \text{ in}}{12 \ \frac{\text{in}}{\text{ft}}}\right)}{1.41 \times 10^{-5} \ \frac{\text{ft}^2}{\text{sec}}}$$

$$= 4.7 \times 10^5$$

$$\text{Re}_{DC} = \frac{v_{DC} D_{DC}}{\nu} = \frac{\left(15.0 \ \frac{\text{ft}}{\text{sec}}\right)\left(\frac{8 \text{ in}}{12 \ \frac{\text{in}}{\text{ft}}}\right)}{1.41 \times 10^{-5} \ \frac{\text{ft}^2}{\text{sec}}}$$

$$= 7.1 \times 10^5$$

The friction values obtained from the Moody diagram are close to the initial estimates. The flow in branch ADC is

$$Q_{ADC} = \left(\frac{\pi D_{AD}^2}{4}\right) v_{AD}$$

$$= \left(\frac{\pi \left(\frac{12 \text{ in}}{12 \ \frac{\text{in}}{\text{ft}}}\right)^2}{4}\right)\left(6.6 \ \frac{\text{ft}}{\text{sec}}\right)$$

$$\times \left(7.4805 \ \frac{\text{gal}}{\text{ft}^3}\right)\left(60 \ \frac{\text{sec}}{\text{min}}\right)$$

$$= 2327 \text{ gal/min}$$

The total flow in the hydrant is the sum of both components.

$$Q = Q_{ABC} + Q_{ADC}$$

$$= 1053 \ \frac{\text{gal}}{\text{min}} + 2327 \ \frac{\text{gal}}{\text{min}}$$

$$= \boxed{3380 \text{ gal/min}}$$

SOLUTION 3

3.1. Neglecting the velocity heads because of the pipe lengths involved, the energy equation between the tank and point D is

$$1160 \text{ ft} + 122 \text{ ft} = 890 \text{ ft} + \frac{p_D}{\gamma} + 3.022\left(\frac{v_{AB}^{1.85} L_{AB}}{C_{AB}^{1.85} D_{AB}^{1.167}}\right)$$

$$+ 3.022\left(\frac{v_{BC}^{1.85} L_{BC}}{C_{BC}^{1.85} D_{BC}^{1.167}}\right)$$

$$+ 3.022\left(\frac{v_{CD}^{1.85} L_{CD}}{C_{CD}^{1.85} D_{CD}^{1.167}}\right)$$

The continuity equation for each pipe produces the following flows and velocities.

$$Q_{CD} = \dfrac{70 \ \dfrac{\text{gal}}{\text{min}}}{\left(7.4805 \ \dfrac{\text{gal}}{\text{ft}^3}\right)\left(60 \ \dfrac{\text{sec}}{\text{min}}\right)} = 0.156 \ \text{ft}^3/\text{sec}$$

$$v_{CD} = \dfrac{Q_{CD}}{\dfrac{\pi D_{CD}^2}{4}} = \dfrac{0.156 \ \dfrac{\text{ft}^3}{\text{sec}}}{\dfrac{\pi(0.3333 \ \text{ft})^2}{4}} = 1.8 \ \text{ft}/\text{sec}$$

$$Q_{BC} = 130 \ \dfrac{\text{gal}}{\text{min}} + Q_{CD} = \dfrac{130 \ \dfrac{\text{gal}}{\text{min}} + 70 \ \dfrac{\text{gal}}{\text{min}}}{\left(7.4805 \ \dfrac{\text{gal}}{\text{ft}^3}\right)\left(60 \ \dfrac{\text{sec}}{\text{min}}\right)}$$

$$= 0.446 \ \text{ft}^3/\text{sec}$$

$$v_{BC} = \dfrac{Q_{BC}}{\dfrac{\pi D_{BC}^2}{4}} = \dfrac{0.446 \ \dfrac{\text{ft}^3}{\text{sec}}}{\dfrac{\pi(0.50 \ \text{ft})^2}{4}} = 2.3 \ \text{ft}/\text{sec}$$

$$Q_{AB} = 140 \ \dfrac{\text{gal}}{\text{min}} + Q_{BC} = \dfrac{140 \ \dfrac{\text{gal}}{\text{min}} + 200 \ \dfrac{\text{gal}}{\text{min}}}{\left(7.4805 \ \dfrac{\text{gal}}{\text{ft}^3}\right)\left(60 \ \dfrac{\text{sec}}{\text{min}}\right)}$$

$$= 0.758 \ \text{ft}^3/\text{sec}$$

$$v_{AB} = \dfrac{Q_{AB}}{\dfrac{\pi D_{AB}^2}{4}} = \dfrac{0.758 \ \dfrac{\text{ft}^3}{\text{sec}}}{\dfrac{\pi(0.5 \ \text{ft})^2}{4}} = 3.9 \ \text{ft}/\text{sec}$$

Replacing the known values in the energy equation yields

$$392 \ \text{ft} = \dfrac{p_D}{\gamma} + \left(\dfrac{3.022}{(150)^{1.85}}\right)\left(\dfrac{\left(3.9 \ \dfrac{\text{ft}}{\text{sec}}\right)^{1.85}(16{,}000 \ \text{ft})}{(0.5 \ \text{ft})^{1.167}}\right.$$

$$+ \dfrac{\left(2.3 \ \dfrac{\text{ft}}{\text{sec}}\right)^{1.85}(14{,}000 \ \text{ft})}{(0.5 \ \text{ft})^{1.167}}$$

$$\left. + \dfrac{\left(1.8 \ \dfrac{\text{ft}}{\text{sec}}\right)^{1.85}(8000 \ \text{ft})}{(0.3333 \ \text{ft})^{1.167}}\right)$$

Solving for the pressure head,

$$\dfrac{p_D}{\gamma} = 199 \ \text{ft}$$

The pressure is

$$p_D = \dfrac{(199 \ \text{ft})\left(62.4 \ \dfrac{\text{lbf}}{\text{ft}^3}\right)}{\left(12 \ \dfrac{\text{in}}{\text{ft}}\right)^2}$$

$$= \boxed{86 \ \text{psig}}$$

3.2. If this pressure is insufficient for the demand at point D, most likely the pipes do not have enough capacity or the water level in the tank is too low.

3.3. Assuming that the peak demand is 1.5 times the average, the flows and velocities through the pipe are

$$Q_{CD} = (1.5)\left(0.156 \ \dfrac{\text{ft}^3}{\text{sec}}\right) = 0.234 \ \text{ft}^3/\text{sec}$$

$$v_{CD} = (1.5)\left(1.8 \ \dfrac{\text{ft}}{\text{sec}}\right) = 2.7 \ \text{ft}/\text{sec}$$

$$Q_{BC} = (1.5)\left(0.446 \ \dfrac{\text{ft}^3}{\text{sec}}\right) = 0.669 \ \text{ft}^3/\text{sec}$$

$$v_{BC} = (1.5)\left(2.3 \ \dfrac{\text{ft}}{\text{sec}}\right) = 3.4 \ \text{ft}/\text{sec}$$

$$Q_{AB} = (1.5)\left(0.758 \ \dfrac{\text{ft}^3}{\text{sec}}\right) = 1.137 \ \text{ft}^3/\text{sec}$$

$$v_{AB} = (1.5)\left(3.9 \ \dfrac{\text{ft}}{\text{sec}}\right) = 5.8 \ \text{ft}/\text{sec}$$

With these new values, the energy equation is

$$392 \ \text{ft} = \dfrac{p_D}{\gamma} + \left(\dfrac{3.022}{(150)^{1.85}}\right)\left(\dfrac{\left(5.8 \ \dfrac{\text{ft}}{\text{sec}}\right)^{1.85}(16{,}000 \ \text{ft})}{(0.5 \ \text{ft})^{1.167}}\right.$$

$$+ \dfrac{\left(3.4 \ \dfrac{\text{ft}}{\text{sec}}\right)^{1.85}(14{,}000 \ \text{ft})}{(0.5 \ \text{ft})^{1.167}}$$

$$\left. + \dfrac{\left(2.7 \ \dfrac{\text{ft}}{\text{sec}}\right)^{1.85}(8000 \ \text{ft})}{(0.3333 \ \text{ft})^{1.167}}\right)$$

The new pressure head is

$$\dfrac{p_D}{\gamma} = -10.1 \ \text{ft}$$

The pressure at D is

$$p_D = \frac{(-10.1 \text{ ft})\left(62.4 \ \frac{\text{lbf}}{\text{ft}^3}\right)}{\left(12 \ \frac{\text{in}}{\text{ft}}\right)^2}$$

$$= \boxed{-4.4 \text{ psig}}$$

The negative pressure indicates a problem at this flow volume.

3.4. For this configuration, it appears that the pipes are too small for the required pressure. Nevertheless, changing the existing pipeline would be costly. Possible alternatives include increasing the water level in the tank, pressurizing the tank, adding a pump somewhere in the pipeline, or replacing only a section of the pipeline. The solution chosen must be adopted based on cost.

SOLUTION 4

Assume the flow directions shown.

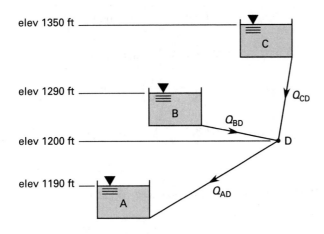

4.1. Disregarding velocity head, the energy equations between the reservoirs and point D are

CD: $\quad 1350 \text{ ft} = 1200 \text{ ft} + \dfrac{p_D}{\gamma} + h_{f,\text{CD}}$

BD: $\quad 1290 \text{ ft} = 1200 \text{ ft} + \dfrac{p_D}{\gamma} + h_{f,\text{BD}}$

AD: $\quad 1190 \text{ ft} = 1200 \text{ ft} + \dfrac{p_D}{\gamma} - h_{f,\text{AD}}$

The velocity heads and the local losses have been neglected. For this set of units (feet and ft³/sec), the friction losses are calculated according to the Hazen-Williams equation as

$$h_f = 4.73 \left(\frac{Q^{1.85} L}{C^{1.85} D^{4.867}} \right)$$

The friction losses are

$$h_{f,\text{CD}} = (4.73)\left(\frac{Q_{\text{CD}}^{1.85}(10{,}000 \text{ ft})}{(150)^{1.85}(0.8333 \text{ ft})^{4.867}}\right) = 10.82 Q_{\text{CD}}^{1.85}$$

$$h_{f,\text{BD}} = (4.73)\left(\frac{Q_{\text{BD}}^{1.85}(4000 \text{ ft})}{(150)^{1.85}(0.6667 \text{ ft})^{4.867}}\right) = 12.82 Q_{\text{BD}}^{1.85}$$

$$h_{f,\text{AD}} = (4.73)\left(\frac{Q_{\text{AD}}^{1.85}(5000 \text{ ft})}{(150)^{1.85}(1 \text{ ft})^{4.867}}\right) = 2.23 Q_{\text{AD}}^{1.85}$$

From the preceding equations, the flows are

$$Q_{\text{CD}} = 0.276 h_{f,\text{CD}}^{0.54}$$

$$Q_{\text{BD}} = 0.252 h_{f,\text{BD}}^{0.54}$$

$$Q_{\text{AD}} = 0.649 h_{f,\text{AD}}^{0.54}$$

Rewrite the energy equations as

$$h_{f,\text{CD}} = 150 \text{ ft} - \frac{p_D}{\gamma}$$

$$h_{f,\text{BD}} = 90 \text{ ft} - \frac{p_D}{\gamma}$$

$$h_{f,\text{AD}} = 10 \text{ ft} + \frac{p_D}{\gamma}$$

Rewrite the continuity equation at point D as

$$Q_{\text{CD}} + Q_{\text{BD}} - Q_{\text{AD}} = 0$$

This produces a system of seven unknowns: three flows, three head losses, and one pressure head at D. These equations are nonlinear and can be best solved iteratively using the following table.

$\dfrac{p_D}{\gamma}$ (assumed) (ft)	$h_{f,\text{CD}}$ (ft)	Q_{CD} (ft³/sec)	$h_{f,\text{BD}}$ (ft)
60.0	90.0	3.13	30.0
50.0	100.0	3.22	40.0
40.0	110.0	3.49	50.0
42.2	107.8	3.46	47.8

Q_{BD} (ft³/sec)	$h_{f,\text{AD}}$ (ft)	Q_{AD} (ft³/sec)	$Q_{\text{CD}} + Q_{\text{BD}} - Q_{\text{AD}}$ (ft³/sec)
1.58	70.0	6.43	−1.73
1.85	60.0	5.92	−0.76
2.08	50.0	5.37	0.20
2.03	52.2	5.49	−0.003

In the table, a value of p_D/γ is assumed. Inserting this value in the energy equations yields the head losses in each pipe. The flows are calculated from the energy loss expressions. Finally, the flows are checked for continuity. The last line in the table uses an estimate of p_D/γ interpolated from the previous two; the continuity closure error is very small, so the choice is correct and the flows are

$$Q_{CD} = \left(3.46 \ \frac{\text{ft}^3}{\text{sec}}\right)\left(7.4805 \ \frac{\text{gal}}{\text{ft}^3}\right)\left(60 \ \frac{\text{sec}}{\text{min}}\right)$$

$$= \boxed{1553 \ \text{gal/min into D}}$$

$$Q_{BD} = \left(2.03 \ \frac{\text{ft}^3}{\text{sec}}\right)\left(7.4805 \ \frac{\text{gal}}{\text{ft}^3}\right)\left(60 \ \frac{\text{sec}}{\text{min}}\right)$$

$$= \boxed{911 \ \text{gal/min into D}}$$

$$Q_{AD} = \left(5.49 \ \frac{\text{ft}^3}{\text{sec}}\right)\left(7.4805 \ \frac{\text{gal}}{\text{ft}^3}\right)\left(60 \ \frac{\text{sec}}{\text{min}}\right)$$

$$= \boxed{2464 \ \text{gal/min into D}}$$

4.2. The table includes the following head losses.

$$h_{f,CD} = \boxed{107.8 \ \text{ft}}$$

$$h_{f,BD} = \boxed{47.8 \ \text{ft}}$$

$$h_{f,AD} = \boxed{52.2 \ \text{ft}}$$

SOLUTION 5

The Hardy Cross method is used to solve the flow loop. The friction losses are given by

$$h_f = K Q^{1.85}$$

$$K = 10.6 \left(\frac{L}{C^{1.85} D^{4.867}}\right)$$

Q is in MGD, and L and D are in feet. Using the data provided produces the K values shown in the following computational tables.

The initial estimates for the directions and magnitudes of the flows are

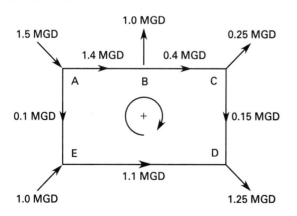

The method converges after two iterations as shown in the tables.

For iteration 1,

pipe	K	Q	h_f	$\dfrac{h_f}{Q}$	corrected Q
AB	6.53	1.40	12.17	8.70	1.26
BC	8.17	0.40	1.50	3.75	0.26
CD	52.86	0.15	1.58	10.54	0.01
DE	4.31	−1.10	−5.14	4.67	−1.24
EA	61.67	−0.10	−0.87	8.71	−0.24
			9.25	36.36	

$$\text{correction} = \delta = \frac{-9.25}{(1.85)(36.36)} = -0.14$$

For iteration 2,

pipe	K	Q	h_f	$\dfrac{h_f}{Q}$	corrected Q
AB	6.53	1.26	10.06	7.96	1.26
BC	8.17	0.26	0.69	2.62	0.26
CD	52.86	0.01	0.02	1.28	0.01
DE	4.31	−1.24	−6.39	5.16	−1.24
EA	61.67	−0.24	−4.31	18.17	−0.24
			0.07	35.19	

$$\text{correction} = \delta = \frac{-0.07}{(1.85)(35.19)} = 0.00$$

The last table contains the solution to the network.

5.1. $Q_{AB} = \boxed{1.26 \ \text{MGD}}$

5.2. Water flows from A to B.

5.3. The pressure drop is equal to the friction losses.

$$\frac{\Delta p_{AB}}{\gamma} = K_{AB} Q_{AB}^{1.85}$$

$$= (6.53)(1.26 \ \text{MGD})^{1.85}$$

$$= 10.01 \ \text{ft}$$

$$\Delta p_{AB} = \frac{(10.01 \ \text{ft})\left(62.4 \ \frac{\text{lbf}}{\text{ft}^3}\right)}{\left(12 \ \frac{\text{in}}{\text{ft}}\right)^2}$$

$$= \boxed{4.34 \ \text{psig}}$$

5.4. $Q_{BC} = \boxed{0.26 \ \text{MGD}}$

5.5. $Q_{AE} = \boxed{0.24 \ \text{MGD}}$

5.6. $Q_{ED} = \boxed{1.24 \ \text{MGD}}$

5.7. $Q_{CD} = \boxed{0.01 \ \text{MGD}}$

5.8. The pressure drop is equal to the friction losses.

$$\frac{\Delta p_{CD}}{\gamma} = K_{CD}\, Q_{CD}^{1.85}$$

$$= (52.86)(0.01 \text{ MGD})^{1.85}$$

$$= 0.0105 \text{ ft}$$

$$\Delta p_{CD} = \frac{(0.0105 \text{ ft})\left(62.4\ \dfrac{\text{lbf}}{\text{ft}^3}\right)}{\left(12\ \dfrac{\text{in}}{\text{ft}}\right)^2}$$

$$= \boxed{0.0046 \text{ psig}}$$

5.9. Water flows from C to D.

5.10. The speed of wave propagation, assuming a rigid conduit (so that pipe compressibility does not affect the wave velocity), is

$$c = \sqrt{\frac{E}{\rho}} = \sqrt{\frac{Eg}{\gamma}}$$

E is the bulk modulus of elasticity for the fluid, and γ is the specific weight. For water at 50°F,

$$E = 305{,}000 \text{ lbf/in}^2$$

$$\gamma = 62.4 \text{ lbf/ft}^3$$

The wave speed is

$$c = \sqrt{\frac{\left(305{,}000\ \dfrac{\text{lbf}}{\text{in}^2}\right)\left(12\ \dfrac{\text{in}}{\text{ft}}\right)^2\left(32.2\ \dfrac{\text{ft}}{\text{sec}^2}\right)}{62.4\ \dfrac{\text{lbf}}{\text{ft}^3}}}$$

$$= 4761 \text{ ft/sec}$$

The pressure surge can be minimized with a very slow closure of the valve. If t_c is the closure time, this objective can be achieved by ensuring that the valve closure time is at least 20 times the return time.

$$t_c > \frac{20L}{c}$$

L is the pipe length.

$$t_c > \frac{(20)(2000 \text{ ft})}{4761\ \dfrac{\text{ft}}{\text{sec}}} = \boxed{8.4 \text{ sec}}$$

SOLUTION 6

6.1. Call Q_1 the flow through branch AD, and call Q_2 the flow through ABCD. The energy equations in branches AD and ABCD are as follows.

In AD,

$$\frac{p_A}{\gamma} = \frac{p_D}{\gamma} + f_{AD}\left(\frac{L_{AD}}{D_{AD}}\right)$$

$$\times\left(\frac{Q_1^2}{(2)\left(32.2\ \dfrac{\text{ft}}{\text{sec}^2}\right)\left(\dfrac{\pi D_{AD}^2}{4}\right)^2}\right)$$

In ABCD,

$$\frac{p_A}{\gamma} = \frac{p_D}{\gamma} + f_{AC}\left(\frac{L_{AC}}{D_{AC}}\right)$$

$$\times\left(\frac{Q_2^2}{(2)\left(32.2\ \dfrac{\text{ft}}{\text{sec}^2}\right)\left(\dfrac{\pi D_{AC}^2}{4}\right)^2}\right)$$

$$+ f_{CD}\left(\frac{L_{CD}}{D_{CD}}\right)\left(\frac{Q_2^2}{(2)\left(32.2\ \dfrac{\text{ft}}{\text{sec}^2}\right)\left(\dfrac{\pi D_{CD}^2}{4}\right)^2}\right)$$

For 5 yr old cast iron pipes, the specific roughness is $\epsilon = 0.001$ ft. The relative roughnesses are

$$\frac{\epsilon}{D_{AD}} = \frac{0.001 \text{ ft}}{0.8333 \text{ ft}} = 0.0012$$

$$\frac{\epsilon}{D_{AC}} = \frac{0.001 \text{ ft}}{0.6667 \text{ ft}} = 0.0015$$

$$\frac{\epsilon}{D_{CD}} = \frac{0.001 \text{ ft}}{1 \text{ ft}} = 0.001$$

Using the Moody diagram, initially assume that $f_{AD} = f_{DC} = 0.021$ and $f_{AC} = 0.023$. Given that $p_A - p_D = 10 \text{ lbf/in}^2 = 1440 \text{ lbf/ft}^2$, the energy equations become

In AD,

$$\frac{1440\ \dfrac{\text{lbf}}{\text{ft}^2}}{62.4\ \dfrac{\text{lbf}}{\text{ft}^3}} = (0.021)\left(\frac{2000 \text{ ft}}{0.8333 \text{ ft}}\right)$$

$$\times\left(\frac{Q_1^2}{(2)\left(32.2\ \dfrac{\text{ft}}{\text{sec}^2}\right)\left(\dfrac{\pi(0.8333 \text{ ft})^2}{4}\right)^2}\right)$$

In ABCD,

$$\frac{1440 \ \frac{\text{lbf}}{\text{ft}^2}}{62.4 \ \frac{\text{lbf}}{\text{ft}^3}} = \left(\frac{Q_2^2}{(2)\left(32.2 \ \frac{\text{ft}}{\text{sec}^2}\right)} \right)$$

$$\times \left((0.023)\left(\frac{2100 \ \text{ft}}{0.6667 \ \text{ft}}\right)\left(\frac{1}{\left(\frac{\pi(0.6667 \ \text{ft})^2}{4}\right)^2}\right) \right.$$

$$\left. + (0.021)\left(\frac{900 \ \text{ft}}{1 \ \text{ft}}\right)\left(\frac{1}{\left(\frac{\pi(1 \ \text{ft})^2}{4}\right)^2}\right) \right)$$

The resulting system of equations is

$$23.08 = 2.631 Q_1^2$$
$$23.08 = 9.709 Q_2^2$$

Solving for Q_1 and Q_2 yields

$$Q_1 = 2.96 \ \text{ft}^3/\text{sec}$$
$$Q_2 = 1.54 \ \text{ft}^3/\text{sec}$$

The velocities in the pipes are

$$v_{AD} = \frac{Q_1}{\dfrac{\pi D_{AD}^2}{4}} = \frac{2.96 \ \frac{\text{ft}^3}{\text{sec}}}{\dfrac{\pi(0.8333 \ \text{ft})^2}{4}}$$
$$= 5.4 \ \text{ft/sec}$$

$$v_{AC} = \frac{Q_2}{\dfrac{\pi D_{AC}^2}{4}} = \frac{1.54 \ \frac{\text{ft}^3}{\text{sec}}}{\dfrac{\pi(0.6667 \ \text{ft})^2}{4}}$$
$$= 4.4 \ \text{ft/sec}$$

$$v_{CD} = \frac{Q_3}{\dfrac{\pi D_{CD}^2}{4}} = \frac{1.54 \ \frac{\text{ft}^3}{\text{sec}}}{\dfrac{\pi(1 \ \text{ft})^2}{4}}$$
$$= 2.0 \ \text{ft/sec}$$

The Reynolds numbers are

$$\text{Re}_{AD} = \frac{v_{AD} D_{AD}}{\nu} = \frac{\left(5.4 \ \frac{\text{ft}}{\text{sec}}\right)(0.8333 \ \text{ft})}{1.41 \times 10^{-5} \ \frac{\text{ft}^2}{\text{sec}}}$$
$$= 3.2 \times 10^5$$

$$\text{Re}_{AC} = \frac{v_{AC} D_{AC}}{\nu} = \frac{\left(4.4 \ \frac{\text{ft}}{\text{sec}}\right)(0.6667 \ \text{ft})}{1.41 \times 10^{-5} \ \frac{\text{ft}^2}{\text{sec}}}$$
$$= 2.1 \times 10^5$$

$$\text{Re}_{CD} = \frac{v_{CD} D_{CD}}{\nu} = \frac{\left(2.0 \ \frac{\text{ft}}{\text{sec}}\right)(1 \ \text{ft})}{1.41 \times 10^{-5} \ \frac{\text{ft}^2}{\text{sec}}}$$
$$= 1.4 \times 10^5$$

With these values, the revised friction factors from the Moody diagram are

$$f_{AD} = 0.021$$
$$f_{AC} = 0.023$$
$$f_{CD} = 0.0215$$

Inserting these new values in the energy equations and recalculating the flows produces no substantial change.

$$Q_1 = 2.96 \ \text{ft}^3/\text{sec}$$
$$Q_2 = 1.54 \ \text{ft}^3/\text{sec}$$

The flow through the pump is

$$Q_{\text{pump}} = Q_1 + Q_2$$
$$= \left(2.96 \ \frac{\text{ft}^3}{\text{sec}} + 1.54 \ \frac{\text{ft}^3}{\text{sec}}\right)\left(7.4805 \ \frac{\text{gal}}{\text{ft}^3}\right)\left(60 \ \frac{\text{sec}}{\text{min}}\right)$$
$$= \boxed{2020 \ \text{gal/min}}$$

6.2. The flow velocity in leg AD is

$$v_{AD} = \frac{Q_1}{\dfrac{\pi D_{AD}^2}{4}} = \frac{2.96 \ \frac{\text{ft}^3}{\text{sec}}}{\dfrac{\pi(0.8333 \ \text{ft})^2}{4}}$$
$$= \boxed{5.5 \ \text{ft/sec}}$$

6.3. From the problem statement,

$$p_A - p_D = \boxed{10 \ \text{psig}}$$

6.4. The pressure drop is the same regardless of the path, so

$$p_A - p_D = \boxed{10 \ \text{psig}}$$

6.5. Assuming that outlet D discharges to the atmosphere, the head provided by the pump is equal to the pressure drop.

$$h_{pump} = \frac{\left(10 \frac{lbf}{in^2}\right)\left(12 \frac{in}{ft}\right)^2}{62.4 \frac{lbf}{ft^3}}$$

$$= \boxed{23.1 \text{ ft}}$$

SOLUTION 7

7.1. The energy equation between the reservoirs is

$$12 \text{ ft} = f\left(\frac{L}{D}\right)\left(\frac{Q^2}{2g\left(\frac{\pi D^2}{4}\right)^2}\right)$$

$$+ 2K\left(\frac{Q^2}{2g\left(\frac{\pi D^2}{4}\right)^2}\right)$$

The last term represents the local losses caused by the two elbows. The loss coefficient for a single elbow of unknown size and bend radius is assumed to be approximately $K = 0.9$. (This may be 2–3 times higher than the actual value, but is chosen to be conservative.)

$$Q = \frac{6 \text{ MGD}}{0.64632 \frac{\text{MGD}}{\frac{ft^3}{sec}}} = 9.3 \text{ ft}^3/\text{sec}$$

For the maximum flow of $Q = 6$ MGD $= 9.3$ ft^3/sec and assuming initially that $f = 0.02$, the energy equation becomes

$$12 \text{ ft} = (0.02)\left(\frac{240 \text{ ft}}{D}\right)\left(\frac{\left(9.3 \frac{ft^3}{sec}\right)^2}{(2)\left(32.2 \frac{ft}{sec^2}\right)\left(\frac{\pi D^2}{4}\right)^2}\right)$$

$$+ (2)(0.9)\left(\frac{\left(9.3 \frac{ft^3}{sec^2}\right)^2}{(2)\left(32.2 \frac{ft}{sec^2}\right)\left(\frac{\pi D^2}{4}\right)^2}\right)$$

$$12 = \frac{10.45}{D^5} + \frac{3.90}{D^4}$$

Solve this nonlinear equation by trial and error for D.

$$D \approx (1.04 \text{ ft})\left(12 \frac{in}{ft}\right) = 12.5 \text{ in}$$

To check the friction factor, the following quantities are needed. Assume that the physical properties of the sludge are similar to those of water.

$$v = \frac{Q}{\frac{\pi D^2}{4}} = \frac{9.3 \frac{ft^3}{sec}}{\frac{\pi(1.04 \text{ ft})^2}{4}} = 10.9 \text{ ft/sec}$$

$$Re = \frac{vD}{\nu} = \frac{\left(10.9 \frac{ft}{sec}\right)(1.04 \text{ ft})}{1.54 \times 10^{-5} \frac{ft^2}{sec}} = 7.4 \times 10^5$$

The viscosity is selected for the worst case of $T = 45°F$. The specific roughness for ductile iron is $\epsilon = 0.0008$ ft. Therefore, the relative roughness is

$$\frac{\epsilon}{D} = \frac{0.0008 \text{ ft}}{1.04 \text{ ft}} = 0.0008$$

From the Moody diagram, $f = 0.019$, which is close to the value used.

Choose a standard-size pipe of

$$D = \boxed{14 \text{ in}}$$

7.2. The system must perform adequately under the worst possible conditions, that is, maximum flow and cold weather. A slightly oversized pipe would compensate for increased roughness due to the corrosion caused by the sludge.

2 Hydraulic Machines

PROBLEM 1

A tank supplies water to a small town during the day. At night, when the demand is low, a pump is used to refill the tank. New PVC pipe connects the pump and tank. All minor losses are insignificant. The water temperature is 50°F.

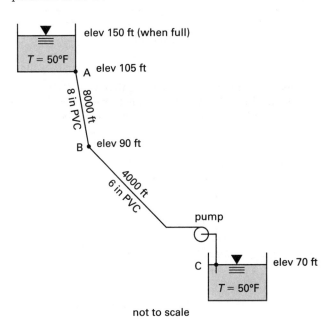

not to scale

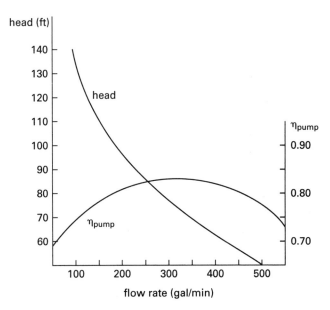

1.1. What is the minimum rate at which the tank will be refilled?

1.2. If a minimum pressure of 20 psig is needed at a hydrant located at point B, what is the maximum flow rate that can be drawn from the hydrant? Assume that the tank is full and the pump is not in use during fire fighting.

PROBLEM 2

The water surface elevations of two reservoirs differ by 350 ft. The reservoirs are connected by 2200 ft of 12 in steel schedule-40 pipe (Darcy friction factor, $f = 0.02$). The transfer rate (from the lower to the higher reservoir) is 9.5 ft³/sec. An 83% efficient centrifugal pump is located at the elevation of the lower reservoir drain. Two fully open gate valves provide for pump maintenance and replacement. The water temperature is 50°F. Disregard all other minor losses.

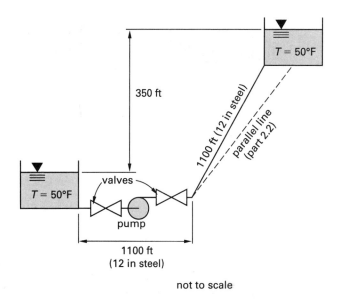

not to scale

2.1. What total motor horsepower is required to pump the water?

2.2. If the pump(s) and motor(s) are unchanged, what will be the flow rate if a second 12 in line is placed in parallel with the last half (1100 ft) of the original pipe?

SOLUTION 1

1.1. Neglecting the velocity head, and representing the head added by the pump as h_A, the energy equation is

$$70 \text{ ft} + h_A = 150 \text{ ft} + f_{AB}\left(\frac{L_{AB}}{D_{AB}}\right)\left(\frac{Q^2}{2g\left(\frac{\pi D_{AB}^2}{4}\right)^2}\right)$$

$$+ f_{BC}\left(\frac{L_{BC}}{D_{BC}}\right)\left(\frac{Q^2}{2g\left(\frac{\pi D_{BC}^2}{4}\right)^2}\right)$$

Assume initially that $f_{AB} = f_{BC} = 0.02$. The energy equation becomes

$$h_A = 80 \text{ ft} + \left(\!(0.02)\left(\frac{8000 \text{ ft}}{0.6667 \text{ ft}}\right)\left(\frac{1}{\left(\frac{\pi(0.6667 \text{ ft})^2}{4}\right)^2}\right)\right.$$

$$+ (0.02)\left(\frac{4000 \text{ ft}}{0.5 \text{ ft}}\right)\left(\frac{1}{\left(\frac{\pi(0.5 \text{ ft})^2}{4}\right)^2}\right)\!\!\bigg)$$

$$\times\left(\frac{Q^2}{(2)\left(32.2 \dfrac{\text{ft}}{\text{sec}^2}\right)}\right)$$

The equation of the system curve is

$$h_A = 80 + 95.02 Q^2 \quad [Q \text{ is in ft}^3/\text{sec}]$$

The operating point is the intersection of this curve with the pump curve, as shown in the following graph.

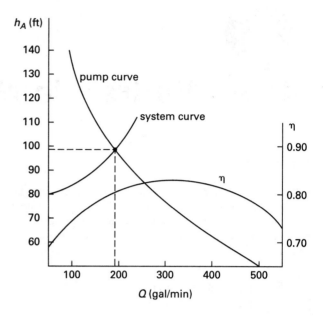

The operating point is at $Q = 190$ gal/min (0.42 ft^3/sec), $h_A = 97$ ft. In order to check the friction factors, the velocities and Reynolds numbers are calculated.

$$v_{AB} = \frac{Q}{\dfrac{\pi D_{AB}^2}{4}} = \frac{0.42 \dfrac{\text{ft}^3}{\text{sec}}}{\dfrac{\pi(0.6667 \text{ ft})^2}{4}}$$

$$= 1.2 \text{ ft/sec}$$

$$\text{Re}_{AB} = \frac{v_{AB} D_{AB}}{\nu} = \frac{\left(1.2 \dfrac{\text{ft}}{\text{sec}}\right)(0.6667 \text{ ft})}{1.41 \times 10^{-5} \dfrac{\text{ft}^2}{\text{sec}}}$$

$$= 5.7 \times 10^4$$

$$v_{BC} = \frac{Q}{\dfrac{\pi D_{BC}^2}{4}} = \frac{0.42 \dfrac{\text{ft}^3}{\text{sec}}}{\dfrac{\pi(0.5 \text{ ft})^2}{4}}$$

$$= 2.1 \text{ ft/sec}$$

$$\text{Re}_{BC} = \frac{v_{BC} D_{BC}}{\nu} = \frac{\left(2.1 \dfrac{\text{ft}}{\text{sec}}\right)(0.5 \text{ ft})}{1.41 \times 10^{-5} \dfrac{\text{ft}^2}{\text{sec}}}$$

$$= 7.4 \times 10^4$$

For PVC pipe, the specific roughness is $\epsilon = 5 \times 10^{-6}$ ft, so the relative roughness is

$$\frac{\epsilon}{D_{AB}} = \frac{5 \times 10^{-6} \text{ ft}}{0.6667 \text{ ft}} = 0.0000075$$

$$\frac{\epsilon}{D_{BC}} = \frac{5 \times 10^{-6} \text{ ft}}{0.5 \text{ ft}} = 0.00001$$

From the Moody diagram, $f_{AB} = 0.020$ and $f_{BC} = 0.019$, which are close to the initial estimate. The flow through the system is

$$Q = \boxed{190 \text{ gal/min}}$$

1.2. Neglecting the velocity heads, the energy equation between the upper tank and point B is

$$150 \text{ ft} = 90 \text{ ft} + \frac{p_B}{\gamma} + f_{AB}\left(\frac{L_{AB}}{D_{AB}}\right)\left(\frac{Q^2}{2g\left(\frac{\pi D_{AB}^2}{4}\right)^2}\right)$$

Assume initially that $f_{AB} = 0.018$. This equation becomes

$$60 \text{ ft} = \frac{\left(20 \frac{\text{lbf}}{\text{in}^2}\right)\left(12 \frac{\text{in}}{\text{ft}}\right)^2}{62.4 \frac{\text{lbf}}{\text{ft}^3}} + (0.018)\left(\frac{8000 \text{ ft}}{0.6667 \text{ ft}}\right)$$

$$\times \left(\frac{Q^2}{(2)\left(32.2 \frac{\text{ft}}{\text{sec}^2}\right)\left(\frac{\pi(0.6667 \text{ ft})^2}{4}\right)^2}\right)$$

Solving for Q yields

$$Q = 0.71 \text{ ft}^3/\text{sec}$$

The velocity is

$$v_{AB} = \frac{Q}{\frac{\pi D_{AB}^2}{4}} = \frac{0.71 \frac{\text{ft}^3}{\text{sec}}}{\frac{\pi(0.6667 \text{ ft})^2}{4}} = 2.0 \text{ ft/sec}$$

The Reynolds number is

$$Re_{AB} = \frac{v_{AB} D_{AB}}{\nu} = \frac{\left(2.0 \frac{\text{ft}}{\text{sec}}\right)(0.6667 \text{ ft})}{1.41 \times 10^{-5} \frac{\text{ft}^2}{\text{sec}}}$$

$$= 9.5 \times 10^4$$

With the value ϵ/D_{AB} calculated in Sol. 1.1, the Moody diagram indicates $f_{AB} = 0.018$, which is the value used. The flow is

$$Q = \left(0.71 \frac{\text{ft}^3}{\text{sec}}\right)\left(7.4805 \frac{\text{gal}}{\text{ft}^3}\right)\left(60 \frac{\text{sec}}{\text{min}}\right)$$

$$= \boxed{319 \text{ gal/min}}$$

SOLUTION 2

2.1. The energy equation between the two reservoirs is

$$h_A = 350 \text{ ft} + f\left(\frac{L}{D}\right)\left(\frac{v^2}{2g}\right) + 2K\left(\frac{v^2}{2g}\right)$$

The local loss coefficient for one fully open gate valve is $K = 0.2$. The velocity is

$$v = \frac{Q}{\frac{\pi D^2}{4}} = \frac{9.5 \frac{\text{ft}^3}{\text{sec}}}{\frac{\pi(1 \text{ ft})^2}{4}}$$

$$= 12.1 \text{ ft/sec}$$

$$h_A = 350 \text{ ft} + (0.02)\left(\frac{2200 \text{ ft}}{1 \text{ ft}}\right)$$

$$\times \left(\frac{\left(12.1 \frac{\text{ft}}{\text{sec}}\right)^2}{(2)\left(32.2 \frac{\text{ft}}{\text{sec}^2}\right)}\right)$$

$$+ (2)(0.2)\left(\frac{\left(12.1 \frac{\text{ft}}{\text{sec}}\right)^2}{(2)\left(32.2 \frac{\text{ft}}{\text{sec}^2}\right)}\right)$$

$$= 450.9 \text{ ft}$$

The motor brake horsepower required by the pump is

$$\text{bhp} = \frac{\gamma h_A Q}{550\eta}$$

$$= \frac{\left(62.4 \frac{\text{lbf}}{\text{ft}^3}\right)(450.9 \text{ ft})\left(9.5 \frac{\text{ft}^3}{\text{sec}}\right)}{\left(550 \frac{\text{ft-lbf}}{\text{hp-sec}}\right)(0.83)}$$

$$= 586 \text{ hp}$$

The power requirement can be met with a set of three pumps driven by three 200 hp motors.

2.2. Assume that the power is 600 hp, then

$$600 \text{ hp} = \frac{\left(62.4 \ \frac{\text{lbf}}{\text{ft}^3}\right) h_A Q}{(550)(0.83)}$$

$$h_A = \frac{4389.4}{Q}$$

Since the additional pipe is identical to the existing one, the flow through it will be half of the total, and the losses through each branch are the same. The energy equation for the new system is

$$h_A = 350 \text{ ft} + (0.02)\left(\frac{1100 \text{ ft}}{1 \text{ ft}}\right)$$

$$\times \left(\frac{Q^2}{(2)\left(32.2 \ \frac{\text{ft}}{\text{sec}^2}\right)\left(\frac{\pi(1 \text{ ft})^2}{4}\right)^2}\right)$$

$$+ (0.02)\left(\frac{1100 \text{ ft}}{1 \text{ ft}}\right)$$

$$\times \left(\frac{\left(\frac{Q}{2}\right)^2}{(2)\left(32.2 \ \frac{\text{ft}}{\text{sec}^2}\right)\left(\frac{\pi(1 \text{ ft})^2}{4}\right)^2}\right)$$

$$+ (2)(0.2)$$

$$\times \left(\frac{Q^2}{(2)\left(32.2 \ \frac{\text{ft}}{\text{sec}^2}\right)\left(\frac{\pi(1 \text{ ft})^2}{4}\right)^2}\right)$$

$$= 350 + 0.702 Q^2$$

Substituting the expression for the power produces

$$\frac{4389.4}{Q} = 350 + 0.702 Q^2$$

Solving for the flow by trial and error (or by other means) yields a value of 10.3 ft³/sec.

$$Q = \left(10.3 \ \frac{\text{ft}^3}{\text{sec}}\right)\left(7.4805 \ \frac{\text{gal}}{\text{ft}^3}\right)\left(60 \ \frac{\text{sec}}{\text{min}}\right)$$

$$= \boxed{4623 \text{ gal/min}}$$

3 Open Channel Flow

PROBLEM 1

The cross section of a very long channel with an average geometric slope of 0.0004 ft/ft is shown. At a particular location, the elevations of various points across the channel and flood plain are: A, 910.0 ft; B, 905.0 ft; C, 900.0 ft; D, 909 ft; E, 910.0 ft.

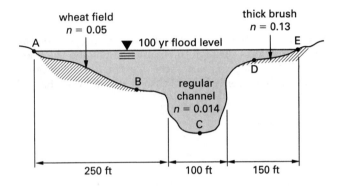

1.1. Find the discharge (in cubic feet per second) from a 100 yr flood.

1.2. Find the average velocity from a 100 yr flood.

PROBLEM 2

A trapezoidal channel (slope = 0.006 ft/ft; Manning's $n = 0.014$) has a cross section with a 12 ft wide base and 1:2 (vertical to horizontal) sides. The flow rate is 2300 ft³/sec. At a particular point, the channel meets (i.e., joins) a 10 ft deep stream. The invert elevations of the channel and stream are the same.

2.1. What is the normal depth of the channel flow?

2.2. What is the critical depth of the channel flow?

2.3. Is the flow subcritical or supercritical?

2.4. What type of control does the stream represent to the channel?

PROBLEM 3

The following questions are not necessarily related.

3.1. What is the most efficient cross section?

3.2. What is the geometry of a most efficient trapezoidal channel? Draw a figure and label all sides and angles.

3.3. What is the meaning of the term *upstream control?*

3.4. What is the meaning of the term *downstream control?*

PROBLEM 4

An existing 22 in diameter PVC pipe (Manning's $n = 0.01$) is installed on a 2% slope. The flow rate that needs to be carried between the two points separated by 1000 ft is 35 ft³/sec. However, the pipe does not appear to have such a capacity.

4.1. What is the capacity of the 22 in pipe?

4.2. What diameter pipe should be installed parallel to the first pipe to carry the excess?

4.3. What will be the diameter if the second pipe is installed over a different route with an average slope of 0.025?

4.4. If the second pipe starts at elevation 4257 ft, what will be the elevation of the second pipe's invert at the end of the run? Assume the two pipes have the same length.

PROBLEM 5

A section of an existing natural channel (Manning's $n = 0.030$) drops 470 ft as it meanders 8.4 mi through an area scheduled to be developed commercially. The existing channel is basically rectangular in shape, is 25 ft wide, and flows (at maximum) with a depth of 12 ft. The developers want to install an artificial channel 3.7 mi shorter over a more direct route, starting and ending at the same points as the natural channel. The artificial channel will be trapezoidal in shape with a base of 20 ft and sides inclined at 1:1, and it will be earth lined.

5.1. Will the new channel have the capacity to replace the existing channel?

5.2. Will erosion be more or less of a problem with the new channel than with the old channel? Why?

PROBLEM 6

A contracted measurement weir is to be installed in a 22 ft wide channel. The depth of the weir opening is 16 in, and the bottom of the weir opening is at elevation 479 ft. The bottom of the channel is at elevation 470 ft. For proper measurement and metering, a minimum of 6 in of hydraulic head over the weir lip is required. The average flow is 6 MGD but varies from a minimum of 3 MGD to a maximum of 12 MGD.

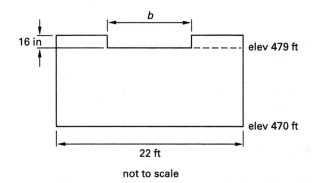

not to scale

6.1. What should be the length, b, of the weir opening?

6.2. What is the hydraulic head over the weir lip at the maximum flow?

6.3. How far back from the weir would you place a gauge station in order to measure the true elevation of the water surface in the channel? Explain the reason for your answer.

SOLUTION 1

Subdivide the channel into three sections as shown.

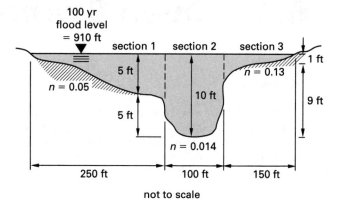

not to scale

1.1. The total flow is assumed to be the sum of the individual flows through the main channel and the flood plains. The slope is the same for the three sections. Uniform flow has been established due to the channel length.

For section 1,

$$A_1 = wd \approx (250 \text{ ft})(5 \text{ ft}) = 1250 \text{ ft}^2$$

Since this section can be considered to be a wide channel, the hydraulic radius is

$$R_1 \approx 5 \text{ ft}$$

Manning's formula yields

$$Q_1 = \left(\frac{1.49}{n_1}\right) A_1 R_1^{2/3} \sqrt{S}$$
$$= \left(\frac{1.49}{0.05}\right)(1250 \text{ ft}^2)(5 \text{ ft})^{2/3}\sqrt{0.0004}$$
$$= 2178 \text{ ft}^3/\text{sec}$$

For section 2,

$$A_2 = wd \approx (100 \text{ ft})(10 \text{ ft}) = 1000 \text{ ft}^2$$

From the illustration, the wetted perimeter is

$$P_2 = 5 \text{ ft} + 100 \text{ ft} + 9 \text{ ft} = 114 \text{ ft}$$

The hydraulic radius is

$$R_2 = \frac{A_2}{P_2} = \frac{1000 \text{ ft}^2}{114 \text{ ft}}$$
$$= 8.8 \text{ ft}$$

Manning's formula for this section is

$$Q_2 = \left(\frac{1.49}{n_2}\right) A_2 R_2^{2/3} \sqrt{S}$$

$$= \left(\frac{1.49}{0.014}\right)(1000 \text{ ft}^2)(8.8 \text{ ft})^{2/3}\sqrt{0.0004}$$

$$= 9073 \text{ ft}^3/\text{sec}$$

For section 3,

$$A_3 = wd \approx (150 \text{ ft})(1 \text{ ft}) = 150 \text{ ft}^2$$

Like section 1, this section can also be considered wide.

$$R_3 \approx 1 \text{ ft}$$

The flow is given by Manning's formula.

$$Q_3 = \left(\frac{1.49}{n_3}\right) A_3 R_3^{2/3} \sqrt{S}$$

$$= \left(\frac{1.49}{0.13}\right)(150 \text{ ft}^2)(1 \text{ ft})^{2/3}\sqrt{0.0004}$$

$$= 34.4 \text{ ft}^3/\text{sec}$$

The total flow is

$$Q = Q_1 + Q_2 + Q_3$$

$$= 2178 \frac{\text{ft}^3}{\text{sec}} + 9073 \frac{\text{ft}^3}{\text{sec}} + 34.4 \frac{\text{ft}^3}{\text{sec}}$$

$$= \boxed{11{,}285 \text{ ft}^3/\text{sec}}$$

1.2. The average velocity is

$$v = \frac{Q}{A} = \frac{11{,}285 \frac{\text{ft}^3}{\text{sec}}}{1250 \text{ ft}^2 + 1000 \text{ ft}^2 + 150 \text{ ft}^2}$$

$$= \boxed{4.7 \text{ ft/sec}}$$

SOLUTION 2

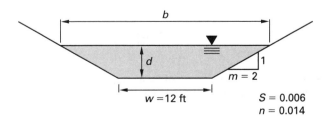

Use the following formulas for geometric properties of a trapezoidal channel. (m is the horizontal-to-vertical side slope.)

$$\text{area } A = d(w + md)$$

$$\text{wetted perimeter } P = w + 2d\sqrt{1 + m^2}$$

$$\text{top width } b = w + 2md$$

$$\text{hydraulic radius } R = \frac{A}{P}$$

2.1. Manning's formula holds for uniform flow—that is, when d equals the normal depth, d_o.

$$Q = \left(\frac{1.49}{n}\right) A R^{2/3}\sqrt{S}$$

$$m = \frac{2}{1} = 2$$

Inserting the known values yields

$$2300 \frac{\text{ft}^3}{\text{sec}} = \left(\frac{1.49}{0.014}\right)d_o(12 \text{ ft} + 2d_o)$$

$$\times \left(\frac{d_o(12 \text{ ft} + 2d_o)}{12 \text{ ft} + 2\sqrt{5}d_o}\right)^{2/3}\sqrt{0.006}$$

Solving by trial and error for d_o yields

$$d_o = \boxed{5.4 \text{ ft}}$$

2.2. The condition for critical flow is

$$\frac{Q^2}{g} = \frac{A^3}{b}$$

Replacing the known quantities yields

$$\frac{\left(2300 \frac{\text{ft}^3}{\text{sec}}\right)^2}{32.2 \frac{\text{ft}}{\text{sec}^2}} = \frac{\left(d_c(12 \text{ ft} + 2d_c)\right)^3}{12 \text{ ft} + (2)(2)d_c}$$

d_c is the critical depth. Solving for d_c by trial and error yields

$$d_c = \boxed{7.2 \text{ ft}}$$

2.3. The flow is supercritical since $d_c > d_o$.

2.4. The flow profile at the junction is shown.

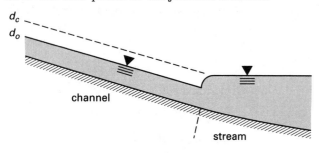

Since the flow is supercritical, the control section must be somewhere upstream, so the stream does not control the flow. A hydraulic jump takes place below the junction.

SOLUTION 3

3.1. The most efficient cross section is one that has the minimum wetted perimeter for a given flow area. The most efficient of all sections is a semicircular channel.

3.2. The most efficient trapezoidal section is half a hexagon.

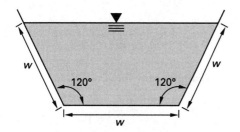

3.3. A channel section is controlled upstream if the discharge is given by the flow depth at the upper end of the section, regardless of downstream features.

3.4. A channel section is controlled downstream if the discharge is given by the flow depth at the lower end of the section.

SOLUTION 4

Assume that the pipe operates as an open channel while flowing full.

4.1. For a circular pipe with diameter $D = 22$ in,

$$D = \frac{22 \text{ in}}{12 \frac{\text{in}}{\text{ft}}} = 1.8333 \text{ ft}$$

$$A = \tfrac{1}{4}\pi D^2 = \tfrac{1}{4}\pi(1.8333 \text{ ft})^2$$
$$= 2.64 \text{ ft}^2$$

$$R = \tfrac{1}{4}D = \left(\tfrac{1}{4}\right)(1.8333 \text{ ft})$$
$$= 0.46 \text{ ft}$$

If the pipe is long enough to allow development of uniform flow, Manning's formula is applicable. (This equation is dimensionally inconsistent.)

$$Q = \left(\frac{1.49}{n}\right) A R^{2/3} \sqrt{S}$$
$$= \left(\frac{1.49}{0.01}\right)(2.64 \text{ ft}^2)(0.46 \text{ ft})^{2/3}\sqrt{0.02}$$
$$= \boxed{33 \text{ ft}^3/\text{sec}}$$

4.2. To meet the 35 ft³/sec demand, an additional PVC pipe should be installed to carry 2 ft³/sec. The diameter of a circular pipe flowing full is

$$D = 1.33\left(\frac{nQ}{\sqrt{S}}\right)^{3/8}$$
$$= (1.33)\left(\frac{(0.01)\left(2 \frac{\text{ft}^3}{\text{sec}}\right)}{\sqrt{0.02}}\right)^{3/8}\left(12 \frac{\text{in}}{\text{ft}}\right)$$
$$= \boxed{7.7 \text{ in}}$$

The excess water can be carried by a pipe with a nominal diameter of

$$D = \boxed{8 \text{ in}}$$

4.3. The diameter is

$$D = 1.33\left(\frac{nQ}{\sqrt{S}}\right)^{3/8}$$
$$= (1.33)\left(\frac{(0.01)\left(2 \frac{\text{ft}^3}{\text{sec}}\right)}{\sqrt{0.025}}\right)^{3/8}\left(12 \frac{\text{in}}{\text{ft}}\right)$$
$$= \boxed{7.3 \text{ in}}$$

The new route will also require a nominal pipe size of

$$D = \boxed{8 \text{ in}}$$

4.4. If 4257 ft is the upstream invert elevation, the downstream invert will be at elevation

$$4257 \text{ ft} - (0.025)(1000 \text{ ft}) = \boxed{4232 \text{ ft}}$$

SOLUTION 5

The capacity of the section can be calculated assuming uniform flow.

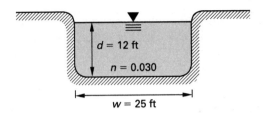

5.1. The bottom slope is

$$S = \frac{470 \text{ ft}}{(8.4 \text{ mi})\left(5280 \frac{\text{ft}}{\text{mi}}\right)} = 0.0106$$

The geometric properties are

$$A = wd = (25 \text{ ft})(12 \text{ ft}) = 300 \text{ ft}^2$$

$$P = w + 2d$$
$$= 25 \text{ ft} + (2)(12 \text{ ft})$$
$$= 49 \text{ ft}$$

$$R = \frac{A}{P} = \frac{300 \text{ ft}^2}{49 \text{ ft}}$$
$$= 6.122 \text{ ft}$$

Manning's formula gives the capacity of the natural channel. (This equation is dimensionally inconsistent.)

$$Q = \left(\frac{1.49}{n}\right) A R^{2/3} \sqrt{S}$$
$$= \left(\frac{1.49}{0.03}\right)(300 \text{ ft}^2)(6.122 \text{ ft})^{2/3}\sqrt{0.0106}$$
$$= 5134 \text{ ft}^3/\text{sec}$$

The nature of the flow regime is determined from the critical depth. For a rectangular channel,

$$d_c = \left(\frac{Q^2}{gw^2}\right)^{1/3}$$
$$= \left(\frac{\left(5134 \frac{\text{ft}^3}{\text{sec}}\right)^2}{\left(32.2 \frac{\text{ft}}{\text{sec}^2}\right)(25 \text{ ft})^2}\right)^{1/3}$$
$$= 10.9 \text{ ft}$$

Since $d > d_c$, the flow is subcritical. The artificial channel will have a smaller roughness coefficient because the effects of meandering and brush growth have been eliminated.

The new length is 8.4 mi - 3.7 mi = 4.7 mi. Since the new length is 4.7 mi, the ratio of meander length to straight length is 8.4 mi/4.7 mi = 1.8, which indicates a severe degree of meandering. The roughness coefficient for the new channel is approximately $n = 0.023$.

The bottom slope is

$$S = \frac{470 \text{ ft}}{(4.7 \text{ mi})\left(5280 \frac{\text{ft}}{\text{mi}}\right)} = 0.0189$$

The following illustration gives data for the artificial channel.

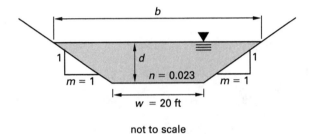

not to scale

The geometric properties of the section are

$$A = d(w + md)$$
$$P = w + 2d\sqrt{1 + m^2}$$
$$b = w + 2md$$
$$R = \frac{A}{P}$$
$$m = \frac{1}{1} = 1$$

In order to determine if the channel will be able to carry the maximum flow of 5134 ft³/sec, the hydraulic profile must be defined first. The normal depth, d_o, is calculated using Manning's formula.

$$Q = \left(\frac{1.49}{n}\right) A R^{2/3} \sqrt{S}$$

$$5134 \frac{\text{ft}^3}{\text{sec}} = \left(\frac{1.49}{0.023}\right) d_o(20 \text{ ft} + d_o)$$
$$\times \left(\frac{d_o(20 \text{ ft} + (1)d_o)}{20 \text{ ft} + (2\sqrt{2}d_o)}\right)^{2/3}\sqrt{0.0189}$$

Solving by trial and error for d_o yields

$$d_o = 7.3 \text{ ft}$$

The critical depth, d_c, is calculated from

$$\frac{Q^2}{g} = \frac{A^3}{b}$$

$$\frac{\left(5134 \ \frac{\text{ft}^3}{\text{sec}}\right)^2}{32.2 \ \frac{\text{ft}}{\text{sec}^2}} = \frac{\left(d_c(20 \text{ ft} + d_c)\right)^3}{20 \text{ ft} + (2)(1)d_c}$$

Solving by trial and error for d_c yields

$$d_c = 10.6 \text{ ft}$$

The artificial channel has a steep slope. The likely flow profile is shown in the illustration.

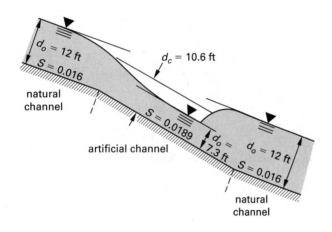

Since the flow depths in the artificial channel are within the range of natural depths, the new channel will be able to carry the maximum flow.

5.2. The erosion potential of the flow can be assessed from the mean velocity.

For the natural channel,

$$\text{v} = \frac{Q}{A} = \frac{5134 \ \frac{\text{ft}^3}{\text{sec}}}{300 \text{ ft}^2} = 17.1 \text{ ft/sec}$$

In the artificial channel, the maximum velocity occurs when the flow becomes uniform. At this point,

$$
\begin{aligned}
A &= d_o(w + md_o) \\
&= (7.3 \text{ ft})\left(20 \text{ ft} + (1)(7.3 \text{ ft})\right) \\
&= 199.3 \text{ ft}^2
\end{aligned}
$$

$$\text{v} = \frac{Q}{A} = \frac{5134 \ \frac{\text{ft}^3}{\text{sec}}}{199.3 \text{ ft}^2} = 25.8 \text{ ft/sec}$$

Erosion problems will be severe in the artificial channel. Both channels are well above maximum permissible velocities for natural bed material.

SOLUTION 6

6.1. In order to guarantee the minimum freeboard, the configuration shown must hold for the minimum flow.

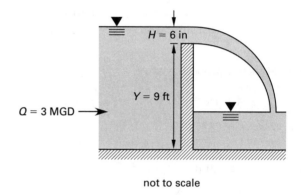

not to scale

The flow area is

$$A = (22 \text{ ft})(9.5 \text{ ft}) = 209 \text{ ft}^2$$

For a flow of 3 MGD $= 4.6 \text{ ft}^3/\text{sec}$, the approach velocity is

$$\text{v}_1 = \frac{Q}{A} = \frac{4.6 \ \frac{\text{ft}^3}{\text{sec}}}{209 \text{ ft}^2} = 0.022 \text{ ft/sec}$$

Since the velocity head is negligible, the following formula can be used.

$$Q = \tfrac{2}{3}C_1(b - 0.2H)\sqrt{2g}H^{3/2}$$

The coefficient C_1 is given by

$$
\begin{aligned}
C_1 &= \left(0.6035 + 0.0813\left(\frac{H}{Y}\right) + \frac{0.000295}{Y}\right) \\
&\quad \times \left(1 + \frac{0.00361}{H}\right)^{3/2} \\
&= \left(0.6035 + (0.0813)\left(\frac{0.5 \text{ ft}}{9 \text{ ft}}\right) + \frac{0.000295}{9 \text{ ft}}\right) \\
&\quad \times \left(1 + \frac{0.00361}{0.5 \text{ ft}}\right)^{3/2} \\
&= 0.615
\end{aligned}
$$

Substituting this and the other numerical values in the discharge equation yields

$$4.6 \ \frac{ft^3}{sec} = \left(\frac{2}{3}\right)(0.615)\big(b - (0.2)(0.5 \ ft)\big)$$
$$\times \sqrt{(2)\left(32.2 \ \frac{ft}{sec^2}\right)}(0.5 \ ft)^{3/2}$$

Solving for b results in

$$b = \boxed{4.1 \ ft}$$

6.2. For the maximum flow at $12 \ MGD = 18.6 \ ft^3/sec$, the approach velocity is

$$v_1 = \frac{18.6 \ \dfrac{ft^3}{sec}}{209 \ ft^2} = 0.09 \ ft/sec$$

This is also negligible, so the following equation is valid and must be solved iteratively for the unknown H.

$$Q = \tfrac{2}{3}C_1(b - 0.2H)\sqrt{2g}H^{3/2}$$

As a first estimate, assume $H = 16 \ in \ (1.33 \ ft)$. The coefficient C_1 is

$$C_1 = \left(0.6035 + 0.0813\left(\frac{H}{Y}\right) + \frac{0.000295}{Y}\right)$$
$$\times \left(1 + \frac{0.00361}{H}\right)^{3/2}$$
$$= \left(0.6035 + (0.0813)\left(\frac{1.33 \ ft}{9 \ ft}\right) + \frac{0.000295}{9 \ ft}\right)$$
$$\times \left(1 + \frac{0.00361}{1.33 \ ft}\right)^{3/2}$$
$$= 0.618$$

A second estimate for the freeboard can be obtained by solving for H in the discharge equation. (The calculation is not very sensitive to the correction for contraction. Therefore, the previous estimate of H is used for computational convenience.)

$$18.6 \ \frac{ft^3}{sec} = \left(\frac{2}{3}\right)(0.618)\big(4.1 \ ft - (0.2)(1.33 \ ft)\big)$$
$$\times \sqrt{(2)\left(32.2 \ \frac{ft}{sec^2}\right)}H^{3/2}$$

This yields

$$H = 1.29 \ ft$$

With this new value, C_1 remains unchanged. An improved estimate of H comes from solving

$$18.6 \ \frac{ft^3}{sec} = \left(\frac{2}{3}\right)(0.618)\big(4.1 \ ft - (0.2)(1.29 \ ft)\big)$$
$$\times \sqrt{(2)\left(32.2 \ \frac{ft}{sec^2}\right)}H^{3/2}$$

This yields the same value of H, $1.29 \ ft$.

$$H = (1.29 \ ft)\left(12 \ \frac{in}{ft}\right)$$
$$= \boxed{15.5 \ in}$$

6.3. The usual rule is to measure the approach depth at a distance from the crest greater than $4H$.

> The gauge station should be placed at least 6 ft upstream from the weir.

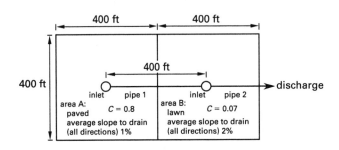
4 Hydrology

PROBLEM 1

An unconfined coarse-sand aquifer is 65 ft thick. Its porosity is 30%, and its transmissivity is 120,000 gal/ft-day. A pump draws water for 24 hr from a fully penetrating production well in the aquifer, after which the pump is shut off. Eight hours after the pump is shut off, an observation well 30 ft from the production well shows a 2.0 ft drawdown.

1.1. Estimate the pumping volume immediately before the production well is shut down.

1.2. Assume the pumping volume is 800 gal/min immediately before the production well is shut down. What is the steady-state drawdown at the production well?

PROBLEM 2

The following equations define the parameters of a hydrograph for a 30 min storm over a 200 ac collection basin.

$$Q = 0.0132 A^{0.9} L^{0.3} C^{1.1} E^{0.08}$$

$$T_d = 0.44 A^{1.1} L^{0.04} C^{1.2} E^{0.17}$$

$$T_p = 0.0024 A^{1.7} L^{0.04} C^{0.7} E^{0.14}$$

$$T_{50} = 0.0367 A^{1.3} L^{0.01} C^{0.7} E^{0.12}$$

$$T_{75} = 0.0671 A^{1.2} L^{0.05} C^{0.7} E^{0.11}$$

Q = peak discharge (ft^3/sec)

A = basin area = 200 ac

L = length of channel = 3000 ft

C = average runoff coefficient = 0.35

S = average channel slope = 0.04

E = average imperviousness = 25%

T_d = time from beginning of runoff

to end of channel discharge (hours)

T_p = time from beginning of runoff

to peak discharge (hours)

T_{50} = time from beginning of runoff

to 50% cumulative runoff (hours)

T_{75} = time from beginning of runoff

to 75% cumulative runoff (hours)

2.1. Draw the unit hydrograph. Indicate Q, T_d, T_p, T_{50}, and T_{75}.

2.2. During a 90 min storm in the same basin, 0.3 in fell in the first half hour, 1.5 in fell during the next half hour, and 0.4 in fell during the last half hour. What are Q, T_d, T_p, T_{50}, and T_{75}? (Make any reasonable assumptions.)

PROBLEM 3

The entrance to a theme park consists of two adjacent areas, A and B, that are both drained by drop inlets located in the geometric centers of the areas. The drop inlets feed a common collector designed for flow at 2 ft/sec. Area A is an asphalt parking area with an average slope of 1% from all directions to the drop inlet. Area B is primarily a well maintained lawn with an average slope of 2% from all directions to the drop inlet.

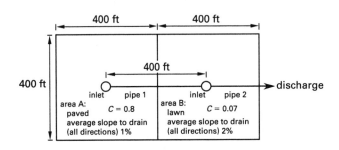

3.1. What is the design flow (in ft^3/sec) in pipe 1 for a 10 yr storm?

3.2. What is the design flow (in ft^3/sec) in pipe 2 for a 10 yr storm?

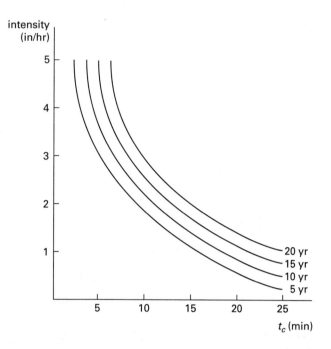

the water under the roadway. The average slope of the channel and culverts is 0.75% (i.e., 0.0075 ft/ft).

The time for runoff from the farthest part of the watershed to begin contributing to the flow is 35 min.

4.1. If curves for 5, 10, 25, 50, and 100 yr floods are available, which would you recommend using? Why?

4.2. What rational-method runoff coefficient would you recommend? Why?

4.3. Using the rational method and assuming the intensity after 35 min is 2 in/hr, what is the runoff?

4.4. Are the culverts under inlet or outlet control?

PROBLEM 5

Data on intensity and duration have been recorded on storms over the past 45 yr. Draw the 10 yr intensity-duration-frequency curve.

year	duration (min)	intensity (in/hr)	year	duration (min)	intensity (in/hr)
1	10	3.0	7	20	2.15
2	10	4.1	8	20	3.05
3	10	4.9	9	20	3.55
4	10	5.1	10	20	3.85
5	10	5.5	11	20	4.07
6	10	6.0	12	20	4.4

year	duration (min)	intensity (in/hr)	year	duration (min)	intensity (in/hr)
13	30	1.9	19	40	1.5
14	30	2.4	20	40	2.0
15	30	2.9	21	40	2.3
16	30	3.1	22	40	2.6
17	30	3.3	23	40	2.9
18	30	3.6	24	40	3.1

year	duration (min)	intensity (in/hr)	year	duration (min)	intensity (in/hr)
25	60	1.1	31	10	2.07
26	60	1.5	32	10	3.5
27	60	1.9	33	10	6.8
28	60	2.0	34	20	1.6
29	60	2.15	35	20	2.6
30	60	2.30	36	20	4.9

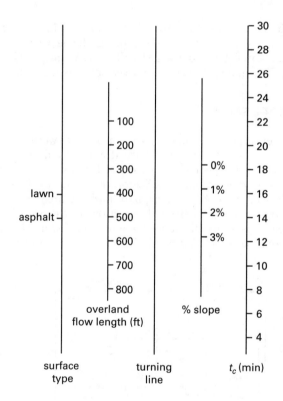

PROBLEM 4

Water from a 175 ac light industrial watershed is collected and drained by a trapezoidal open channel. The channel (Manning's roughness coefficient, $n = 0.02$) has a 4.5 ft wide bottom and 1:1 sides. The channel direction is perpendicular to a road where twin, side-by-side 54 in diameter corrugated metal pipe (CMP) culverts take

year	duration (min)	intensity (in/hr)	year	duration (min)	intensity (in/hr)
37	30	1.2	42	40	3.4
38	30	2.1	43	60	0.8
39	30	4.0	44	60	1.25
40	40	1.05	45	60	2.6
41	40	1.75			

PROBLEM 6

Two square areas, A and B, drain through pipes into a collection manhole as shown. All pipes have a Manning roughness constant of $n = 0.012$. All slopes are 0.035 ft/ft. Area A is 4 ac and has a rational coefficient of $C = 0.9$. The time to concentration, t_c, for area A is 5 min. Area B is 10 ac and has a rational coefficient of $C = 0.25$. The time to concentration for area B is 18 min. Assume the pipes flow full. Determine the diameters of standard pipe sections 1, 2, and 3 for a 10 yr storm.

		intensity (in/hr)		
t_c	5 yr	10 yr	20 yr	50 yr
5 min	6.1	7.0	7.9	8.8
10 min	4.6	5.5	6.4	7.3
15 min	3.5	4.4	5.3	6.2
18 min	2.9	3.8	4.7	5.6
20 min	2.7	3.6	4.5	5.4

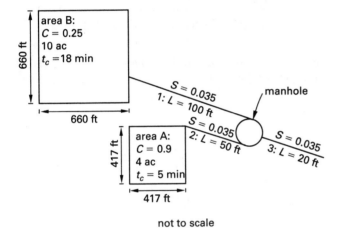

not to scale

PROBLEM 7

A 20 ac pasture (rational coefficient $C = 0.3$, time to concentration $t_c = 15$ min) was originally used for cattle grazing. The pasture drains into a creek that runs along one edge and has barely enough capacity to carry away the peak flow from a 10 yr storm. The creek passes through a nearby residential area. When the creek overflows, significant property damage occurs.

The intensity-duration curve for a 10 yr storm in this area is given by the following table.

time (min)	intensity (in/hr)
5	7.7
10	6.9
15	5.8
20	4.7
30	3.5

The pasture has been purchased by a foreign car manufacturer that wants to pave over all 20 ac (rational coefficient $C = 0.90$, $t_c = 10$ min) to provide storage for newly imported cars. The car manufacturer has proposed construction of a detention basin to capture and hold runoff such that the creek's pre-existing capacity is not exceeded during any storm with a frequency of 10 years or less. What should be the capacity in ac-ft of the detention basin?

SOLUTION 1

1.1. For a small drawdown compared with the initial thickness of the aquifer, the equation governing unsteady flow toward a fully penetrating well in an unconfined aquifer is

$$Y - y = \left(\frac{Q}{4\pi K_p Y}\right) W(u)$$

$W(u)$ is the well function and

$$u = \frac{r^2 S_y}{4 K_p Y t}$$

Q is the flow rate, S_y is the yield, K_p is the hydraulic conductivity, t is the time, and the other variables are depicted in the illustration shown.

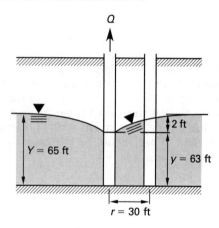

not to scale

The pumping schedule is

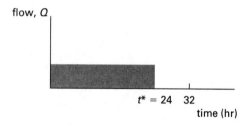

The principle of superposition indicates that the position of the water table is given by

$$Y - y = \left(\frac{Q}{4\pi K_p Y}\right)(W(u_1) - W(u_2))$$

$$u_1 = \frac{r^2 S_y}{4 K_p Y t}$$

$$u_2 = \frac{r^2 S_y}{4 K_p Y (t - t^*)}$$

For $u < 0.1$, the well function is approximately

$$W(u) \approx 0.5772 - \ln(u)$$

The equation becomes

$$Y - y = \left(\frac{Q}{4\pi K_p Y}\right)\ln\frac{t}{t - t^*}$$

The hydraulic conductivity is

$$K_p = \frac{120,000 \ \frac{\text{gal}}{\text{ft-day}}}{65 \ \text{ft}} = 1846 \ \text{gal/ft}^2\text{-day}$$

For coarse sand, assume the yield is 85% of the porosity.

$$S_y = (0.85)(0.30) = 0.26$$

$$u_1 = \frac{(30 \ \text{ft})^2(0.26)}{(4)\left(10.3 \ \frac{\text{ft}}{\text{hr}}\right)(65 \ \text{ft})(32 \ \text{hr})} = 0.0027$$

$$u_2 = \frac{(30 \ \text{ft})^2(0.26)}{(4)\left(10.3 \ \frac{\text{ft}}{\text{hr}}\right)(65 \ \text{ft})(32 \ \text{hr} - 24 \ \text{hr})} = 0.011$$

Since both u_1 and u_2 are less than 0.1, the simplified equation can be used.

$$65 \ \text{ft} - 63 \ \text{ft} = \left(\frac{Q}{(4\pi)\left(1846 \ \frac{\text{gal}}{\text{ft}^2\text{-day}}\right)(65 \ \text{ft})}\right)$$
$$\times \ln\frac{32 \ \text{hr}}{32 \ \text{hr} - 24 \ \text{hr}}$$

Solving, the flow rate is 2.175×10^6 gal/day.

$$Q = \frac{2.175 \times 10^6 \ \frac{\text{gal}}{\text{day}}}{1440 \ \frac{\text{min}}{\text{day}}}$$
$$= \boxed{1511 \ \text{gal/min}}$$

1.2. For steady flow from a fully penetrating well into an unconfined aquifer, the following empirical expression is valid at the production well.

$$\frac{Q}{d_o} = \frac{T}{2000 \ \frac{\text{min}}{\text{day}}}$$

Q is in gal/min, d_o is in feet, and T is in gal/ft-day.

$$d_o = \frac{\left(2000 \ \frac{\text{min}}{\text{day}}\right)\left(800 \ \frac{\text{gal}}{\text{min}}\right)}{120,000 \ \frac{\text{gal}}{\text{ft-day}}} = \boxed{13.3 \ \text{ft}}$$

SOLUTION 2

2.1. Substituting the known data into the hydrograph equation yields

$$Q = (0.0132)(200 \text{ ac})^{0.9}(3000 \text{ ft})^{0.3}(0.35)^{1.1}(0.25)^{0.08}$$
$$= 4.8 \text{ ft}^3/\text{sec}$$

$$T_d = (0.44)(200 \text{ ac})^{1.1}(3000 \text{ ft})^{0.04}(0.35)^{1.2}(0.25)^{0.17}$$
$$= 46.2 \text{ hr}$$

$$T_p = (0.0024)(200 \text{ ac})^{1.7}(3000 \text{ ft})^{0.04}(0.35)^{0.7}(0.25)^{0.14}$$
$$= 10.7 \text{ hr}$$

$$T_{50} = (0.0367)(200 \text{ ac})^{1.3}(3000 \text{ ft})^{0.01}(0.35)^{0.7}(0.25)^{0.12}$$
$$= 15.8 \text{ hr}$$

$$T_{75} = (0.0671)(200 \text{ ac})^{1.2}(3000 \text{ ft})^{0.05}(0.35)^{0.7}(0.25)^{0.11}$$
$$= 23.8 \text{ hr}$$

The hydrograph variables are shown in the graph.

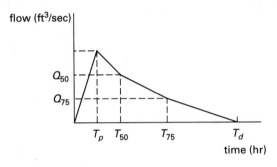

The total runoff is equal to the area below the hydrograph.

$$R = \left(\tfrac{1}{2}\right)(10.7 \text{ hr})\left(4.8 \ \frac{\text{ft}^3}{\text{sec}}\right)$$
$$+ \left(\tfrac{1}{2}\right)\left(4.8 \ \frac{\text{ft}^3}{\text{sec}} + Q_{50}\right)(15.8 \text{ hr} - 10.7 \text{ hr})$$
$$+ \left(\tfrac{1}{2}\right)(Q_{50} + Q_{75})(23.8 \text{ hr} - 15.8 \text{ hr})$$
$$+ \tfrac{1}{2}Q_{75}(46.2 \text{ hr} - 23.8 \text{ hr})$$
$$= 37.9 + 6.55Q_{50} + 15.2Q_{75}$$

According to the definitions given, the area up to T_{50} must be 50% of the total runoff, and the area up to T_{75} must be 75% of the total runoff.

$$0.5R = \left(\tfrac{1}{2}\right)(10.7 \text{ hr})\left(4.8 \ \frac{\text{ft}^3}{\text{sec}}\right)$$
$$+ \left(\tfrac{1}{2}\right)\left(4.8 \ \frac{\text{ft}^3}{\text{sec}} + Q_{50}\right)(15.8 \text{ hr} - 10.7 \text{ hr})$$

$$0.75R = 0.5R + \left(\tfrac{1}{2}\right)(Q_{50} + Q_{75})(23.8 \text{ hr} - 15.8 \text{ hr})$$

Substituting R into these last two equations results in a system of two equations.

$$1.45Q_{50} + 15.2Q_{75} = 37.94 \text{ ft}^3\text{-hr}/\text{sec}$$
$$9.445Q_{50} + 0.8Q_{75} = 37.9 \text{ ft}^3\text{-hr}/\text{sec}$$

Solving for Q_{50} and Q_{75} yields

$$Q_{50} = 3.83 \text{ ft}^3/\text{sec}$$
$$Q_{75} = 2.13 \text{ ft}^3/\text{sec}$$

Substituting these values in the expression for the runoff volume,

$$R = 37.9 \ \frac{\text{ft}^3\text{-hr}}{\text{sec}} + 6.55Q_{50} + 15.2Q_{75}$$
$$= 37.9 \ \frac{\text{ft}^3\text{-hr}}{\text{sec}} + (6.55)\left(3.83 \ \frac{\text{ft}^3}{\text{sec}}\right) + (15.2)\left(2.13 \ \frac{\text{ft}^3}{\text{sec}}\right)$$
$$= 95.36 \text{ ft}^3\text{-hr}/\text{sec}$$
$$R = \frac{\left(95.36 \ \frac{\text{ft}^3\text{-hr}}{\text{sec}}\right)\left(3600 \ \frac{\text{sec}}{\text{hr}}\right)}{43{,}560 \ \frac{\text{ft}^2}{\text{ac}}} = 7.88 \text{ ac-ft}$$

In terms of depth,

$$R = \left(\frac{7.88 \text{ ac-ft}}{200 \text{ ac}}\right)\left(12 \ \frac{\text{in}}{\text{ft}}\right)$$
$$= 0.47 \text{ in}$$

The unit hydrograph is obtained by scaling the ordinates so that the area under the curve is 1 in.

$$Q = \left(4.8 \ \frac{\text{ft}^3}{\text{sec}}\right)\left(\frac{1}{0.47}\right) = 10.2 \text{ ft}^3/\text{sec}$$
$$Q_{50} = \left(3.83 \ \frac{\text{ft}^3}{\text{sec}}\right)\left(\frac{1}{0.47}\right) = 8.15 \text{ ft}^3/\text{sec}$$
$$Q_{75} = \left(2.13 \ \frac{\text{ft}^3}{\text{sec}}\right)\left(\frac{1}{0.47}\right) = 4.53 \text{ ft}^3/\text{sec}$$

The resulting unit hydrograph is shown.

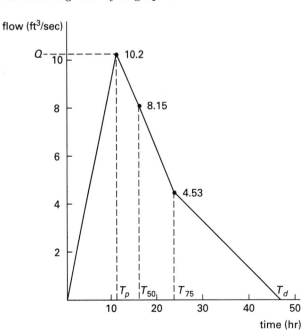

2.2. The hydrograph for this storm is

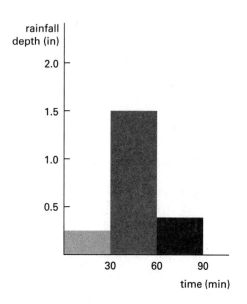

The duration of the storm is relatively short compared with the base of the unit hydrograph. Also, most of the precipitation occurs between 30 min and 60 min. This storm can be approximated by the following hydrograph.

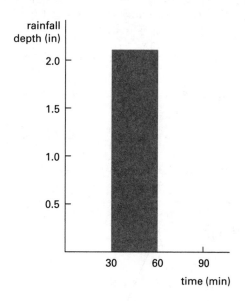

The total depth is equal to

$$0.3 \text{ in} + 1.5 \text{ in} + 0.4 \text{ in} = 2.2 \text{ in}$$

With this assumption, the peak will be

$$Q = (2.2)\left(10.2 \ \frac{\text{ft}^3}{\text{sec}}\right) = \boxed{22.4 \ \text{ft}^3/\text{sec}}$$

The characteristic times will be those of the unit hydrograph displaced by 30 min (0.5 hr).

$$
\begin{array}{l}
T_d = 46.2 \text{ hr} + 0.5 \text{ hr} = 46.7 \text{ hr} \\
T_p = 10.7 \text{ hr} + 0.5 \text{ hr} = 11.2 \text{ hr} \\
T_{50} = 15.8 \text{ hr} + 0.5 \text{ hr} = 16.3 \text{ hr} \\
T_{75} = 23.8 \text{ hr} + 0.5 \text{ hr} = 24.3 \text{ hr}
\end{array}
$$

SOLUTION 3

The time of concentration is taken as the travel time along a diagonal toward the inlet. This distance is equal to $(200 \text{ ft})\sqrt{2} = 283 \text{ ft}$.

3.1. For area A, the nomograph yields $t_c = 9$ min.

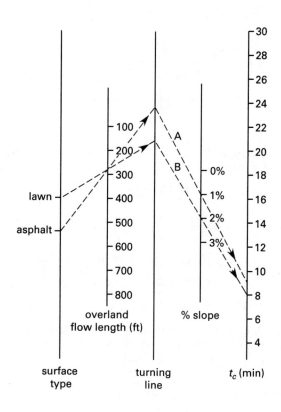

For a storm duration of 9 min, the 10 yr IDF curve indicates an intensity of 2.3 in/hr.

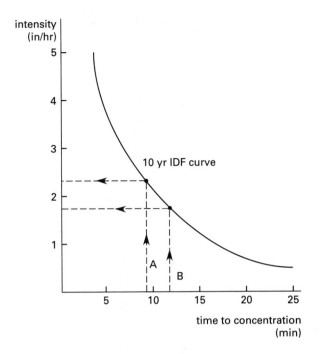

intensity (in/hr) vs. time to concentration (min); 10 yr IDF curve

The surface areas of A and B are 3.7 ac. The runoff coefficient for asphalt area A is $C=0.8$. The peak flow in pipe 1 is

$$Q_p = CIA_d$$

$$= \frac{(0.8)\left(2.3 \frac{\text{in}}{\text{hr}}\right)(3.7 \text{ ac})\left(43{,}560 \frac{\text{ft}^2}{\text{ac}}\right)}{\left(12 \frac{\text{in}}{\text{ft}}\right)\left(3600 \frac{\text{sec}}{\text{hr}}\right)}$$

$$= \boxed{6.8 \text{ ft}^3/\text{sec}}$$

3.2. For area B, the nomograph yields a time of concentration (inlet time) of 8 min. The travel time in pipe 1 is

$$t = \frac{L}{v} = \frac{\dfrac{400 \text{ ft}}{2 \frac{\text{ft}}{\text{sec}}}}{60 \frac{\text{sec}}{\text{min}}} = 3.3 \text{ min}$$

Area A's flow will reach the second inlet at

$$t = 9 \text{ min} + 3.3 \text{ min}$$
$$= 12.3 \text{ min}$$

Pipe 2 has to be able to carry the maximum flow it will see. There are two possiblities for peak flow. The first occurs at area B's time of concentration, 8 min. The runoff from area B will never be higher than it is at 8 min.

Another possibility for peak flow in pipe 2 is when the peak flow from area A reaches pipe 2. This will occur at 12.3 min. The runoff from area B at 12.3 min will combine with the peak runoff from area A.

The maximum of these two flows determines pipe 2's size.

The intensity at 8 min is about 2.5 in/hr. The peak runoff from area B is

$$Q_p(8 \text{ min}, \text{B}) = CIA_d$$

$$= \frac{(0.07)\left(2.5 \frac{\text{in}}{\text{hr}}\right)(3.7 \text{ ac})\left(43{,}560 \frac{\text{ft}^2}{\text{ac}}\right)}{\left(12 \frac{\text{in}}{\text{ft}}\right)\left(3600 \frac{\text{sec}}{\text{hr}}\right)}$$

$$= 0.65 \text{ ft}^3/\text{sec}$$

This is less than $Q_p(\text{peak}, \text{A})$, so it doesn't control.

But at $t = 12.3$ min, the storm intensity is 1.7 in/hr. The runoff from area B at this moment is

$$Q_p(12.3 \text{ min}, \text{B}) = CIA_d$$

$$= \frac{(0.07)\left(1.7 \frac{\text{in}}{\text{hr}}\right)(3.7 \text{ ac})\left(43{,}560 \frac{\text{ft}^2}{\text{ac}}\right)}{\left(12 \frac{\text{in}}{\text{ft}}\right)\left(3600 \frac{\text{sec}}{\text{hr}}\right)}$$

$$= 0.44 \text{ ft}^3/\text{sec}$$

At the moment that area A's peak reaches the second inlet, area B will be producing 0.44 ft³/sec. This will combine with the peak flow from area A.

$$Q_p(\text{peak, pipe 2}) = 0.44 \frac{\text{ft}^3}{\text{sec}} + 6.8 \frac{\text{ft}^3}{\text{sec}}$$

$$= \boxed{7.2 \text{ ft}^3/\text{sec}}$$

SOLUTION 4

4.1. The return period for culvert design depends mainly on land and road use in the surrounding area. Federal or state guidelines for highway design recommend the return period.

4.2. A runoff coefficient for light industrial areas is between 0.5 and 0.8. For this problem, an average value of $\boxed{0.65}$ will be used.

4.3. The runoff is given by the rational formula.

$$Q_p = CIA_d$$

$$= \frac{(0.65)\left(2 \frac{\text{in}}{\text{hr}}\right)(175 \text{ ac})\left(43{,}560 \frac{\text{ft}^2}{\text{ac}}\right)}{\left(12 \frac{\text{in}}{\text{ft}}\right)\left(3600 \frac{\text{sec}}{\text{hr}}\right)}$$

$$= \boxed{229 \text{ ft}^3/\text{sec}}$$

4.4. For a corrugated metal pipe,

$$n = 0.022$$

The hydraulic radius for a 54 in culvert flowing full is

$$R = \frac{D}{4} = \frac{\dfrac{54 \text{ in}}{4}}{12 \dfrac{\text{in}}{\text{ft}}}$$

$$= 1.13 \text{ ft}$$

The discharge through one culvert flowing full is

$$Q_{\text{full}} = \left(\frac{1.49}{n}\right) A R^{2/3} \sqrt{S}$$

$$= \left(\frac{1.49}{0.022}\right)\left(\frac{\pi(4.5 \text{ ft})^2}{4}\right)(1.13 \text{ ft})^{2/3}\sqrt{0.0075}$$

$$= 101.2 \text{ ft}^3/\text{sec}$$

The maximum flow that can be carried by a circular pipe operating as an open channel is

$$Q_{\text{max}} = 1.08 Q_{\text{full}}$$

$$= (1.08)\left(101.2 \frac{\text{ft}^2}{\text{sec}}\right)$$

$$= 109.3 \text{ ft}^3/\text{sec}$$

Each culvert must carry $(228 \text{ ft}^3/\text{sec})/2 = 114 \text{ ft}^3/\text{sec}$. So, the pipe must be under pressure to meet this flow. This could occur as shown.

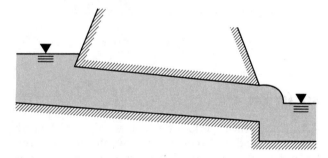

In this case, the culverts are controlled at the outlet.

SOLUTION 5

Each storm duration has nine events as shown in the following tables.

10 min storm

rank	i (in/hr)
1	6.80
2	6.00
3	5.50
4	5.10
5	4.90
6	4.10
7	3.50
8	3.00
9	2.07

20 min storm

rank	i (in/hr)
1	4.90
2	4.40
3	4.07
4	3.85
5	3.55
6	3.05
7	2.60
8	2.15
9	1.60

30 min storm

rank	i (in/hr)
1	4.00
2	3.60
3	3.30
4	3.10
5	2.90
6	2.40
7	2.10
8	1.90
9	1.20

40 min storm

rank	i (in/hr)
1	3.40
2	3.10
3	2.90
4	2.60
5	2.30
6	2.00
7	1.75
8	1.50
9	1.05

60 min storm

rank	i (in/hr)
1	2.60
2	2.30
3	2.15
4	2.00
5	1.90
6	1.50
7	1.25
8	1.10
9	0.80

The probability of exceedance is calculated as

$$p = \frac{r}{N+1}$$

r is the rank when the storms are sorted from the highest intensity to the lowest, and N is the total number of events (nine in this case).

The return period is $T = 1/p$. For the 10 yr IDF, $T = 10$ yr, or $p = 0.1$. The curve is made up of all the events with rank

$$r = (0.1)(9 + 1) = 1$$

These storms are

duration (min)	i (in/hr)
10	6.80
20	4.90
30	4.00
40	3.40
60	2.60

The plot of the 10 yr IDF curve is shown.

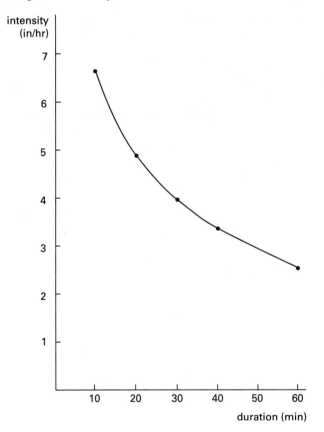

SOLUTION 6

For area A, the 10 yr IDF curve yields $I = 7.0$ in/hr for $t_c = 5$ min. The rational formula produces

$$Q_p = CIA_d$$

$$= \frac{(0.9)\left(7\ \frac{\text{in}}{\text{hr}}\right)(4\ \text{ac})\left(43{,}560\ \frac{\text{ft}^2}{\text{ac}}\right)}{\left(12\ \frac{\text{in}}{\text{ft}}\right)\left(3600\ \frac{\text{sec}}{\text{hr}}\right)}$$

$$= 25.2\ \text{ft}^3/\text{sec}$$

Pipe 2 must carry 25.2 ft³/sec while flowing full. Using the geometric properties of the full section,

$$R = \frac{D}{4}$$

$$A = \frac{\pi D^2}{4}$$

Manning's formula is used to size the pipe.

$$Q = A\text{v} = A\left(\frac{1.49}{n}\right)R^{2/3}\sqrt{S}$$

Inserting the geometric properties, this equation becomes

$$Q = \left(\frac{\pi D^2}{4}\right)\left(\frac{1.49}{n}\right)\left(\frac{D}{4}\right)^{2/3}\sqrt{S}$$

Solving for the diameter of the pipe,

$$D = 1.33\left(\frac{nQ}{\sqrt{S}}\right)^{3/8}$$

Inserting the known values yields

$$D = (1.33)\left(\frac{(0.012)\left(25.2\ \frac{\text{ft}^3}{\text{sec}}\right)}{\sqrt{0.035}}\right)^{3/8}\left(12\ \frac{\text{in}}{\text{ft}}\right)$$

$$= 19.1\ \text{in}$$

A standard size pipe of 20 in diameter should be used for pipe 2.

For area B, the intensity for $t_c = 18$ min is $I = 3.8$ in/hr, and the peak flow is

$$Q_p = CIA_d$$

$$= \frac{(0.25)\left(3.8\ \frac{\text{in}}{\text{hr}}\right)(10\ \text{ac})\left(43{,}560\ \frac{\text{ft}^2}{\text{ac}}\right)}{\left(12\ \frac{\text{in}}{\text{ft}}\right)\left(3600\ \frac{\text{sec}}{\text{hr}}\right)}$$

$$= 9.5\ \text{ft}^3/\text{sec}$$

The diameter is found from Manning's equation for a full pipe.

$$D = 1.33\left(\frac{nQ}{\sqrt{S}}\right)^{3/8}$$

$$= (1.33)\left(\frac{(0.012)\left(9.5\ \frac{\text{ft}^3}{\text{sec}}\right)}{\sqrt{0.035}}\right)^{3/8}\left(12\ \frac{\text{in}}{\text{ft}}\right)$$

$$= 13\ \text{in}$$

A standard size pipe of 14 in diameter should be used for pipe 1.

The flow through pipe 3 is given by the most severe of the following two storms.

storm 1: duration $= 5$ min, $I = 7$ in/hr

storm 2: duration $= 18$ min, $I = 3.8$ in/hr

The travel times in pipes 1 and 2 have been neglected since the pipes are short.

Storm 1:

After 5 min, all of area A responds to the storm, but only a portion of area B contributes to the flow. Assuming constant overland flow velocities, area B can be linearly divided into isochrones.

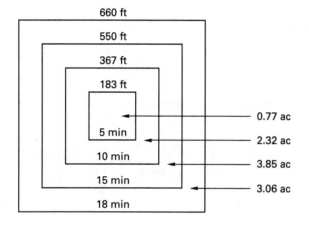

A square area of 10 ac has a dimension of 660 ft on the side. The isochrones are also square, and their dimensions are interpolated from the total concentration time. For instance, for the 5 min isochrone, the length of the side is

$$(660 \text{ ft})\left(\frac{5 \text{ min}}{18 \text{ min}}\right) = 183 \text{ ft}$$

At the end of the storm, only the 183 ft region will contribute to the flow. Consequently, the peak discharge during this storm will be

$$Q_p = \frac{(0.25)\left(7 \frac{\text{in}}{\text{hr}}\right)(0.77 \text{ ac})\left(43{,}560 \frac{\text{ft}^2}{\text{ac}}\right)}{\left(12 \frac{\text{in}}{\text{ft}}\right)\left(3600 \frac{\text{sec}}{\text{hr}}\right)}$$
$$+ \frac{(0.9)\left(7 \frac{\text{in}}{\text{hr}}\right)(4 \text{ ac})\left(43{,}560 \frac{\text{ft}^2}{\text{ac}}\right)}{\left(12 \frac{\text{in}}{\text{ft}}\right)\left(3600 \frac{\text{sec}}{\text{hr}}\right)}$$
$$= 26.5 \text{ ft}^3/\text{sec}$$

Storm 2:

After 18 min, both areas A and B contribute fully to the flow in pipe 3. The peak discharge is

$$Q_p = \frac{(0.25)\left(3.8 \frac{\text{in}}{\text{hr}}\right)(10 \text{ ac})\left(43{,}560 \frac{\text{ft}^2}{\text{ac}}\right)}{\left(12 \frac{\text{in}}{\text{ft}}\right)\left(3600 \frac{\text{sec}}{\text{hr}}\right)}$$
$$+ \frac{(0.9)\left(3.8 \frac{\text{in}}{\text{hr}}\right)(4 \text{ ac})\left(43{,}560 \frac{\text{ft}^2}{\text{ac}}\right)}{\left(12 \frac{\text{in}}{\text{ft}}\right)\left(3600 \frac{\text{sec}}{\text{hr}}\right)}$$
$$= 23.2 \text{ ft}^3/\text{sec}$$

The design flow is the greatest of the two peaks—that is, 26.5 ft³/sec.

The diameter of pipe 3 is given by Manning's equation for a full pipe.

$$D = 1.33\left(\frac{nQ}{\sqrt{S}}\right)^{3/8}$$
$$= (1.33)\left(\frac{(0.012)\left(26.5 \frac{\text{ft}^3}{\text{sec}}\right)}{\sqrt{0.035}}\right)^{3/8}\left(12 \frac{\text{in}}{\text{ft}}\right)$$
$$= 19.2 \text{ in}$$

A standard size pipe of 20 in diameter should be used for pipe 3.

SOLUTION 7

From the IDF table, the 10 yr storm for the grassy area has an intensity of 5.8 in/hr at 15 min. The peak flow is

$$Q_{p,\text{grass}} = CIA_d$$
$$= \frac{(0.3)\left(5.8 \frac{\text{in}}{\text{hr}}\right)(20 \text{ ac})\left(43{,}560 \frac{\text{ft}^2}{\text{ac}}\right)}{\left(12 \frac{\text{in}}{\text{ft}}\right)\left(3600 \frac{\text{sec}}{\text{hr}}\right)}$$
$$= 34.8 \text{ ft}^3/\text{sec}$$

This is the maximum flow handled by the creek. If the area is paved, the time of concentration decreases to 10 min, and the IDF table shows an intensity of $I = 6.9$ in/hr. The new peak flow is

$$Q_{p,\text{paved}} = CIA_d$$
$$= \frac{(0.9)\left(6.9 \frac{\text{in}}{\text{hr}}\right)(20 \text{ ac})\left(43{,}560 \frac{\text{ft}^2}{\text{ac}}\right)}{\left(12 \frac{\text{in}}{\text{ft}}\right)\left(3600 \frac{\text{sec}}{\text{hr}}\right)}$$
$$= 124.2 \text{ ft}^3/\text{sec}$$

The approximate hydrograph for a 10 min storm is

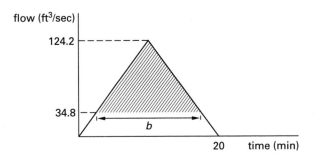

Since flows above 34.8 ft^3/sec are unacceptable, the detention basin must hold the volume in the shaded area. The base, b, of the shaded triangle is interpolated as

$$b = 20 - (20 \text{ min}) \left(\frac{34.8 \ \frac{\text{ft}^3}{\text{sec}}}{124.2 \ \frac{\text{ft}^3}{\text{sec}}} \right) = 14.4 \text{ min}$$

The required volume of the basin is

$$V = \frac{\left(\frac{1}{2}\right)\left(124.2 \ \frac{\text{ft}^3}{\text{sec}} - 34.8 \ \frac{\text{ft}^3}{\text{sec}}\right)(14.4 \text{ min})\left(60 \ \frac{\text{sec}}{\text{min}}\right)}{43{,}560 \ \frac{\text{ft}^3}{\text{ac-ft}}}$$

$$= 0.89 \text{ ac-ft}$$

The required capacity is approximately $\boxed{0.89 \text{ ac-ft.}}$

5 Water Supply

PROBLEM 1

A 28 ft × 20 ft (plan dimensions) sand filter (not shown) feeds a clearwell that has plan dimensions of 200 ft × 250 ft. The elevation of the surface of the clearwell is 150 ft and is constant. A pump at elevation 165 ft is driven by a 95% efficient electric motor and removes 3.5 MGD. The pressure gauge at the exit (located at elevation 175 ft) reads 80 psig (static pressure). All pipes are plain cast iron. Minor losses should be ignored.

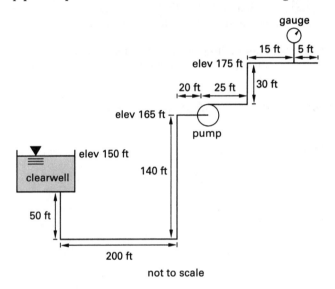

not to scale

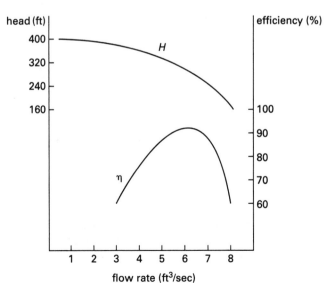

1.1. What is the total dynamic head supplied by the pump?

1.2. Draw the system curve on the accompanying pump curve.

1.3. What is the pump efficiency at the operating point?

1.4. What motor horsepower would you recommend?

1.5. If the sand filter is overloaded by 25% above its design flow, what will be the depletion or accumulation in the clearwell during an 8 hr period?

PROBLEM 2

Due to the presence of organics, a water treatment plant receives water with excessive color (much higher than 100 CU). The water is taken through the following processes.

(a) prechlorination

(b) flash mixing

(c) flocculation

(d) sedimentation

(e) filtration

(f) storage

(g) chlorination

(h) distribution

2.1. Briefly explain the purpose of each of the first seven steps.

2.2. What problems might the plant have in meeting the EPA requirements for chlorinated organics?

2.3. How are trihalomethanes (THMs) introduced into water supplies?

2.4. How are THMs reduced?

2.5. What are the problems inherent with THM treatments?

PROBLEM 3

A rapid mixing paddle device in a drinking water system has four flat plate rectangular paddles, each with an area of 2.81 ft^2. Each paddle is mounted to arms on a rotating spindle turning at 100 rpm. The maximum diameter of the mixing device is 3.75 ft. The plates are completely submerged in 55°F water, and each plate addresses the water flat-on. The mixing paddle unit is located in a 185.7 ft^3 tank through which 2 MGD of water flows.

3.1. What power is required to drive the paddle mixer?

3.2. What is the mean velocity gradient?

3.3. Is the value of Gt_d (i.e., the product of the velocity gradient and the mean detention time) within the normal range for such mixing units?

PROBLEM 4

A water treatment plant receives 2 MGD of well water with a turbidity of 300 NTU.

4.1. Design a treatment sequence.

4.2. Draw a profile diagram showing the locations and elevations of the treatment processes.

4.3. Size all settling basins in your design.

PROBLEM 5

In order to build an underground parking lot for a large office building, a 400 ft × 400 ft square excavation site 35 ft deep needs to be made in a sandy soil (permeability = 190 ft/day) in a location where the water table is only 25 ft below the surface. An impervious clay stratum exists 120 ft down. In order for the soil to develop enough bearing strength to support the heavy excavation machinery, the water surface must be lowered 10 ft below the excavation floor. To accomplish this, a decision has been made to drop one or more dewatering wells 1 ft in diameter in the middle of the excavation site.

5.1. What amount of water must the well(s) withdraw to completely dewater the site to a depth of 10 ft below the excavation floor? Assume the well's radius of influence will be 800 ft.

5.2. Are dewatering wells a good method to use in this situation?

5.3. What other methods could be used to provide a dry work site?

PROBLEM 6

The population of a small residential area with its own water supply system varies between 24,500 during the day and 35,000 at night. The area is relatively flat. The town is approximately rectangular in shape with length and width of 4.4 mi and 2.9 mi, respectively. The water district is under a mandate to provide all water customers with minimum and maximum water pressures of 50 psig and 80 psig, respectively. Disregard minor losses, and state all other assumptions.

6.1. For what average demand should the water district design its water treatment facility?

6.2. Considering hourly, daily, and seasonal variations, what is the maximum instantaneous demand the water district could expect to see?

6.3. Using the National Board of Fire Underwriters' equation, what additional requirements should the water district add for firefighting?

6.4. In general, how would you recommend meeting the minimum and maximum water pressure requirements?

6.5. What total volume of water storage would you recommend?

6.6. Assuming that houses are uniformly distributed in the town, sketch a reasonable layout for the water distribution system, including all towers, mains, and submains.

6.7. Design the water storage system, indicating the volume and elevation of each tank. Assume there are a total of seven water tanks, evenly distributed.

Assume that the distribution path from tower to house is made up of the following pipe lengths, types, and roughness coefficients, in series.

type	diameter (in)	length (ft)	roughness coefficient
steel	16	4000	100
steel	12	4000	100
steel	6	1000	100
copper tube	$^3/_4$	80	130

6.8. How often and at what time of day should the water tanks be refilled?

6.9. Assuming that the water tank surface elevation is 80 ft above the ground, will the water pressure requirements be satisfied?

PROBLEM 7

The composition of water as delivered from an aquifer is as follows.

Ca^{++}	80 mg/L
Cl^-	18 mg/L
CO_2	15 mg/L
Fe^{++}	3 mg/L
HCO_3^-	336 mg/L
Mg^{++}	30 mg/L
SO_4^{--}	64 mg/L
TDS	650 mg/L

7.1. What is the hardness?

7.2. What is the pH of this water?

7.3. What is the acidity or alkalinity?

7.4. Is the water acidic or alkaline? Why?

7.5. What is the most likely color of this water?

7.6. Should this water be softened?

7.7. What amounts of lime and soda ash are required to obtain a final hardness of 80 mg/L?

SOLUTION 1

1.1. The total dynamic head (TDH) can be found from the pump curve in the illustration using the flow given.

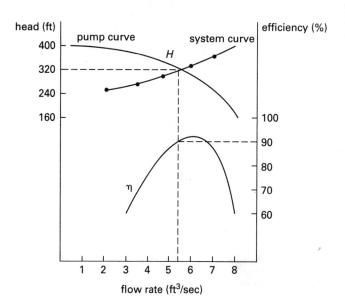

The flow, Q, is

$$Q = (3.5 \text{ MGD})\left(1.55 \frac{\frac{ft^3}{sec}}{MGD}\right) = 5.42 \text{ ft}^3/\text{sec}$$

TDH at 5.42 ft³/sec is $\boxed{320 \text{ ft.}}$

1.2. Static lift is the difference between the water elevation at the outlet (175 ft) and at the inlet (150 ft).

$$\text{static lift} = 175 \text{ ft} - 150 \text{ ft} = 25 \text{ ft}$$

$$\text{static exit pressure} = \frac{p}{\gamma} = \frac{\left(80 \frac{lbf}{in^2}\right)\left(12 \frac{in}{ft}\right)^2}{62.4 \frac{lbf}{in^3}}$$

$$= 184.6 \text{ ft}$$

$$\text{static head} = \text{lift} + \text{exit pressure}$$

$$= 25 \text{ ft} + 184.6 \text{ ft}$$

$$= 209.6 \text{ ft}$$

For this problem, minor losses are small and can be neglected.

$$\text{TDH} = \text{static head} + \text{friction head} + \text{minor losses}$$

$$320 \text{ ft} = 209.6 \text{ ft} + \text{friction head} + 0 \text{ ft}$$

The friction head is the difference between the TDH and the static head.

$$h_f = 320 \text{ ft} - 209.6 \text{ ft} = 110.4 \text{ ft}$$

Use the Hazen-Williams equation to determine the pipe size, and calculate the system curve.

$$h_f = \frac{3.022 v^{1.85} L}{C^{1.85} D^{1.165}} \qquad \text{[Eq. 1]}$$

This system uses plain cast-iron pipe. The roughness coefficient, C, is 100.

The pipe length can be determined from the illustration in the problem statement.

$$L = 50 \text{ ft} + 200 \text{ ft} + 140 \text{ ft} + 20 \text{ ft} + 25 \text{ ft}$$
$$+ 30 \text{ ft} + 15 \text{ ft} + 5 \text{ ft}$$
$$= 485 \text{ ft}$$

The velocity is calculated from the continuity equation.

$$Q = A v$$

The area in the continuity equation can be expressed as

$$A = \frac{\pi D^2}{4}$$

Substitute the equation for area in the continuity equation and solve for v.

$$v = \frac{4Q}{\pi D^2} \qquad \text{[Eq. 2]}$$

Substitute values for h_f, C, L, v, and Q in Eq. 1.

$$110.4 \text{ ft} = \frac{(3.022)\left(\dfrac{(4)\left(5.42 \,\frac{\text{ft}^3}{\text{sec}}\right)}{\pi D^2}\right)^{1.85}(485 \text{ ft})}{(100)^{1.85} D^{1.165}}$$

$$= \frac{(0.292)\left(\dfrac{6.9}{D^2}\right)^{1.85}}{D^{1.165}}$$

$$= \frac{(0.292)(6.9)^{1.85}}{D^{3.7} D^{1.165}}$$

Rearrange and simplify.

$$D^{4.865} = \frac{(0.292)(6.9)^{1.85}}{110.4 \text{ ft}} = 0.0944$$

$$D = 0.616 \text{ ft}$$

$$A = \frac{\pi (0.616 \text{ ft})^2}{4}$$

$$= 0.298 \text{ ft}^2$$

Next, compute the system curve, flow versus TDH. Use values of Q from 3 ft³/sec to 7 ft³/sec and $D = 0.616$ ft to solve for v with Eq. 2. Use values of v, $L = 485$ ft, $C = 100$, and $D = 0.616$ ft to calculate values of h_f with Eq. 1. Finally, TDH $= h_f +$ static head (209.8 ft). These calculations are shown in tabular form.

Q (ft³/sec)	v (ft/sec)	h_f (ft)	static head (ft)	TDH (ft)
3	10.1	37.15	209.8	247.0
4	13.5	63.5	209.8	273.3
5	16.8	95.2	209.8	305.0
6	20.2	133.9	209.8	343.7
7	23.6	178.9	209.8	388.4

This system curve, Q versus TDH, is shown in the previous illustration.

1.3. At the operating point (5.42 ft³/sec) on the efficiency curve in the illustration, the pump efficiency, η_{pump}, is $\boxed{90\%.}$

1.4. The required motor horsepower, P, can be found from the equation

$$P = \frac{h_A \dot{m}}{550 \eta_{\text{pump}}} \times \left(\frac{g}{g_c}\right)$$

The head added, h_A, equals the TDH found in Sol. 1.1.

$$h_A = 320 \text{ ft}$$

The water mass flow rate is

$$\dot{m} = Q\rho$$
$$= \left(5.42 \,\frac{\text{ft}^3}{\text{sec}}\right)\left(62.4 \,\frac{\text{lbm}}{\text{ft}^3}\right)$$
$$= 338.2 \text{ lbm/sec}$$

$$P = \frac{(320 \text{ ft})\left(338.2 \,\frac{\text{lbm}}{\text{sec}}\right)}{\left(550 \,\frac{\text{ft-lbf}}{\text{sec}}}{\text{hp}}\right)(0.90)}\left(\frac{32.2 \,\frac{\text{ft}}{\text{sec}^2}}{32.2 \,\frac{\text{ft-lbm}}{\text{lbf-sec}^2}}\right)$$

$$= \boxed{218.6 \text{ hp} \quad \text{[use a 225 hp motor]}}$$

Motors are rated by their output power, and the motor efficiency is not needed.

1.5. A reasonable assumption is that the inflow to the sand filter equals the outflow, Q. The volume, V, is the product of the flow, the time, and the overload rate.

$$V = \left(5.42 \,\frac{\text{ft}^3}{\text{sec}}\right)\left(3600 \,\frac{\text{sec}}{\text{hr}}\right)(8 \text{ hr})(1.25)$$
$$= 195{,}120 \text{ ft}^3$$

$$\text{clearwell area} = (200 \text{ ft})(250 \text{ ft}) = 50{,}000 \text{ ft}^2$$

$$\text{increased clearwell level} = \frac{195{,}120 \text{ ft}^3}{50{,}000 \text{ ft}^2} = 3.90 \text{ ft}$$

The accumulation in the clearwell should be no more than 3.9 ft because it will decrease the pump head and increase the flow.

SOLUTION 2

2.1. (a) *Prechlorination* is the process of adding chlorine to water at the beginning of the treatment process to disinfect water with high coliform levels, remove certain tastes and odors, and pretreat water for removal of high concentrations of iron and manganese.

(b) *Flash mixing* is the immediate and complete mixing of chemicals (such as coagulants) with water for 2 min or less to achieve maximum process efficiency.

(c) *Flocculation* is the slow, gentle mixing of water with a coagulant to remove colloidal material. This allows the destabilized colloidal particles to agglomerate and form flocs. The mixing time is usually in the range of 20–60 min.

(d) *Sedimentation* is usually 2–8 hr of low (or no) flow when flocs and other discrete particles settle out by gravity in a clarifier.

(e) *Filtration* is the process of passing clarified water through a bed of granular media such as sand, anthracite, or garnet to remove particles that have not been removed by sedimentation.

(f) *Storage* is provided by reservoirs or by water tanks/towers to account for daily and seasonal variations in water demand as well as for extra water demand for firefighting purposes. This excess storage capacity allows for the design capacity of the water treatment plant to be less than the peak water demand.

(g) *Chlorination* is the final disinfection of water with chlorine before distribution and consumption. The chlorine dose is determined by local requirements to provide either a low free-chlorine residual or chloramines only at the points of use.

2.2. Chlorinated organics would probably be present in the treated water because chlorine is added first—before any organic compounds are removed by flocculation, sedimentation, or filtration. Chlorinated organics are difficult to remove by these treatment processes.

2.3. *Trihalomethanes* (THMs) are organic chemicals produced during water treatment when organic compounds (precursors) in the water react with disinfectants such as chlorine, bromine, and iodine. These organic compounds occur naturally from decaying plants.

2.4. THMs can be reduced by

- using a raw water source with less organic THM precursors
- disinfecting with chlorine, bromine, or iodine only at the end of the treatment process after most THM precursors have been removed
- removing precursors before chlorination—for example, by treatment with granular activated carbon (GAC)
- using ozone, chlorine dioxide, or potassium permanganate for disinfection
- adding ammonia to form chloramines, which do not form as many THMs as free chlorine

2.5. Trihalomethane treatments add more expense to the treatment process. The cost of alternate disinfectants is also greater. Alternate disinfectants should be evaluated for disinfecting power, residual stability, and toxicity.

SOLUTION 3

3.1. Power is

$$P = D\text{v} \qquad \text{[Eq. 1]}$$

Drag is

$$D = \frac{C_D A \rho \text{v}^2}{2g_c} \qquad \text{[Eq. 2]}$$

$C_D =$ drag coefficient $= 1.8$ for flat plates

$\rho =$ density of water $= 62.4 \text{ lbm/ft}^3$

$g_c =$ gravitational constant $= 32.2 \text{ ft-lbm/lbf-sec}^2$

$A =$ total paddle area

$$= \left(\frac{2.81 \text{ ft}^2}{\text{paddle}}\right)(4 \text{ paddles})$$

$$= 11.25 \text{ ft}^2$$

$\text{v} =$ mixing velocity $= 0.75 \text{ (tip speed)}$

$\text{tip speed} = (\text{rpm})(\text{perimeter circumference of paddle})$

$$= (\text{rpm})\pi D$$

$$= \left(\frac{100 \frac{\text{rev}}{\text{min}}}{60 \frac{\text{sec}}{\text{min}}}\right)\pi(3.75 \text{ ft})$$

$$= 19.63 \text{ ft/sec}$$

$$\text{v} = (0.75)\left(19.63 \frac{\text{ft}}{\text{sec}}\right)$$

$$= 14.73 \text{ ft/sec}$$

Substitute Eq. 2 for drag into Eq. 1 for power.

$$P = \frac{C_D A \rho v^3}{2 g_c}$$

$$= \frac{(1.8)(11.25 \text{ ft}^2)\left(62.4 \ \dfrac{\text{lbm}}{\text{ft}^3}\right)\left(14.73 \ \dfrac{\text{ft}}{\text{sec}}\right)^3}{(2)\left(32.2 \ \dfrac{\text{ft-lbm}}{\text{lbf-sec}^2}\right)}$$

$$= \frac{62{,}709 \ \dfrac{\text{ft-lbf}}{\text{sec}}}{550 \ \dfrac{\text{ft-lbf}}{\text{sec-hp}}}$$

$$= \boxed{114.0 \text{ hp}}$$

3.2. The mean velocity gradient can be found from the following equation.

$$G = \sqrt{\frac{P}{\mu V_{\text{tank}}}}$$

$$P = 62{,}709 \text{ ft-lbf/sec}$$

$$\mu = 2.55 \times 10^{-5} \text{ lbf-sec/ft}^2 \text{ at } 55°F$$

$$V_{\text{tank}} = 185.7 \text{ ft}^3$$

$$G = \sqrt{\frac{62{,}709 \ \dfrac{\text{ft-lbf}}{\text{sec}}}{\left(2.55 \times 10^{-5} \ \dfrac{\text{lbf-sec}}{\text{ft}^2}\right)(185.7 \text{ ft}^3)}}$$

$$= \boxed{3639 \ 1/\text{sec}}$$

This value of the mean velocity gradient is high.

3.3.

$$Q = (2 \text{ MGD})\left(1.55 \ \frac{\frac{\text{ft}^3}{\text{sec}}}{\text{MGD}}\right)$$

$$= 3.1 \text{ ft}^3/\text{sec}$$

$$t_d = \frac{V}{Q} = \frac{185.7 \text{ ft}^3}{3.1 \ \dfrac{\text{ft}^3}{\text{sec}}}$$

$$= 59.9 \text{ sec} \quad [\text{high end}]$$

$$Gt_d = \left(3639 \ \frac{1}{\text{sec}}\right)(59.9 \text{ sec})$$

$$= 2.17 \times 10^5$$

The Gt_d value is high. The typical range of Gt_d is 10^4–10^5.

SOLUTION 4

4.1. When designing a treatment sequence for this water supply, the following items must be taken into account. The turbidity of 300 NTU is very high and should be reduced to 5 NTU or less. Assume that the water does not require treatment to remove hardness or strong tastes or odors. Iron (Fe) and managanese (Mn) are present in the water and require treatment to be removed. The bacteria count in the raw water is relatively low.

The treatment sequence is

(a) intake

(b) aeration to oxidize Fe and Mn

(c) presedimentation to settle out large solids

(d) coagulant addition and rapid mix

(e) coagulation/flocculation

(f) clarification to settle out flocs

(g) filtration with dual media filters to remove non-settleable particles

(h) disinfection with chlorine

(i) fluoridation

4.2. Treatment process profile:

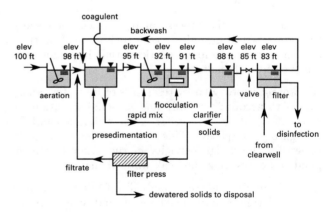

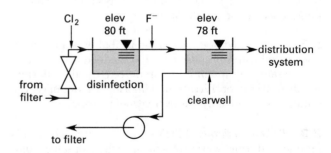

(All elevations are approximate to provide gravity flow between the units.)

4.3. When sizing the settling basins, the following assumptions should be made. The particles are fine sand and silt with specific gravities ranging from 1.2–2.65 and diameters ranging from 0.01–0.1 mm. For spherical particles with these characteristics, the settling velocity, v_s, is typically 2×10^{-3} ft/sec. In this problem, the particles are not assumed to be spherical, so v_s will be reduced by a factor of 0.7.

$$v_s = (0.7)\left(2 \times 10^{-3} \; \frac{\text{ft}}{\text{sec}}\right) = 1.4 \times 10^{-3} \text{ ft/sec}$$

Select the overflow rate, v^*, to be less than v_s.

$$v^* = 9.3 \times 10^{-4} \; \frac{\text{ft}}{\text{sec}} = 600 \text{ gal/day-ft}^2$$

Use two basins, each with half of the flow.

$$\frac{Q}{2} = 1 \times 10^6 \text{ gal/day}$$

By the continuity equation,

$$\begin{aligned} A_{\text{surface}} &= \frac{Q}{v^*} \\ &= \frac{1 \times 10^6 \; \dfrac{\text{gal}}{\text{day}}}{600 \; \dfrac{\text{gal}}{\text{day-ft}^2}} \\ &= 1667 \text{ ft}^2 \end{aligned}$$

Use a rectangular basin with a length-to-width ratio of 5:1.

$$\begin{aligned} A_{\text{surface}} &= (5\,W)\,W \\ &= 1667 \text{ ft}^2 \end{aligned}$$

$$\boxed{\begin{aligned} W &= 18.25 \text{ ft} \\ L &= 91.33 \text{ ft} \\ \text{water depth} &= 10 \text{ ft} \end{aligned}}$$

$$\begin{aligned} \text{tank volume} &= (18.25 \text{ ft})(91.33 \text{ ft})(10 \text{ ft}) \\ &= (16{,}668 \text{ ft}^3)\left(7.48 \; \frac{\text{gal}}{\text{ft}^3}\right) \\ &= \boxed{124{,}675 \text{ gal}} \end{aligned}$$

$$\begin{aligned} \text{detention time} = t &= \frac{V}{Q} \\ &= \left(\frac{124{,}675 \text{ gal}}{1 \times 10^6 \; \dfrac{\text{gal}}{\text{day}}}\right)\left(24 \; \frac{\text{hr}}{\text{day}}\right) \\ &= 3.0 \text{ hr} \quad [\text{OK}] \end{aligned}$$

SOLUTION 5

5.1. The excavation may be represented as

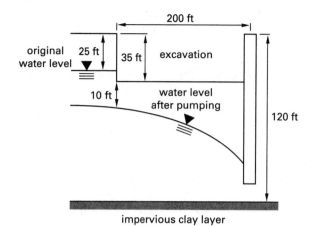

impervious clay layer

not to scale

$$\text{well flow,} \quad Q = \frac{\pi K_p (y_1^2 - y_2^2)}{\ln \dfrac{r_1}{r_2}}$$

K_p = permeability = 190 ft/day

y_1 = original thickness of saturated layer

$\quad = 120 \text{ ft} - 25 \text{ ft}$

$\quad = 95 \text{ ft}$

y_2 = thickness of aquifer at edge of excavation

$\quad = 120 \text{ ft} - 35 \text{ ft} - 10 \text{ ft}$

$\quad = 75 \text{ ft}$

r_1 = radius of influence of the well, 800 ft

r_2 = radius of excavation (distance from center of square to corner)

$\quad = (200 \text{ ft})\sqrt{2}$

$\quad = 283 \text{ ft}$

$$Q = \frac{\pi \left(190 \; \dfrac{\text{ft}}{\text{day}}\right)\left((95 \text{ ft})^2 - (75 \text{ ft})^2\right)}{\left(86{,}400 \; \dfrac{\text{sec}}{\text{day}}\right)\ln \dfrac{800 \text{ ft}}{283 \text{ ft}}}$$

$$\quad = 22.60 \text{ ft}^3/\text{sec}$$

$$Q = \left(22.60 \; \frac{\text{ft}^3}{\text{sec}}\right)\left(448.83 \; \frac{\dfrac{\text{gal}}{\text{min}}}{\dfrac{\text{ft}^3}{\text{sec}}}\right)$$

$$\quad = \boxed{10{,}144 \text{ gal/min}}$$

5.2. 10,144 gal/min is too much water to draw from wells in the center of this excavation site. More than one well would be required. A 20 ft drawdown almost 300 ft from a well requires pumping a great deal of water. Wells in the center of a 400 ft × 400 ft excavation do not appear to be the best way to lower the water table 20 ft at this site.

5.3. Other methods that could be used to provide a dry working site include the following.

(a) Wells spaced around the perimeter of the site, as well as one or two wells in the middle, would also dewater the site.

(b) A bentonite slurry or concrete cutoff wall could be built around the site to a depth greater than the desired dewatered depth (such as 55 ft) to slow the flow of groundwater into the site. The excavation could then be dewatered by wells or trenches.

(c) Dewatering trenches could be used around the site.

SOLUTION 6

6.1. The assumed average water demands are

residential	100 gpcd
commercial	80 gpcd
public	20 gpcd
waste	20 gpcd
total demand	220 gpcd

Population records from the past 20–40 yr for this town, as well as for surrounding towns, should be analyzed to help estimate future population growth.

Use the night population of 35,000, and assume a constant population growth of 20% per decade. Design a water treatment plant for a 30 yr life. The multiplier for the design population is $1 + 0.2 = 1.2$. The exponent for the design population is 30 yr/10 yr = 3.0.

$$\text{design population} = (35{,}000)(1.2)^3$$
$$= 60{,}480 \quad [\text{use } 60{,}500]$$
$$\text{average design flow} = \frac{(220 \text{ gpcd})(60{,}500)}{1{,}000{,}000 \ \frac{\text{gal}}{\text{MG}}}$$
$$= \boxed{13.31 \text{ MGD}}$$

6.2. The maximum hourly flow can be calculated using the average flow and a peak flow multiplier of 3.0.

$$\frac{\left(13.31 \times 10^6 \ \frac{\text{gal}}{\text{day}}\right)(3.0)}{24 \ \frac{\text{hr}}{\text{day}}} = \boxed{1.66 \times 10^6 \text{ gal/hr}}$$

The maximum daily flow is 1.8 times the average flow.

$$\left(13.31 \times 10^6 \ \frac{\text{gal}}{\text{day}}\right)(1.8) = 23.96 \times 10^6 \text{ gal/day}$$

6.3. Fire-fighting water requirements are

$$Q_{\text{gpm}} = (1020\sqrt{P})(1 - 0.01\sqrt{P})$$
$$P = \text{population in thousands of people}$$
$$= \frac{60{,}500}{1000}$$
$$= 60.5$$
$$Q_{\text{gpm}} = (1020\sqrt{60.5})(1 - 0.01\sqrt{60.5})$$
$$= \left(7317 \ \frac{\text{gal}}{\text{min}}\right)\left(60 \ \frac{\text{min}}{\text{hr}}\right)\left(24 \ \frac{\text{hr}}{\text{day}}\right)$$
$$= \boxed{10.54 \times 10^6 \text{ gal/day}}$$

6.4. The minimum water pressure requirement is 50 lbf/in². This can be maintained by pumping water to storage tanks that are high enough to provide adequate pressure, such as

$$\left(50 \ \frac{\text{lbf}}{\text{in}^2}\right)\left(2.31 \ \frac{\text{ft}}{\frac{\text{lbf}}{\text{in}^2}}\right) = 115.5 \text{ ft} + \text{head losses}$$

The maximum water pressure requirement is 80 lbf/in² = 184.7 ft. This can be maintained (not exceeded) by proper design of the storage tanks and distribution system and by using pressure-regulating or altitude valves where appropriate.

6.5. The storage flow to equalize the pumping rate is 0.25 times the maximum daily flow.

$$\text{storage flow} = (0.25)\left(23.96 \times 10^6 \ \frac{\text{gal}}{\text{day}}\right)(1 \text{ day})$$
$$= 6.0 \times 10^6 \text{ gal}$$
$$\text{fire-flow storage} = (7 \text{ hr})\left(7317 \ \frac{\text{gal}}{\text{min}}\right)\left(60 \ \frac{\text{min}}{\text{hr}}\right)$$
$$= 3.07 \times 10^6 \text{ gal}$$
$$\text{emergency storage} = (3 \text{ days})\left(13.31 \times 10^6 \ \frac{\text{gal}}{\text{day}}\right)$$
$$= 39.93 \times 10^6 \text{ gal}$$
$$\text{total storage} = 6.0 \times 10^6 \text{ gal} + 3.07 \times 10^6 \text{ gal}$$
$$+ 39.93 \times 10^6 \text{ gal}$$
$$= \boxed{49 \times 10^6 \text{ gal} \quad (49 \text{ MG})}$$

6.6. The system layout is

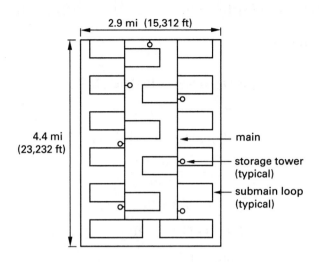

6.7. This answer is subject to variations in assumed layout of the network. Using seven tanks,

$$\text{storage in each tank} = \frac{49 \times 10^6 \text{ gal}}{7 \text{ tanks}}$$
$$= 7 \times 10^6 \text{ gal/tank}$$

$$\text{maximum flow} = \frac{1.66 \times 10^6 \dfrac{\text{gal}}{\text{hr}}}{60 \dfrac{\text{min}}{\text{hr}}}$$
$$= 27{,}730 \text{ gal/min}$$

There are seven towers, each feeding two directions.

$$\text{flow in 16 in (main) pipe} = \frac{27{,}730 \dfrac{\text{gal}}{\text{min}}}{(2)(7)}$$
$$= 1980 \text{ gal/min}$$

Assume each feed direction has three submain take-offs.

$$\text{flow in 12 in pipe} = \frac{1980 \dfrac{\text{gal}}{\text{min}}}{3}$$
$$= 660 \text{ gal/min}$$

Assume reasonable values for the remaining pipes.

$$\text{flow in 6 in pipe} = 200 \text{ gal/min}$$
$$\text{flow in 3/4 in tube} = 8 \text{ gal/min}$$

Calculate the head losses with the Hazen-Williams equation.

$$h_f = \frac{10.44 L Q^{1.85}}{C^{1.85} d^{4.87}}$$

pipe diameter (in)	length (ft)	flow, Q (gal/min)	C	h_f (ft)
16	4000	1980	100	14.5
12	4000	660	100	7.7
6	1000	200	100	6.2
$\frac{3}{4}$	80	8	130	19.5

$$\text{total friction losses} = 47.9 \text{ ft}$$

$$\text{minimum pressure at house} = 50 \text{ lbf/in}^2$$

$$\text{static head} = \left(50 \ \frac{\text{lbf}}{\text{in}^2}\right)\left(2.31 \ \frac{\text{ft}}{\text{lbf-in}^2}\right)$$
$$= 115.5 \text{ ft}$$

Add static head and friction losses. The working tank elevation is

$$115.5 \text{ ft} + 47.9 \text{ ft} = \boxed{163.4 \text{ ft}}$$

6.8. The water tanks should be filled at periods of low demand, such as early morning (2:00–5:00 a.m.) and afternoon (2:00–4:00 p.m.). In addition to being filled twice a day, the tanks may need to have water pumped in during peak demand periods, such as morning (6:00–9:00 a.m.) and evening (4:30–8:30 p.m.).

6.9. The minimum water pressure requirement is

$$\left(50 \ \frac{\text{lbf}}{\text{in}^2}\right)\left(2.31 \ \frac{\text{ft}}{\text{lbf-in}^2}\right) = 115.5 \text{ ft}$$

The working static elevation in a tank 80 ft above the ground is the static head. Subtract the friction losses to get the total available head: 80 ft < 115.5 ft.

$$\boxed{\text{The water pressure requirements will not be met.}}$$

SOLUTION 7

7.1. Hardness is contributed by Ca^{++}, Fe^{++}, and Mg^{++}. Convert mg/L of substance to mg/L as $CaCO_3$.

$$\text{multiplier} = \frac{\text{equivalent weight } CaCO_3}{\text{equivalent weight substance}}$$
$$= \frac{50}{\dfrac{\text{MW}}{\text{charge}}}$$

For Ca^{++}, MW = 40.

$$\text{multiplier} = \frac{50}{\dfrac{40}{2}} = 2.5$$

compound	concentration (mg/L)	multiplier	hardness (mg/L as $CaCO_3$)
Ca^{++}	80	2.5	200.0
Fe^{++}	3	1.79	5.4
Mg^{++}	30	4.1	123.0

$$\text{total hardness} = 200.0 \; \frac{mg}{L} + 5.4 \; \frac{mg}{L} + 123.0 \; \frac{mg}{L}$$

$$= \boxed{328.4 \;\; mg/L \text{ as } CaCO_3}$$

7.2. Calculate $[CO_2]$ and $[HCO_3^-]$.

$$[CO_2] = \frac{15 \times 10^{-3} \; \frac{g}{L}}{44 \; \frac{g}{mol}}$$

$$= 3.41 \times 10^{-4} \; mol/L$$

$$[HCO_3^-] = \frac{336 \times 10^{-3} \; \frac{g}{L}}{61 \; \frac{g}{mol}}$$

$$= 5.51 \times 10^{-3} \; mol/L$$

Assume

$$[H_2CO_3^*] = [CO_{2 \; aq}] = 3.41 \times 10^{-4} \; mol/L$$

$$H_2CO_3^* \rightleftharpoons HCO_3^- + H^+$$

$$pK_1 = 6.3$$

$$\frac{[H^+][HCO_3^-]}{[H_2CO_3^*]} = 10^{-6.3}$$

$$[H^+] = \frac{(10^{-6.3})[H_2CO_3^*]}{[HCO_3^-]}$$

$$= \frac{(10^{-6.3})\left(3.41 \times 10^{-4} \; \frac{mol}{L}\right)}{5.51 \times 10^{-3} \; \frac{mol}{L}}$$

$$= 3.10 \times 10^{-8} \; mol/L$$

$$pH = -\log[H^*]$$

$$= -\log\left(3.10 \times 10^{-8} \; \frac{mol}{L}\right)$$

$$= \boxed{7.5}$$

7.3. The alkalinity is found by

$$\text{alkalinity (eq/L)} = [HCO_3^-] + 2[CO_3^{--}] + [OH^-] - [H^+]$$

From Sol. 7.2,

$$[HCO_3^-] = 5.51 \times 10^{-3} \; mol/L$$

$$[H^+] = 3.1 \times 10^{-8} \; mol/L$$

$$[H^+][OH^-] = (1 \times 10^{-14})[OH^-]$$

$$= \frac{1 \times 10^{-14}}{3.1 \times 10^{-8}}$$

$$= 3.23 \times 10^{-7} \; mol/L$$

$$HCO_3^- \rightleftharpoons CO_3^{--} + H^+$$

$$pK_2 = 10.3$$

$$\frac{[H^+][CO_3^{--}]}{[HCO_3^-]} = 10^{-10.3}$$

$$[CO_3^{--}] = \frac{(10^{-10.3})\left(5.51 \times 10^{-3} \; \frac{mol}{L}\right)}{3.1 \times 10^{-8} \; \frac{mol}{L}}$$

$$= 8.91 \times 10^{-6} \; mol/L$$

$$\text{alkalinity} = 5.51 \times 10^{-3} \; \frac{mol}{L}$$

$$+ (2)\left(8.91 \times 10^{-6} \; \frac{mol}{L}\right)$$

$$+ 3.23 \times 10^{-7} \; \frac{mol}{L}$$

$$- 3.1 \times 10^{-8} \; \frac{mol}{L}$$

$$= 5.528 \times 10^{-3} \; \frac{eq}{L}$$

$$\begin{aligned} \text{alkalinity} \\ \text{as } CaCO_3 \end{aligned} = \left(5.528 \times 10^{-3} \; \frac{eq}{L}\right)$$

$$\times \left(50 \times 10^3 \; \frac{mg \text{ as } CaCO_3}{eq}\right)$$

$$= \boxed{276.4 \; mg/L \text{ as } CaCO_3}$$

$$\text{acidity (eq/L)} = 2[H_2CO_3^*] + [HCO_3^-] + [H^+] - [OH^-]$$

From Sol. 7.2,

$$[H_2CO_3^*] = 3.41 \times 10^{-4} \ \frac{mol}{L}$$

$$acidity = (2)\left(3.41 \times 10^{-4} \ \frac{mol}{L}\right)$$
$$+ 5.51 \times 10^{-3} \ \frac{mol}{L}$$
$$+ 3.1 \times 10^{-8} \ \frac{mol}{L}$$
$$+ 3.23 \times 10^{-7} \ \frac{mol}{L}$$
$$= 6.19 \times 10^{-3} \ eq/L$$

$$\begin{aligned} \text{acidity} \\ \text{as } CaCO_3 \end{aligned} = \left(6.19 \times 10^{-3} \ \frac{eq}{L}\right)$$
$$\times \left(50 \times 10^3 \ \frac{mg \ as \ CaCO_3}{eq}\right)$$
$$= \boxed{309.6 \ mg/L \ as \ CaCO_3}$$

7.4. Alkalinity is mainly caused by HCO_3^-. Acidity is mainly caused by $H_2CO_3^*$ (i.e., the combination of H_2O and CO_2). Assume the water is alkaline—check in Sol. 7.3. From Sol. 7.2, pH $= 7.5$.

$$\boxed{\text{The water is alkaline.}}$$

7.5. This water would probably be the color of rust or rusty tea. The iron would contribute a rust color. Any organics (such as TDS) may also contribute a tea color.

7.6. This water should be softened to reduce the hardness to 60 mg/L or less (which is considered soft water).

7.7. From Sol. 7.3,

$$total \ hardness = 328.4 \ mg/L \ as \ CaCO_3$$
$$alkalinity = 276.4 \ mg/L \ as \ CaCO_3$$

$$CO_2 = \left(15 \ \frac{mg}{L}\right)\left(\frac{50}{\frac{44}{2}}\right)$$
$$= 34.1 \ mg/L \ as \ CaCO_3$$

In the first stage, add hydrated lime (93% pure $Ca(OH)_2$) to remove CO_2, alkalinity $Ca(HCO_3)_2$, and $Mg(HCO_3)_2$).

Hydrated lime must be converted to $CaCO_3$ equivalents.

$$\text{formula weight, } Ca(OH)_2 = 74$$
$$\text{valence} = 2$$
$$\text{equivalence, } Ca(OH)_2 = \frac{50}{\frac{74}{2}}$$
$$= 1.35$$

The amount of lime required to remove CO_2 and HCO_3^- is

$$\frac{34.1 \ \frac{mg}{L} + 276.4 \ \frac{mg}{L}}{(1.35)(0.93)} = 247.3 \ mg/L \ of \ Ca(OH)_2$$

Add another 50 mg/L of lime to raise the pH above 10.8 to precipitate $Mg(OH)_2$.

$$\text{extra } Ca(OH)_2 = \frac{50 \ \frac{mg}{L}}{0.93} = 53.8 \ mg/L$$
$$\text{total } Ca(OH)_2 = 247.3 \ \frac{mg}{L} + 53.8 \ \frac{mg}{L}$$
$$= \boxed{301.1 \ mg/L}$$

In the second stage, add soda ash (98% Na_2CO_3) to remove $CaSO_4$ and $MgSO_4$.

$$\text{noncarbonate hardness} = \text{total hardness} - \text{alkalinity}$$
$$= 328.4 \ \frac{mg}{L} - 276.4 \ \frac{mg}{L}$$
$$= 52.0 \ mg/L \ as \ CaCO_3$$

The hardness needs to be reduced to 80 mg/L as $CaCO_3$. Remove

$$52.0 \ \frac{mg}{L} - 80 \ \frac{mg}{L} = -28 \ mg/L \ as \ CaCO_3$$

$$CaCO_3 \ \text{equivalence of } Na_2CO_3 = \frac{50}{\frac{106}{2}}$$
$$= 0.94$$

$$\boxed{\begin{aligned}\text{No soda ash is required because hardness is less than} \\ \text{80 mg/L after lime treatment.}\end{aligned}}$$

Topic II: Geotechnical

Geotechnical

6 Soils

PROBLEM 1

Soil is taken from a borrow site and used as fill at a construction site. A standard Proctor test indicates that the dry specific weight of a borrow soil is 105 lbf/ft^3 at an optimum moisture content of 20%. The borrow soil has a water content of 26%. The specific gravity of the soil solids is 2.65. Per contract, the fill must be compacted to 95% of the standard Proctor maximum, but the earthwork contractor complains that the fill pumps and ruts badly before reaching 95% compaction.

1.1. Plot three points on the zero air void curve for moisture contents of 20%, 25%, and 30%.

1.2. What is the maximum water content when the fill is compacted to 95% of the standard Proctor maximum?

1.3. What is the maximum density without pumping or rutting at the original water content of 26%?

1.4. Does the earthwork contractor have a valid argument?

1.5. Briefly give your recommendations.

PROBLEM 2

Samples of five different soils, A, B, C, D, and E, are taken. The properties of the five soils are given.

soil	sieve analysis (% passing)			
	no. 10	no. 40	no. 100	no. 200
A	45	15	5	–
B	75	40	20	15
C	85	62	35	7
D	89	72	48	39
E	95	87	72	68

soil	liquid limit, w_l	plastic limit, w_p	plasticity index, I_p
A	8	–	–
B	27	15	12
C	37	29	8
D	35	22	13
E	52	25	27

2.1. What are the soils' classifications according to the AASHTO system?

2.2. What are the soils' classifications according to the Unified Soils system?

2.3. Which soil(s) would require undercutting if exposed after excavation and grading (i.e., would not be suitable for roadway base material)?

2.4. Which soil(s) would not make suitable roadway embankment material?

2.5. Which soil(s) would not make suitable borrow material for a subbase?

PROBLEM 3

The cut and fill volumes (in cubic yards) of soil for 20 stations along a new highway are given. The soil has a shrinkage factor of 0.25. The contractual free haul distance is 600 ft.

station	cut (yd^3)	fill (yd^3)
1+00	–	1000
2+00	2000	–
3+00	1000	–
4+00	–	1500
5+00	500	–
6+00	1000	–
7+00	1500	–
8+00	–	500
9+00	–	1500
10+00	–	500
11+00	–	1000
12+00	2000	–
13+00	2500	–
14+00	1000	–
15+00	0	0
16+00	–	1000
17+00	–	1500
18+00	–	2000
19+00	–	2500
20+00	–	2000

3.1. Draw the mass diagram.

3.2. Show all balance points.

3.3. Calculate all haul amounts and draw arrows to show all haul directions.

3.4. What is the station yardage of overhaul between 1+00 and the first balance point?

3.5. What is the definition of a bank yard?

3.6. What methods can be used to protect the slope and control erosion from the sides of cuts made along the highway?

PROBLEM 4

Two 20 ft deep cuts are made at an angle of 45°, one in sandy soil (angle of internal friction = 30°, specific weight = 120 lbf/ft³), and another in homogeneous soft clay (specific weight = 110 lbf/ft³; unconfined compressive strength = 1600 lbf/ft²). Answer parts 4.1 through 4.4 for both soils.

4.1. Find the slope stability factor of safety for the material alone.

4.2. Find the factor of safety for translational or lateral block failure.

4.3. Is either of the factors of safety (from parts 4.1 and 4.2) adequate? Why or why not?

4.4. If the slope is inadequate, what slope is required?

PROBLEM 5

An 18 ft high, 40 ft wide braced cut is made in clay (cohesion = 450 lbf/ft²; specific weight = 115 lbf/ft³). The bottom of the cut is well above a hard clay layer.

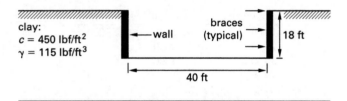

5.1. What is the factor of safety against bottom heave?

5.2. At what depth will the bottom heave?

5.3. If the bottom of the cut is only 10 ft above a hard clay layer, what will be the factor of safety?

PROBLEM 6

A 20 in diameter concrete pipe with a crushing strength of 1000 lbf/ft² is buried 4 ft below grade. The trench is backfilled with granular soil having a dry specific weight of 120 lbf/ft³. The pipe is loaded with an additional dead load of 100 lbf/ft². Determine if the pipe will break under H-20 loading for bedding classes B and D.

PROBLEM 7

Two soils have been proposed for use as a landfill soil liner. Falling head permeability tests have been performed.

permeability test results

test parameter	soil A	soil B
sample length	15 cm	15 cm
sample area	40 cm²	40 cm²
initial head	100 cm	200 cm
final head	75 cm	190 cm
test time	300 min	6000 min
standpipe area	5 cm²	2 cm²

7.1. What are the permeabilities of the two soils in units of ft/yr?

7.2. Based on the permeabilities alone, make a recommendation as to which soil(s) should be used as a liner.

SOLUTION 1

1.1. Plot the zero air voids curve. Use the following equation in which the degree of saturation, s, equals 1. The dry specific weight is

$$\gamma_d = \frac{\gamma_w s}{w + \dfrac{s}{SG}}$$

γ_w = specific weight of water = 62.4 lbf/ft^3

w = water content expressed as a decimal

SG = specific gravity of solids

Choose values for water content and solve for γ_d.

w	γ_d (lbf/ft^3)
15%	118.3
20%	108.1
25%	99.5
30%	92.1

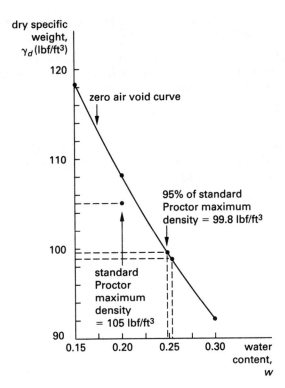

1.2. 95% of the standard Proctor maximum density is

$$\gamma = (0.95)\left(105 \ \frac{\text{lbf}}{\text{ft}^3}\right) = 99.8 \ \text{lbf/ft}^3$$

Soil can be no more than fully saturated ($s = 1$). Find the intersection of the zero air voids curve with 95% of the standard Proctor maximum.

From the plot, the maximum water content at 95% of the standard Proctor maximum is approximately 25%.

1.3. Maximum specific weight without pumping at 26% water content is the point on the zero air voids curve at 26%. From the plot, $\gamma \approx 98$ lbf/ft^3.

The maximum specific weight can also be found by

$$\gamma_d = \frac{62.4 \ \dfrac{\text{lbf}}{\text{ft}^3}}{0.26 + \dfrac{1}{2.65}} = \boxed{97.9 \ \text{lbf/ft}^3}$$

1.4. The earthwork contractor has a valid point. At 26% moisture content, any density over 98 lbf/ft^3 will cause the soil to pump and rut.

1.5. The contractor should dry the soil back to near-optimum water content (20%). Each lift should be spread and then disked until it dries to about 20% water content. Disking should be done prior to compaction.

Drying all the way back to 20% is recommended, but not required. Anything less than or equal to about 24% should be acceptable.

Finally, the contractor should be reminded that the nearer the soil is to optimum water content, the easier it will be to reach 95% of the standard Proctor maximum. Fewer compactor passes are required at 20% water content than would be required at 24% or 25% water content.

SOLUTION 2

2.1. Use the AASHTO soil classification chart. Proceed from left to right in the chart. The correct group will be found by the process of elimination. The first group meeting all the test data is the group classification. Interpolate unknown sieve percentages.

soil	AASHTO group	Unified Soils Classification
A	A-1-a	SW or SP (C_u undetermined)
B	A-2-6	SC (CL fines)
C	A-2-4	SP–SM (ML fines; $C_u < 4$)
D	A-6	SC (CL fines)
E	A-7-6	CH (CH fines)

2.2. Use the Unified Soil Classification flow charts found in ASTM D2487, *Standard Practice for Classification of Soils for Engineering Purposes*. Choose the appropriate flow chart based on 50% or more passing the no. 200 sieve, or 50% or more retained on the no. 200 sieve. Enter the flow chart from the left and follow the path that meets all the test data. Assume all soils have 50% or more passing the no. 4 sieve.

Geotechnical

C_u is not specified and cannot be determined for soil A. Classify fines using ASTM D2487 or any standard plasticity chart.

$$C_u = \frac{D_{60}}{D_{10}}$$

2.3. Refer to the AASHTO soil classification chart. Soils B, D, and E would not be suitable for roadway base material because they are rated A-2-6 or worse.

2.4. Soils D and E are clayey soils rated A-6 or worse. They would not be suitable for a roadway embankment.

2.5. Soils rated A-2-6 or worse (such as soils B, D, and E) would not make good subbase borrow soil.

SOLUTION 3

3.1. The mass diagram is constructed by first modifying the fill values by the shrinkage factor given for the soil. Next, add all the cut values and subtract all the modified fill values, then tabulate the cumulative values for each station as shown in the following table. Plot the cumulative values as shown in *Illustration for Sols. 3.1 and 3.2.*

3.2. Balance points may be found by drawing a horizontal line at any location on the mass diagram. Cut and fill are balanced between any two adjacent stations where the mass diagram is intersected by the horizontal. Consider the points where the mass diagram crosses the zero cut and fill horizontal. Balance points along this line are indicated by vertical arrows on the mass diagram. These points occur approximately at stations 1+50, 4+00, 9+50, 11+70, and 17+40. (See *Table for Sols. 3.1 and 3.2* and *Illustration for Sols. 3.1 and 3.2.*)

Haul directions are indicated on the mass diagram. In general, a positive slope indicates cutting, and a negative slope indicates filling.

3.3. Haul volumes may be calculated as the areas below the mass diagram between balance points. For example, the volume between stations 1+50 and 4+00 is found by

$$\text{sta } (2 - 1.5)\left(\frac{750 \text{ yd}^3}{2}\right)$$
$$+ \text{sta } (3 - 2)\left(\frac{1750 \text{ yd}^3 + 750 \text{ yd}^3}{2}\right)$$
$$+ \text{sta } (4 - 3)\left(\frac{-125 \text{ yd}^3 + 1750 \text{ yd}^3}{2}\right)$$
$$= 188 \text{ sta yd}^3 + 1250 \text{ sta yd}^3 + 813 \text{ sta yd}^3$$
$$= \boxed{2251 \text{ sta yd}^3 \quad (2251 \text{ sta yd})}$$

Other haul volumes for balance points are shown in the preceding table.

3.4. The overhaul volume between station 1+00 and the first balance point is 0 sta yd, since 50 ft is less than the freehand distance.

3.5. A bank yard is defined as cubic yardage in place or in the ground.

3.6. Measures to control slope stability and erosion include reducing slope angles, hydroseeding or vegetating slopes, and using silt fences.

Illustration for Sols. 3.1 and 3.2

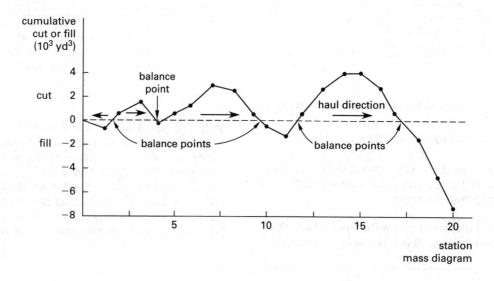

Table for Sols. 3.1 and 3.2

sta	cut (10^3 yd^3)	fill (10^3 yd^3)	fill (10^3 yd^3) $\times 1.25$	cumulative cut or fill 10^3 yd^3	haul volume (sta-10^3 yd)	total volume at balance points (sta-10^3 yd)
0	–	0	–	0	–	–
1	0	1	1.25	−1.25	−0.625	–
1.5	–	–	–	0	−0.3125	−0.9375
2	2	0	0	0.75	0.1875	–
3	1	0	0	1.75	1.25	–
4	0	1.5	1.875	−0.125	0.8125	2.25
5	0.5	0	0	0.375	0.125	–
6	1	0	0	1.375	0.875	–
7	1.5	0	0	2.875	2.125	–
8	0	0.5	0.625	2.25	2.5625	–
9	0	1.5	1.875	0.375	1.3125	–
9.5	–	–	–	0	0.09375	7.09375
10	0	0.5	0.625	−0.25	−0.0625	–
11	0	1	1.25	−1.5	−0.875	–
11.7	–	–	–	0	−0.525	−1.4625
12	2	0	0	0.5	0.075	–
13	2.5	0	0	3	1.75	–
14	1	0	0	4	3.5	–
15	0	0	0	4	4	–
16	0	1	1.25	2.75	3.375	–
17	0	1.5	1.875	0.875	1.8125	–
17.4	–	–	–	0	0.175	14.6875
18	0	2	2.5	−1.625	−0.4875	–
19	0	2.5	3.125	−4.75	−3.1875	–
20	0	2	2.5	−7.25	−6	−9.675

SOLUTION 4

4.1. The factor of safety, FS, for the material alone is calculated as the ratio of soil friction angle, ϕ, to slope angle, β.

$$\text{FS}_{\text{sand}} = \frac{\phi}{\beta} = \frac{30°}{45°}$$
$$= \boxed{0.67}$$

For homogeneous soft clay, the friction angle is taken as zero.

$$\text{FS}_{\text{clay}} = \frac{\phi}{\beta} = \frac{0°}{45°}$$
$$= \boxed{0}$$

4.2. The factor of safety against a soil block translation or rotation is computed by a computer program or taken from a slope stability chart. For the sand, there is no surcharge and no water inside or outside of the slope.

$$\text{FS} = \frac{\tan \phi}{\tan \beta}$$
$$= \frac{\tan 30°}{\tan 45°}$$
$$= \boxed{0.58}$$

For the clay, use the Taylor slope stability chart ($\phi = 0°$). The factor of safety can be computed as

$$\text{FS} = \frac{N_o c}{\gamma H}$$

N_o is the stability number from the chart, c is the cohesion intercept taken as one-half the unconfined compressive strength for $\phi = 0$, γ is the specific weight of the soil, and H is the slope height.

From the Taylor slope stability chart, an average value of N_o is approximately 6 for $\beta = 45°$. The factor of safety is

$$FS = \frac{N_o c}{\gamma H} = \frac{(6)\left(\dfrac{1600 \ \frac{lbf}{ft^2}}{2}\right)}{\left(110 \ \frac{lbf}{ft^3}\right)(20 \ ft)}$$

$$= \boxed{2.2}$$

4.3. Any factor of safety less than 1 is unacceptable because failure is indicated. Generally, a factor of safety of 1.25–1.5 is considered acceptable. The clay slope should be stable as shown in Sol. 4.2, but the sand slope will be unstable.

4.4. The clay slope is stable with a factor of safety of 2.2. The sand slope should be revised to have a factor of safety of approximately 1.5. From the previous equation, a maximum stable slope angle may be determined by

$$\beta = \arctan \frac{\tan \phi}{FS}$$

$$= \arctan \frac{\tan 30°}{1.5}$$

$$= \boxed{21°}$$

SOLUTION 5

5.1. When the hard clay layer is deeper than $0.7B$ (B = excavation width), the factor of safety, FS, against the bottom heave is

$$FS = \frac{Q_u}{Q} = \frac{5.7cB_1}{\gamma H B_1 - cH}$$

$$c = \text{cohesion}$$

$$B_1 = 0.7B$$

$$\gamma = \text{soil specific weight}$$

$$H = \text{excavation height}$$

$$FS = \frac{(5.7)\left(450 \ \frac{lbf}{ft^2}\right)(0.7)(40 \ ft)}{\left(115 \ \frac{lbf}{ft^3}\right)(18 \ ft)(0.7)(40 \ ft) - \left(450 \ \frac{lbf}{ft^2}\right)(18 \ ft)}$$

$$= \boxed{1.44}$$

5.2. If all other parameters remain the same but the excavation is carried deeper, the bottom will heave when the factor of safety equals 1.

$$FS = \left(\frac{1}{H}\right)\left(\frac{5.7c}{\gamma - \dfrac{c}{0.7B}}\right) = 1$$

Solve for H.

$$H = \frac{5.7c}{\gamma - \dfrac{c}{0.7B}}$$

$$= \frac{(5.7)\left(450 \ \frac{lbf}{ft^2}\right)}{115 \ \frac{lbf}{ft^3} - \dfrac{450 \ \frac{lbf}{ft^2}}{(0.7)(40 \ ft)}}$$

$$= \boxed{25.9 \ ft}$$

5.3. Assume that the original conditions apply but that the hard clay is now 10 ft below the bottom of the excavation. For a depth to hard layer, D, of less than $0.7B$, use

$$FS = \left(\frac{1}{H}\right)\left(\frac{5.7c}{\gamma - \dfrac{c}{D}}\right)$$

$$= \left(\frac{1}{18 \ ft}\right)\left(\frac{(5.7)\left(450 \ \frac{lbf}{ft^2}\right)}{115 \ \frac{lbf}{ft^3} - \dfrac{450 \ \frac{lbf}{ft^2}}{10 \ ft}}\right)$$

$$= \boxed{2.03}$$

SOLUTION 6

Compute the dead load per unit length, w, and the dead load pressure, $p_{\text{dead load}}$. C_w is the correction coefficient, and B is the trench width.

$$w = C_w \gamma B^2$$

For a 20 in diameter pipe, a trench of approximately 36 in would be required. The C_w is approximately 1, so

$$w = (1)\left(120 \ \frac{\text{lbf}}{\text{ft}^3}\right)(3 \ \text{ft})^2$$
$$= 1080 \ \text{lbf/ft}$$

$$p_{\text{dead load}} = \frac{1080 \ \frac{\text{lbf}}{\text{ft}}}{3 \ \text{ft}} + 100 \ \frac{\text{lbf}}{\text{ft}^2}$$
$$= 460 \ \text{lbf/ft}^2$$

The live load, $p_{\text{live load}}$, for H-20 loading is determined from UFC 3-220-10N, 2005 (previously NAVFAC DM 7.1 (1982)), Chap. 4, Fig. 19. For 4 ft of cover, the live load on the pipe is approximately 350 lbf/ft². This value includes impact effects.

$$p_{\text{live load}} = 350 \ \text{lbf/ft}^2$$

The allowable load, $p_{\text{allowable}}$, that the pipe should be able to withstand without excessive cracking is

$$p_{\text{allowable}} = (p_{\text{dead load}} + p_{\text{live load}})\left(\frac{\text{FS}}{L_f}\right)$$

L_f is a load factor depending on bedding conditions. For bedding class B, $L_f = 1.9$; for bedding class D, $L_f = 1.1$. For class B,

$$p_{\text{allowable}} = \left(460 \ \frac{\text{lbf}}{\text{ft}^2} + 350 \ \frac{\text{lbf}}{\text{ft}^2}\right)\left(\frac{1.25}{1.9}\right)$$
$$= \boxed{533 \ \text{lbf/ft}^2 < 1000 \ \text{lbf/ft}^2 \quad \text{[safe]}}$$

For class D,

$$p_{\text{allowable}} = \left(460 \ \frac{\text{lbf}}{\text{ft}^2} + 350 \ \frac{\text{lbf}}{\text{ft}^2}\right)\left(\frac{1.25}{1.1}\right)$$
$$= \boxed{920 \ \text{lbf/ft}^2 < 1000 \ \text{lbf/ft}^2 \quad \text{[safe]}}$$

However, if a factor of safety of 1.5 is desired, the pipe would not be safe for class D bedding.

SOLUTION 7

7.1. The hydraulic conductivity of the two soils may be computed from the equation for the falling-head test.

$$k = \frac{A'L}{At} \ln \frac{h_i}{h_f}$$

$$k_{\text{soil A}} = \left(\frac{(5 \ \text{cm}^2)(15 \ \text{cm})}{(40 \ \text{cm}^2)(300 \ \text{min})}\right)\ln \frac{100 \ \text{cm}}{75 \ \text{cm}}$$
$$= 0.0018 \ \text{cm/min}$$

Converting the units,

$$k_{\text{soil A}} = \left(0.0018 \ \frac{\text{cm}}{\text{min}}\right)\left(\frac{525{,}600 \ \frac{\text{min}}{\text{yr}}}{30.5 \ \frac{\text{cm}}{\text{ft}}}\right)$$
$$= \boxed{31.0 \ \text{ft/yr}}$$

$$k_{\text{soil B}} = \left(\frac{(2 \ \text{cm}^2)(15 \ \text{cm})}{(40 \ \text{cm}^2)(6000 \ \text{min})}\right)\ln \frac{200 \ \text{cm}}{190 \ \text{cm}}$$
$$= 6.4 \times 10^{-6} \ \text{cm/min}$$

$$k_{\text{soil B}} = \left(6.4 \times 10^{-6} \ \frac{\text{cm}}{\text{min}}\right)\left(\frac{525{,}600 \ \frac{\text{min}}{\text{yr}}}{30.5 \ \frac{\text{cm}}{\text{ft}}}\right)$$
$$= \boxed{0.11 \ \text{ft/yr}}$$

7.2. Most regulations require a soil liner to have a minimum in-place hydraulic conductivity of 1×10^{-7} cm/sec (or 0.1 ft/yr). Soil B would be acceptable as a liner.

7 Foundations

PROBLEM 1

20 in diameter concrete end-bearing piles placed every 12 ft (each way) carry a 24 in thick, steel-reinforced concrete slab and a 1100 lbf/ft^2 live load. The piles penetrate 10 ft of recently placed granular fill (specific weight = 110 lbf/ft^3) and 24 ft of normally consolidated clay (specific weight = 95 lbf/ft^3; angle of internal friction = 30°; unconfined compressive strength = 900 lbf/ft^2). The piles continue another 16 ft, and pile ends bear on a dense sand layer (ϕ = 35°; γ = 125 lbf/ft^3). Piles have been installed loosely through a steel casing for the first 10 ft.

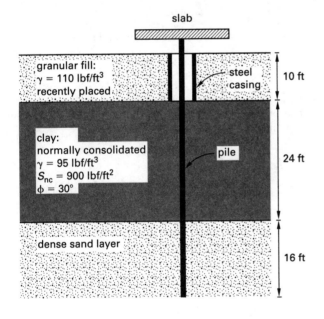

1.1. Is the end capacity sufficient for this installation? Why or why not?

1.2. Would your approach to determining the pile strength be any different if the clay layer was overconsolidated? If so, in what manner would your approach differ?

PROBLEM 2

Two square footings in a structure must have the same settlement even though their loadings and sizes are different. Both footings are located at the same depth in a sand layer 2 ft above the water table. The first footing (A) is 6 ft square. Disregard any interaction between the two footings.

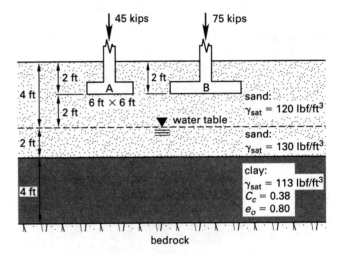

2.1. What is the settlement of footing A?

2.2. For a first approximation, assume a 2:1 (vertical: horizontal) pressure distribution and determine the size of footing B that will produce the same settlement as footing A.

PROBLEM 3

A pipe support in a refinery consists of a single 20 ft long, 20 in diameter concrete pile. A lateral (horizontal to the ground) load is applied at the top of the pile. Six feet of the pile remain above ground. The modulus of elasticity of the concrete is 60,000 lbf/ft^2. The soil consists of sand (effective friction angle = 30°; specific weight = 120 lbf/ft^3; coefficient of modulus variation = 40,000 lbf/ft^3). Assume the water table is at the ground surface. Use a factor of safety of 2.5 to determine the maximum allowable lateral load. State the authority for your method.

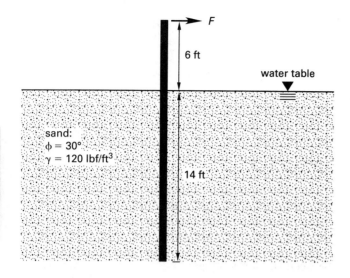

PROBLEM 4

The bay spacing in a reinforced concrete frame is 20 ft. Each bay has a series of 6 ft × 6 ft footings that carry vertical loads of 50,000 lbf. The footings are located over two layers of clay and two layers of sand. The specific gravity of the clay solids is 2.7.

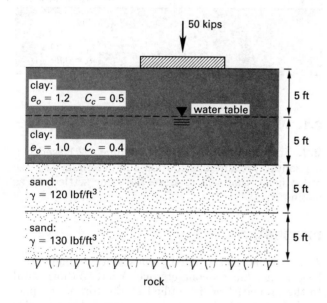

4.1. Find the primary settlement under the center of the footing.

4.2. Is this settlement tolerable? Why or why not?

SOLUTION 1

1.1. Compute the pile load.

$$\text{pile load} = (\text{area})(\text{live load} + \text{dead load})$$
$$= (12 \text{ ft})(12 \text{ ft})\left(1100 \ \frac{\text{lbf}}{\text{ft}^2} + (2 \text{ ft})\left(150 \ \frac{\text{lbf}}{\text{ft}^3}\right)\right)$$
$$= 201{,}600 \text{ lbf}$$

Pile capacity is derived from side (skin) resistance and tip (end) resistance. In this problem, negative skin friction or downdrag will occur as the clay consolidates. Additional load will be applied to the pile. Downdrag will occur between the top of the clay layer and the neutral axis (generally taken as the center of the layer, although a more rigorous solution may be employed). The average lateral force applied by the clay is

$$f_{\text{ave}} = \bar{\sigma}_v k \tan \delta$$
$$\bar{\sigma}_v = \text{effective vertical stress}$$
$$k = \text{earth pressure coefficient}$$
$$= 1 - \sin \phi$$
$$\delta = \text{soil-pile friction angle}$$
$$\approx 0.5\phi \text{ to } 0.7\phi$$

The average vertical stress acting over the top half of the layer is computed at a depth of 6 ft below the top of the clay.

$$f_{\text{ave}} = \bar{\sigma}_v k \tan \delta$$
$$= \left((10 \text{ ft})\left(110 \ \frac{\text{lbf}}{\text{ft}^3}\right) + (6 \text{ ft})\left(95 \ \frac{\text{lbf}}{\text{ft}^3}\right)\right)$$
$$\times (1 - \sin 30°)\tan\left((0.6)(30°)\right)$$
$$= 271 \text{ lbf/ft}^2$$

The downdrag force may be computed as

$$Q_n = f_{\text{ave}} A_{\text{pile surface}}$$
$$= \left(271 \ \frac{\text{lbf}}{\text{ft}^2}\right)(12 \text{ ft})\pi\left(\frac{20 \text{ in}}{12 \ \frac{\text{in}}{\text{ft}}}\right)$$
$$= 17{,}027 \text{ lbf}$$

The load applied to the pile is

$$\text{load applied} = 201{,}600 \text{ lbf} + 17{,}027 \text{ lbf}$$
$$= 218{,}627 \text{ lbf}$$

The critical embedment depth in the dense sand may be determined.

$$\left(\frac{L_b}{D}\right)_{\text{critical}} \approx 20 \text{ for dense sand}$$

$$L_b = \text{embedment length}$$

$$D = \text{pile diameter}$$

The maximum tip capacity is obtained at

$$L_b = 20D$$

$$= (20)\left(\frac{20 \text{ in}}{12\dfrac{\text{in}}{\text{ft}}}\right)$$

$$= 33.3 \text{ ft}$$

For $\phi = 35°$, $N_q = 41.4$. (Value used may vary with other sources.)

For a depth of 16 ft into the dense sand,

$$q' = (10 \text{ ft})\left(110 \ \frac{\text{lbf}}{\text{ft}^3}\right)$$

$$+ (24 \text{ ft})\left(95 \ \frac{\text{lbf}}{\text{ft}^3}\right) + (16 \text{ ft})\left(125 \ \frac{\text{lbf}}{\text{ft}^3}\right)$$

$$= 5380 \text{ lbf/ft}^2$$

The tip capacity of the pile may be computed as

$$Q_p = q_p A_p = A_p q' N_q$$

$$A_p = \text{pile tip cross-sectional area}$$

$$q' = \text{effective vertical stress at pile tip}$$

$$N_q = \text{bearing capacity factor}$$

$$Q_p = A_p q' N_q$$

$$= \left(\frac{20 \text{ in}}{12\dfrac{\text{in}}{\text{ft}}}\right)^2 \left(\frac{\pi}{4}\right)\left(5380 \ \frac{\text{lbf}}{\text{ft}^2}\right)(41.4)$$

$$= 486{,}000 \text{ lbf}$$

$$Q_{\text{allowable}} = \frac{Q_p}{\text{FS}}$$

A factor of safety of about 3 should be used.

$$Q_{\text{allowable}} = \frac{486{,}000 \text{ lbf}}{3}$$

$$= 162{,}000 \text{ lbf}$$

$Q_{\text{allowable}} < \text{load applied, so the pile end capacity is not sufficient.}$

1.2. If the clay was overconsolidated, settlement would be negligible and no downdrag would occur. Additional pile capacity would be derived from friction between the pile and the clay. For highly overconsolidated clays, it would be conservative to ignore increased capacity because the clay may crack, which would limit the strength increase. Potential for cracking may be reduced by augering a hole for the pile before driving.

SOLUTION 2

2.1. Footing A applies 45 kips of load over 36 ft² of footing area. The applied stress on the bottom of the footing, q, may be computed as

$$q = \left(\frac{45 \text{ kips}}{36 \text{ ft}^2}\right)\left(1000 \ \frac{\text{lbf}}{\text{kip}}\right)$$

$$= 1250 \text{ lbf/ft}^2$$

Compute effective overburden pressures, $\bar{\sigma}_{v_o}$, and the change in stress, Δp, at the top of the sand layers and the center of the clay. z is the depth from the bottom of the footing. Use Boussinesq stress contour charts or other suitable means to determine the ratio $\Delta p/q$ (or calculate Δp graphically from an influence chart). (The sand settlement is probably negligible, but check it anyway.)

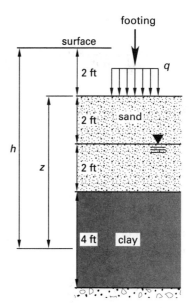

$$\bar{\sigma}_{v_o} = \gamma h$$

At a depth of 2 ft,

$$\bar{\sigma}_{v_o} = \left(120 \ \frac{\text{lbf}}{\text{ft}^3}\right)(2 \text{ ft})$$

$$= 240 \text{ lbf/ft}^2 \quad [\text{at base of footing}]$$

At a depth of 4 ft,

$$\bar{\sigma}_{v_o} = 240 \ \frac{\text{lbf}}{\text{ft}^2} + \left(120 \ \frac{\text{lbf}}{\text{ft}^3}\right)(2 \ \text{ft})$$
$$= 480 \ \text{lbf/ft}^2 \quad [\text{at water table}]$$

At a depth of 8 ft,

$$\bar{\sigma}_{v_o} = 480 \ \frac{\text{lbf}}{\text{ft}^2} + (2 \ \text{ft})\left(130 \ \frac{\text{lbf}}{\text{ft}^3} - 62.4 \ \frac{\text{lbf}}{\text{ft}^3}\right)$$
$$+ (2 \ \text{ft})\left(113 \ \frac{\text{lbf}}{\text{ft}^3} - 62.4 \ \frac{\text{lbf}}{\text{ft}^3}\right)$$
$$= 716.4 \ \text{lbf/ft}^2 \quad [\text{at center of clay}]$$

depth below footing, z (ft)	$\bar{\sigma}_{v_o}$ (lbf/ft²)	$\dfrac{z}{B}$	$\dfrac{\Delta p}{q}$	Δp (lbf/ft²)
0	240	0	1	1250
2	480	0.33	0.9	1125
6	716.4	1.0	0.33	412

Use the illustration in the solution to compute the change in stress below the footing at depths of 1 ft, 3 ft, 5 ft, and 8 ft from the bottom of the footing.

At 1 ft,

$$\frac{z}{B} = \frac{1 \ \text{ft}}{6 \ \text{ft}} = 0.17$$

At 3 ft,

$$\frac{z}{B} = \frac{3 \ \text{ft}}{6 \ \text{ft}} = 0.5$$

At 5 ft,

$$\frac{z}{B} = \frac{5 \ \text{ft}}{6 \ \text{ft}} = 0.8$$

At 8 ft,

$$\frac{z}{B} = \frac{8 \ \text{ft}}{6 \ \text{ft}} = 1.3$$

Settlement in the sand layers may be computed using the following equation.

$$S_{\text{sand}} = C_1 C_2 (q - \bar{\sigma}_{v_o}) \sum \left(\frac{I_z}{E_s}\right) \Delta z$$

$I_z = $ strain influence factor

$C_1 = $ correction for embedment

$$= (1 - 0.5)\left(\frac{\bar{\sigma}_{v_o}}{q - \bar{\sigma}_{v_o}}\right)$$

$C_2 = $ correction for creep

$$= 1 + 0.2 \log(\text{time in yr}/0.1 \ \text{yr})$$

$E_s = $ Young's modulus

For a square footing,

$$I_z = 0.1 \ \text{at} \ z = 0$$
$$I_z = 0.5 \ \text{at} \ z = 0.5B = (0.5)(6 \ \text{ft}) = 3 \ \text{ft}$$

At $z = 1$ ft and 3 ft, $z/B = 0.167$ and 0.5, respectively, and $I_z = 0.23$ and 0.5, respectively.

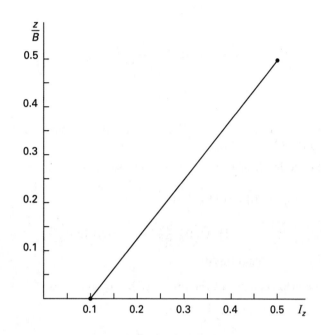

The settlement in the sand one year after the footing is placed is

$$S = \left(1 - (0.5)\left(\frac{240 \ \frac{\text{lbf}}{\text{ft}^2}}{1250 \ \frac{\text{lbf}}{\text{ft}^2} - 240 \ \frac{\text{lbf}}{\text{ft}^2}}\right)\right)$$
$$\times \left(1 + 0.2 \log\left(\frac{1 \ \text{yr}}{0.1 \ \text{yr}}\right)\right)$$
$$\times \left(\left(\frac{0.23}{720{,}000 \ \frac{\text{lbf}}{\text{ft}^2}}\right)(24 \ \text{in})\right.$$
$$\left. + \left(\frac{0.5}{720{,}000 \ \frac{\text{lbf}}{\text{ft}^2}}\right)(24 \ \text{in})\right)$$
$$\times \left(1250 \ \frac{\text{lbf}}{\text{ft}^2} - 240 \ \frac{\text{lbf}}{\text{ft}^2}\right)$$
$$= 0.026 \ \text{in} \quad [\text{negligible}]$$

Settlement for a normally consolidated clay may be computed as

$$S = \left(\frac{\text{thickness}}{1 + e_o}\right) C_c \log\left(\frac{\overline{\sigma}_{v_o} + \Delta p}{\overline{\sigma}_{v_o}}\right)$$

$$= \left(\left(\frac{4 \text{ ft}}{1 + 0.80}\right)(0.38)\log\left(\frac{716.4 \frac{\text{lbf}}{\text{ft}^2} + 412 \frac{\text{lbf}}{\text{ft}^2}}{716.4 \frac{\text{lbf}}{\text{ft}^2}}\right)\right)$$

$$\times \left(12 \frac{\text{in}}{\text{ft}}\right)$$

$$= 2.04 \text{ in} \quad (2.0 \text{ in})$$

> Settlement in the sand layers is negligible, so the total settlement beneath footing A is 2.0 in. Note that this value is excessive for a typical structure.

2.2. Assume, based on the previous calculations, that settlement in the sand is negligible for footing B. Footing B should be sized so that 2 in of settlement is induced in the clay layer. In order for this to occur, the change in stress at the center of the clay layer must be the same as for footing A—that is, 412 lbf/ft². The hypothetical loaded area at the center of the clay layer would be

$$A = \frac{\text{load}}{\text{stress}} = \frac{75,000 \text{ lbf}}{412 \frac{\text{lbf}}{\text{ft}^2}}$$

$$= 182 \text{ ft}^2$$

A square loaded area of 182 ft² would have dimensions of 13.5 ft × 13.5 ft. Assuming the 2:1 (vertical-to-horizontal) stress distribution, project upward to the depth of footing B as shown.

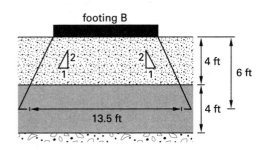

Footing B should have a side dimension of

$$13.5 \text{ ft} - (2)(0.5)(6 \text{ ft}) = \boxed{7.5 \text{ ft}}$$

SOLUTION 3

The first step is to determine if the pile is short and rigid or long and flexible. Compute the stiffness factor, R.

$$R = \sqrt[4]{\frac{EI}{kB}}$$

$$k = \frac{k_1}{1.5}$$

$k_1 = $ Terzaghi's subgrade modulus for a 1 ft² plate

$B = $ pile width

$I = $ moment of inertia

$$= \tfrac{1}{4}\pi r^4$$

$$= 0.0491 D^4$$

For sand with $\phi = 30°$, density may range from loose to medium. k_1 may be estimated to be from 50 tons/ft³ to 100 tons/ft³. Use $k_1 = (50 \text{ tons/ft}^3)(2000 \text{ lbf/ton}) = 100,000 \text{ lbf/ft}^3$. Solve for R.

$$R = \sqrt[4]{\frac{\left(60,000 \frac{\text{lbf}}{\text{ft}^2}\right)(0.0491)(20 \text{ in})^4\left(12 \frac{\text{in}}{\text{ft}}\right)}{\left(\dfrac{100,000 \frac{\text{lbf}}{\text{ft}^3}}{1.5}\right)(20 \text{ in})}}$$

$$= 8 \text{ in}$$

Compute the stiffness factor, T.

$$T = \sqrt[5]{\frac{EI}{n_h}}$$

$n_h = $ coefficient of modulus variation

$$T = \sqrt[5]{\frac{\left(60,000 \frac{\text{lbf}}{\text{ft}^2}\right)(0.0491)(20 \text{ in})^4\left(12 \frac{\text{in}}{\text{ft}}\right)}{40,000 \frac{\text{lbf}}{\text{ft}^3}}}$$

$$= 10.7 \text{ in}$$

Compare the embedded pile length, L, to R and T.

$$L = (14 \text{ ft})\left(12 \frac{\text{in}}{\text{ft}}\right) = 168 \text{ in}$$

Since $L \geq 4T$ and $L \geq 3.5R$, the pile is considered to be long and elastic. An elastic solution must be used.

Find the ultimate moment resistance, M_u, for the pile. A 20 in round (or octagonal) pile will typically have eight no. 5 or no. 6 bars. Such piles will have M_u values of 70 kN·m and 100 kN·m, respectively.

$$k_p = \tan^2\left(45° + \frac{\phi}{2}\right)$$

$$\frac{M_u}{B^4 \gamma k_p} = \frac{(70{,}000 \text{ N·m})\left(3.28 \frac{\text{ft}}{\text{m}}\right)}{\left(\dfrac{20 \text{ in}}{12 \frac{\text{in}}{\text{ft}}}\right)^4 \left(120 \frac{\text{lbf}}{\text{ft}^3} - 62.4 \frac{\text{lbf}}{\text{ft}^3}\right)}$$

$$\times \tan^2\left(45° + \frac{30°}{2}\right)\left(4.45 \frac{\text{N}}{\text{lbf}}\right)$$

$$= 38.7 \quad (40)$$

$$\frac{H_u}{k_p B^3 \gamma} = 8 \text{ for } \frac{e}{B} \text{ ratio} = 3.6$$

$$H_u = \text{ultimate lateral load}$$

$$= 8 k_p B^3 \gamma$$

$$= (8)\left(\tan^2\left(45° + \frac{30°}{2}\right)\left(\frac{20 \text{ in}}{12 \frac{\text{in}}{\text{ft}}}\right)^3\right.$$

$$\left. \times \left(120 \frac{\text{lbf}}{\text{ft}^3} - 62.4 \frac{\text{lbf}}{\text{ft}^3}\right)\right)$$

$$= 6400 \text{ lbf}$$

The allowable lateral load is

$$H_{\text{allowable}} = \frac{H_u}{\text{FS}} = \frac{6400 \text{ lbf}}{2.5}$$

$$= \boxed{2560 \text{ lbf} \quad [\text{minimum}]}$$

If the larger steel is used, then

$$M_u = 100 \text{ kN·m}$$

$$\frac{M_u}{B^4 \gamma k_p} = 55$$

$$\frac{H_u}{k_p B^3 \gamma} = 10$$

$$H_u = 8000 \text{ lbf}$$

$$H_{\text{allowable}} = \frac{8000 \text{ lbf}}{2.5}$$

$$= \boxed{3200 \text{ lbf} \quad [\text{maximum}]}$$

SOLUTION 4

4.1. The applied footing load is

$$q = \frac{\text{load}}{\text{area}} = \frac{50{,}000 \text{ lbf}}{(6 \text{ ft})(6 \text{ ft})}$$

$$= 1389 \text{ lbf/ft}^2$$

Determine the change in stress, Δp, at the center of each soil layer from stress-contour charts or by other means.

center of layer no.	depth (ft)	depth/B (for $B=6$ ft)	$\Delta p/q$ ($L/B=1$)	Δp (lbf/ft²)
1	2.5	0.4	0.8	1111
2	7.5	1.3	0.23	319
3	12.5	2.1	0.095	132
4	17.5	2.9	0.055	76

Compute the specific weight of the clay layers.

$$\gamma_{\text{solids}} = (2.7)\left(62.4 \frac{\text{lbf}}{\text{ft}^3}\right) = 168.5 \text{ lbf/ft}^3$$

$$\gamma_d = \frac{\gamma_{\text{solids}}}{1+e}$$

For clay 1,

$$\gamma_d = \frac{168.5 \frac{\text{lbf}}{\text{ft}^3}}{1 + 1.2} = 76.6 \text{ lbf/ft}^3$$

For clay 2,

$$\gamma_d = \frac{168.5 \frac{\text{lbf}}{\text{ft}^3}}{1 + 1.0} = 84.3 \text{ lbf/ft}^3$$

Clay 2 is below the water table, so compute the water content.

$$w = \frac{Se}{\text{SG}}$$

Below the water table, the degree of saturation, S, is approximately 1.

$$w = \frac{(1)(1.0)}{2.70} = 0.37 \quad (37\%)$$

The saturated density of clay 2 is

$$\gamma_{\text{sat}} = \gamma_d (1 + w)$$

$$= \left(84.3 \frac{\text{lbf}}{\text{ft}^3}\right)(1 + 0.37)$$

$$= 115.5 \text{ lbf/ft}^3$$

The change in stress at a depth of $2B$ below a footing is usually negligible. Only 4–9% of the applied footing load reaches the sand layers. Settlement in the sand layers is negligible.

Geotechnical

The effective overburden pressure at the center of the clay layers is

$$\overline{\sigma}_{v_o,\text{layer 1}} = (2.5 \text{ ft})\left(76.6 \ \frac{\text{lbf}}{\text{ft}^3}\right)$$

$$= 192 \text{ lbf/ft}^2$$

$$\overline{\sigma}_{v_o,\text{layer 2}} = (5 \text{ ft})\left(76.6 \ \frac{\text{lbf}}{\text{ft}^3}\right)$$

$$+ (2.5 \text{ ft})\left(115.5 \ \frac{\text{lbf}}{\text{ft}^3} - 62.4 \ \frac{\text{lbf}}{\text{ft}^3}\right)$$

$$= 516 \text{ lbf/ft}^2$$

Assume the clay is normally consolidated and compute the settlement.

$$S = \left(\frac{\text{layer height}}{1 + e_o}\right)\left(C_c \log\left(\frac{\overline{\sigma}_{v_o} + \Delta p}{\overline{\sigma}_{v_o}}\right)\right)$$

$$S_{\text{layer 1}} = \left(\frac{5 \text{ ft}}{1 + 1.2}\right)\left(0.5 \log\left(\frac{192 \ \frac{\text{lbf}}{\text{ft}^2} + 1111 \ \frac{\text{lbf}}{\text{ft}^2}}{192 \ \frac{\text{lbf}}{\text{ft}^2}}\right)\right)$$

$$= 0.95 \text{ ft}$$

$$S_{\text{layer 2}} = \left(\frac{5 \text{ ft}}{1 + 1.0}\right)\left(0.4 \log\left(\frac{516 \ \frac{\text{lbf}}{\text{ft}^2} + 319 \ \frac{\text{lbf}}{\text{ft}^2}}{516 \ \frac{\text{lbf}}{\text{ft}^2}}\right)\right)$$

$$= 0.21 \text{ ft}$$

$$\text{total settlement} = (0.95 \text{ ft} + 0.21 \text{ ft})\left(12 \ \frac{\text{in}}{\text{ft}}\right)$$

$$= \boxed{13.9 \text{ in}}$$

4.2. This settlement is not tolerable. Generally, a settlement of 1 in or less is acceptable. For a concrete frame, angular distortion, δ/l of 0.003 is considered acceptable. This results in a settlement of approximately $3/4$ in in a column span, l, of 20 ft. The computed settlement would be excessive.

Geotechnical

Topic III: Environmental

Chapter

Environmental

8 Wastewater

PROBLEM 1

A wastewater treatment plant receives 4.5 MGD of 70% wastewater influent with the following characteristics.

- 950 lbm BOD/MG
- 3200 mg/L MLSS
- 2 mg/L dissolved oxygen
- 70% volatile solids

As part of its processing, the wastewater is aerated with diffusion aerators operating with the following characteristics.

BOD constant	1.15 lbm O_2/lbm BOD removed
MLSS constant	0.03 lbm O_2/lbm BOD removed
oxygen transfer efficiency	15%
BOD removal efficiency	85%
mechanical efficiency	80%
outlet pressure	22.7 lbf/in^2

What is the cost of 24 hr of constant aeration if the aerators have been sized with 100% excess capacity and electrical power costs $0.08/kW-hr?

PROBLEM 2

Three covered digesters are used to process 30,000 gal of waste sludge per day. Each digester has an internal diameter of 45 ft, vertical walls of unspecified height, and a 7.5 ft high tapered section at the bottom. Prior to loading, the waste is thickened 5% and enters with the following characteristics.

BOD$_5$ loading	12,000 lbm/day
volatile organics	70%
temperature	61°F
BOD reaction constant	0.2 1/day

Use the following data in calculating methane production.

efficiency of waste utilization, E	70%
yield coefficient, Y	0.06 lbm/lbm
endogenous coefficient, R_d	0.03 1/day
mean cell residence time, θ_c	10 days

The average ambient temperature is 58°F, and the temperature of digestion is 91°F. The average heat loss through the vertical portion of the digester walls is 110,000 Btu/hr per digester. The overall coefficients of heat transfer for the floor and roof of the digester are 0.15 Btu/ft^2-°F-hr and 0.16 Btu/ft^2-°F-hr, respectively.

2.1. How much heat (in Btus) must be supplied to maintain the 91°F digestion temperature?

2.2. Is the methane produced from the digestion process sufficient to provide the energy required?

PROBLEM 3

An existing 24 in (inside diameter) unlined concrete sewer pipe currently serves a small town but is expected to be under capacity when the population peaks at 300,000. The existing line drops 100 ft as it travels 4000 ft to the first pumping station. A second pipe will be installed in parallel with the existing pipe. Use 250 gpcd (gal per capita per day) as an estimate of the future wastewater production.

3.1. What diameter pipe should be installed so that the combined capacity is adequate for the peak population?

3.2. If both pipes are 24 in in diameter, what is the depth of flow in each pipe?

PROBLEM 4

A pretreatment plant processes waste from several local industries prior to discharging it into the municipal system. The following mix of influents must be handled simultaneously.

volume	source
13,000 gal/day	sanitary waste from employees
41,250 gal/day	hospital medical waste
5000 gal/day	plating plant chemical waste
15,000 gal/day	dairy animal waste

4.1. What typical limits of BOD would be expected from each of these four sources?

4.2. What other wastewater quality characteristics must be considered when designing the pretreatment process?

4.3. Design a pretreatment process and briefly explain why each step is required.

PROBLEM 5

Design a stabilization pond to support a large national park campground. The population of the campground increases dramatically during the summer. During winter months, a skeleton crew remains to perform maintenance. The following characteristics must be considered in the pond design.

BOD loading (all sources)	250 mg/L
average annual temperature	70°F
latitude	29°N
skeleton population	300 people
peak summer population	10,000 people
hospital	100 beds
laundromat	20 washers
dining rooms	500 seats
gas stations	8 pumps

The dining room serves three meals a day. During the summer, the dining facility offers a choice of three sittings/servings per meal.

Use the following data in calculating the performance of the stabilization pond.

BOD loading from all sources	250 mg/L
BOD_5 conversion	90%
first-order BOD_5 rate constant, k	0.25 1/day at 20°C
temperature coefficient, θ	1.06 at 20°C
summer pond temperature	32°C (90°F)
winter pond temperature	10°C (50°F)
maximum individual pond area	10 ac
maximum pond depth	2 ft

The product of the first-order reaction constant, k, and the detention time, t, is 5.

PROBLEM 6

Specify five different methods for disposing of solid waste in open areas. For each, indicate the (a) advantages, (b) disadvantages, and (c) relative cost. Also specify (d) the approximate population size for which each method is appropriate.

PROBLEM 7

The present population of a large town is 500,000 and is expected to increase by 3% each year as it has in the past. The population currently disposes of 5.4 lbm of solid waste per day per person, but this amount is expected to increase at the rate of 1% per year as it has in the past.

Through a new recycling program, 3% of all solid waste can be recovered. A solid-waste disposal site has been in use (without recycling) for 16 years and was originally designed with a 20 year life. The site is roughly square in configuration with vertical side slopes. There is a 250 ft buffer zone around the site, and waste can be piled to an average height of 40 ft. Daily, intermittent, and final cover occupies 25% of the disposal site. The existing site has four years of capacity remaining. Solid waste is compacted to a density of 1000 lbm/yd^3.

7.1. How many acres does the current site occupy?

7.2. Specify the size (in acres) of a replacement site that would extend the useful life by an additional five years.

PROBLEM 8

One MGD of wastewater is treated in a lagoon/holding pond arrangement. The effluent from the ponds is subsequently processed by spraying into fields at the rate of 2.5 in/wk. The static head available in the spray lines is 45 ft, but head losses due to friction through the piping system are 30 ft. Each spray nozzle covers a circular area 350 ft in diameter when operating at its rated capacity of 275 gal/min and 75 psig. Pumps increase the head above the static value. To ensure uniform coverage, only 25% of the field nozzles are in use at one time. Pump mechanical efficiency is 82%; motor electrical conversion efficiency is 92%.

8.1. At rated capacity, what land area (in acres) can be in use at any given moment?

8.2. How many nozzles (total) are needed?

8.3. What horsepower motor(s) should be used to drive the pumps?

8.4. What considerations should be taken into account when determining the application rate of effluent to the spray fields?

PROBLEM 9

A typical block in a five block by eight block subdivision is illustrated. Streets are 24 ft wide. The subdivision is constructed on a 160 ac section of land. Individual building sites (parcels, lots, etc.) are 0.25 ac (90 ft × 120 ft) each, and building sites are grouped into blocks containing 12 sites as shown. There are 480 such sites planned. Each site will support, on the average, 4.1 residents. Infiltration to the sewer is 200 gal/in-mi-day. Inflow to the sewer is 2000 gal/in-mi-day. What flow must the wastewater pump station be able to handle?

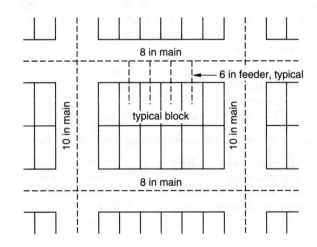

Environmental

PROBLEM 10

A complete mix-activated sludge processing facility will be designed using the theory of kinetics and the following characteristics.

incoming flow, Q	5 MGD
influent soluble BOD_5, S_0	250 mg/L
effluent BOD	20 mg/L
effluent suspended solids	20 mg/L
mixed liquor volatile suspended solids, MLVSS = 0.8 MLSS	3500 mg/L
return sludge suspended solids, SS	10,000 mg/L
return sludge volatile suspended solids, x_r	8000 mg/L
mean cell residence time, θ_c	10 days
kinetic coefficient, y	0.6 lbm VSS/lbm BOD_5
kinetic coefficient, k_d	0.06/day
percent of effluent biodegradable suspended solids	65%
ratio of BOD_5/BOD_L	0.68

Calculate the following quantities.

10.1. biological efficiency (in soluble BOD_5 removal)

10.2. overall efficiency (in BOD_5 removal)

10.3. biomass production (lbm/day)

10.4. increase in mixed liquor suspended solids (lbm/day)

10.5. sludge to be wasted (lbm/day)

10.6. sludge-wasting rate (MGD)

10.7. recirculation ratio

10.8. hydraulic retention time (hr)

10.9. specific substrate utilization rate (mg BOD_5/mg MLVSS-day)

10.10. food-to-microorganism ratio (mg BOD_5/mg MLVSS-day)

10.11. blower capacity (ft^3/min) if the oxygen transfer efficiency is 8% and 100% excess capacity is used

PROBLEM 11

Design an industrial wastewater treatment plant to process wastewater from a beet sugar processing facility. The facility will operate for 3 months during the fall and winter. The facility will have a total capacity of 1000 tons per day. The flows and characteristics of different sources within the plant are as follows.

source	unit volume (gal/ton)	daily flow (gal/day)	BOD (mg/L)	TSS (mg/L)	TDS (mg/L)
flume/wash water	2200	2.2×10^6	200	800	780
process water	660	6.6×10^5	1230	1100	1120
lime drainage	75	7.5×10^4	1420	450	2850

11.1. Suppose the flume/wash water will be recovered and reused. Design a sedimentation basin to treat the flume/wash water. Assume the detention time will be 2.0 hr and the water depth will be 10.0 ft.

11.2. Evaluate the use of an aerobic lagoon to treat the facility's wastewater. Assume a detention time of 2.0 days, a working depth of 1.5 ft, and a target loading rate of 86 lbm BOD/ac-day.

11.3. Briefly describe two alternative methods for treating the wastewater from this facility.

PROBLEM 12

A wastewater treatment plant will be designed according to the *Ten States' Standards* to process incoming wastewater with the following characteristics.

> 2.0 MGD average annual daily flow (AADF)
> 250 mg/L BOD_5
> 180 mg/L suspended solids

Treatment will consist of circular trickling filters using synthetic plastic media, a grit chamber, and rectangular primary and secondary clarifiers.

12.1. Draw the process diagram showing each operation in sequence.

12.2. Size the primary clarifiers with an aspect ratio of 4:1. State your assumptions.

12.3. Size the trickling filters. State your assumptions.

PROBLEM 13

Outline the general procedure and list the general equations you would use to design a three-cell aeration pond based on the following characteristics.

wastewater flow	Q
wastewater organic strength	S_0, BOD_5
wastewater nutrients	P, NH_3-N or TKN
average winter wastewater temperature	T_{ww}
average summer wastewater temperature	T_{ws}
critical ambient winter air temperature	T_{aw}
critical ambient summer air temperature	T_{as}
substrate utilization coefficient	k
theoretical yield coefficient	$Y_{T,20°C}$
microbial decay coefficient	$k_{d,20°C}$
mean cell residence time	θ_c
temperature factor	f
detention time in each cell	t
depth of each cell	D

13.1. What are the cell volume and cell surface area?

13.2. What are the winter and summer lagoon operating temperatures?

13.3. How are k and k_d corrected for the winter and summer operating temperatures?

13.4. What is the soluble BOD_5 from the first cell?

13.5. What is the biomass concentration entering the first cell?

13.6. What is the biomass concentration entering the second cell?

13.7. What is the biomass concentration entering the third cell?

13.8. Estimate the effluent BOD_5 before clarification, assuming the soluble BOD_5 from the third cell is S_e.

13.9. Calculate the minimum mean cell residence time for nitrification. Assume μ_{max} and pH_{opt} are known.

13.10. Calculate the oxygen requirements for each cell of the lagoon, including nitrification where appropriate.

13.11. Determine the motor horsepower required to supply oxygen to each cell, assuming a surface aerator delivers 1.8 lbf of oxygen per horsepower-hour.

13.12. Determine the motor horsepower for complete mixing in each cell.

13.13. Specify a maximum length:width aspect ratio for rectangular cells.

13.14. Compare the nutrient requirements and supply in the first cell. Assume the first cell controls the nutrient requirements.

PROBLEM 14

Effluent from a primary treatment plant discharges from an 8 in diameter pipe into an intermittent stream.

The characteristics of the effluent are

velocity	0.6 ft/sec (flowing full)
BOD_5	150 mg/L (20°C)
dissolved oxygen, DO	1.0 mg/L
temperature	22°C

The characteristics of the intermittent stream are

flow	2 ft^3/sec
velocity	0.2 ft/sec
BOD_5	4 mg/L (20°C)
dissolved oxygen, DO	10.0 mg/L
summer temperature	52°F
winter temperature	45°F
deaeration coefficient, K_D	0.1 1/day
reaeration coefficient, K_R	0.25 1/day
deaeration temperature constant	1.047
aeration temperature constant	1.016

14.1. Find the effluent volumetric flow rate.

14.2. Find the 5 day BOD immediately after mixing.

14.3. Find the dissolved oxygen immediately after mixing.

14.4. Find the temperature in the summer immediately after mixing.

14.5. Find the temperature in the winter immediately after mixing.

14.6. Find the dissolved oxygen concentration at a point 20 mi downstream.

PROBLEM 15

A retirement community supports 2000 people. A single concrete sewer main connects the community to a local treatment plant.

15.1. Determine the required diameter of the waste main assuming a minimum permissible slope. State all of your assumptions.

15.2. Given that the waste main diameter is 8 in and the slope is 0.021 ft/ft, find the depth and velocity of flow.

PROBLEM 16

A large spill of a toxic liquid chemical has occurred. The following steps have already been taken.

(a) The spill has been adequately contained by a make-shift dike.

(b) The identity of the chemical has been determined and no other hazardous or toxic contaminants have been identified.

(c) It has been determined that there are no immediate hazards due to fire or explosion.

(d) It has been determined that there is no serious threat of contamination to municipal water supplies, major groundwater supplies, or surface water.

Specify the steps you would take to neutralize the chemical and clean up the site.

PROBLEM 17

Explain how you would design and use a lagoon as an equalizing basin to alter peak flow. Discuss the merits of using such a method to equalize flow.

PROBLEM 18

18.1. What is the difference (if any) between scouring and self-cleansing velocities? What are typical values?

18.2. What minimum scouring velocity should be maintained given a roughly spherical particle with a diameter of 0.5 in?

PROBLEM 19

The following questions apply to sanitary landfills.

19.1. What agencies in the United States control landfill design?

19.2. What is the average density of shredded waste?

19.3. What is the average density of baled and compacted waste?

19.4. What is a typical depth of daily soil cover?

19.5. What is a typical depth of intermediate soil cover?

19.6. What is a typical depth of final soil cover?

19.7. What is the maximum length of time that intermediate cover can be used?

19.8. What is the purpose of a clay liner?

19.9. What is the primary limitation of clay liners?

19.10. What is the significance (as related to sanitary landfill design) of the acronym NIMBY?

SOLUTION 1

The influent wastewater flow is

$$(4.5 \text{ MGD})\left(950 \ \frac{\text{lbm BOD}}{\text{MG}}\right) = 4275 \text{ lbm BOD/day}$$

85% of the BOD is removed in this process.

$$(0.85)\left(4275 \ \frac{\text{lbm BOD}}{\text{day}}\right) = 3634 \text{ lbm BOD/day}$$

The O_2 required is

$$\begin{array}{r} 1.15 \ O_2/\text{lbm BOD removed} \\ + \ 0.03 \ O_2/\text{lbm BOD removed} \\ \hline 1.18 \ O_2/\text{lbm BOD removed} \end{array}$$

$$\left(1.18 \ \frac{\text{lbm } O_2}{\text{lbm BOD}}\right)\left(3634 \ \frac{\text{lbm BOD}}{\text{day}}\right)$$
$$= 4288 \text{ lbm } O_2/\text{day}$$

Air is 23.2% O_2 by weight and has a density of 0.075 lbm/ft^3. The oxygen transfer efficiency of the diffusers in wastewater is 15%. The airflow required is

$$\frac{4288 \ \dfrac{\text{lbm } O_2}{\text{day}}}{\left(0.075 \ \dfrac{\text{lbm}}{\text{ft}^3}\right)(0.232)(0.15)} = 1.643 \times 10^6 \text{ ft}^3/\text{day}$$

The air blowers have been sized with 100% excess capacity, so double the air flow.

$$\frac{\left(1.643 \times 10^6 \ \dfrac{\text{ft}^3}{\text{day}}\right)(2.0)}{1440 \ \dfrac{\text{min}}{\text{day}}} = 2282 \text{ ft}^3/\text{min}$$

Assuming the compression is steady state, open flow, and isentropic, the theoretical blower power can be calculated from the following equation. (Assumes $k = 1.4$ for air.)

$$P = \left(\frac{\dot{m}RT_1}{(550)(0.283)\eta}\right)\left(\left(\frac{p_2}{p_1}\right)^{0.283} - 1\right)$$

$P =$ power (hp)

$\dot{m} =$ mass flow rate of air (lbm/sec)

$$= \frac{\left(2282 \ \dfrac{\text{ft}^3}{\text{min}}\right)\left(0.075 \ \dfrac{\text{lbm}}{\text{ft}^3}\right)}{60 \ \dfrac{\text{sec}}{\text{min}}}$$

$$= 2.85 \text{ lbm/sec}$$

$R =$ gas constant $= 53.3$ ft-lbf/lbm air-°R

$T =$ temperature (°R) $=$ °F $+ 460° = 70°$F $+ 460°$
$$= 530°\text{R}$$

η = efficiency of blowers

p_1 = inlet pressure = standard atmospheric pressure

$= 14.7 \ \text{lbf/in}^2$

p_2 = outlet pressure

$$P = \left(\frac{\left(2.85 \ \frac{\text{lbm}}{\text{sec}}\right)\left(53.3 \ \frac{\text{ft-lbf}}{\text{lbm-°R}}\right)(530°\text{R})}{\left(550 \ \frac{\text{ft-lbf}}{\text{sec-hp}}\right)(0.283)(0.80)} \right)$$

$$\times \left(\left(\frac{22.7 \ \frac{\text{lbf}}{\text{in}^2}}{14.7 \ \frac{\text{lbf}}{\text{in}^2}} \right)^{0.283} - 1 \right)$$

$= 84.6 \ \text{hp}$ (85 hp)

The power cost for 24 hr of constant aeration is

$$(85 \ \text{hp})\left(0.746 \ \frac{\text{kW}}{\text{hp}}\right)\left(24 \ \frac{\text{hr}}{\text{day}}\right)(1 \ \text{day})\left(\frac{\$0.08}{\text{kW-hr}}\right)$$

$$= \boxed{\$121.75/\text{day}}$$

SOLUTION 2

2.1. The heat required for raw sludge is calculated from the following equation.

$q_s = V c_p (T_2 - T_1)$

$V = 30{,}000 \ \text{gal/day of sludge}$

$c_p = 1.0 \ \text{Btu/lbm-°F (water)}$

$T_1 = 61°\text{F}$

$T_2 = 91°\text{F}$

$$q_s = \left(\frac{30{,}000 \ \frac{\text{gal}}{\text{day}}}{24 \ \frac{\text{hr}}{\text{day}}} \right)\left(8.34 \ \frac{\text{lbm}}{\text{gal}}\right)\left(1.0 \ \frac{\text{Btu}}{\text{lbm-°F}}\right)$$

$$\times (91°\text{F} - 61°\text{F})$$

$= 312{,}750 \ \text{Btu/hr}$

The digester shape is

45 ft

7.5 ft

22.5 ft | 22.5 ft

not to scale

The heat required to make up for heat loss from the walls is

$q_w = q_{\text{digester}}(\text{no. of digesters})$

$$= \left(110{,}000 \ \frac{\text{Btu}}{\text{hr-digester}}\right)(3 \ \text{digesters})$$

$= 330{,}000 \ \text{Btu/hr}$

For the floor,

$A = \pi r \sqrt{r^2 + b^2}$

$= \pi(22.5 \ \text{ft})\sqrt{(22.5 \ \text{ft})^2 + (7.5 \ \text{ft})^2}$

$= 1676 \ \text{ft}^2/\text{digester}$

$q = AU(T_1 - T_2)$

$U = 0.15 \ \text{Btu/ft}^2\text{-°F-hr}$

$T_2 = 91°\text{F}$

$T_1 = 58°\text{F}$

$$q_f = \left(1676 \ \frac{\text{ft}^2}{\text{digester}}\right)(3 \ \text{digesters})\left(0.15 \ \frac{\text{Btu}}{\text{ft}^2\text{-°F-hr}}\right)$$

$$\times (91°\text{F} - 58°\text{F})$$

$= 24{,}889 \ \text{Btu/hr}$

For the roof,

$A = \pi r^2 = \pi(22.5 \ \text{ft})^2$

$= 1590 \ \text{ft}^2/\text{digester}$

$U = 0.16 \ \text{Btu/ft}^2\text{-°F-hr}$

$$q_r = \left(1590 \ \frac{\text{ft}^2}{\text{digester}}\right)(3 \ \text{digesters})\left(0.16 \ \frac{\text{Btu}}{\text{ft}^2\text{-°F-hr}}\right)$$

$$\times (91°\text{F} - 58°\text{F})$$

$= 25{,}186 \ \text{Btu/hr}$

Add the heat values required from the walls, floor, and roof.

$q_t = q_w + q_f + q_r$

$$= 330{,}000 \ \frac{\text{Btu}}{\text{hr}} + 24{,}889 \ \frac{\text{Btu}}{\text{hr}} + 25{,}186 \ \frac{\text{Btu}}{\text{hr}}$$

$= 380{,}075 \ \text{Btu/hr}$

The total heat required is

$$q = 312{,}750 \ \frac{\text{Btu}}{\text{hr}} + 380{,}075 \ \frac{\text{Btu}}{\text{hr}}$$

$$= \boxed{692{,}825 \ \text{Btu/hr}}$$

Environmental

2.2. The volume of methane produced from the sludge is calculated by

$$V_{CH_4} = \left(5.61 \ \frac{ft^3}{lbm}\right)(EQS_0 - 1.42P_x)$$

V_{CH_4} = volume of methane produced (ft³/day)

5.61 = theoretical conversion factor for amount of methane produced from conversion of 1 lbm of BOD_L

E = efficiency of waste utilization = 70%

Q = sludge flow rate = 30,000 gal/day

S_0 = ultimate BOD_L (lbm/gal)

P_x = net mass of cell tissue produced (lbm/day)

$$BOD_L = \frac{BOD_5}{1 - e^{-Rt}}$$

$R = 0.2$ 1/day and $t = 5$ days.

$$BOD_L = \frac{12,000 \ \frac{lbm}{day}}{1 - e^{-(0.2 \ 1/day)(5 \ days)}}$$

$$= 18,984 \ lbm/day$$

$$S_0 = \frac{BOD_L}{Q} = \frac{18,984 \ \frac{lbm}{day}}{30,000 \ \frac{gal}{day}}$$

$$= 0.63 \ lbm/gal$$

$$P_x = \frac{YQES_0}{1 + R_d\theta_c}$$

yield coefficient, $Y = 0.06$ lbm/lbm

endogenous coefficient, $R_d = 0.03$ 1/day

mean cell residence time, $\theta_c = 10$ days

$$P_x = \frac{\left(0.06 \ \frac{lbm}{lbm}\right)\left(30,000 \ \frac{gal}{day}\right)(0.70)\left(0.63 \ \frac{lbm}{gal}\right)}{1 + \left(0.03 \ \frac{1}{day}\right)(10 \ days)}$$

$$= 611 \ lbm/day$$

$$V_{CH_4} = \left(5.61 \ \frac{ft^3}{lbm}\right)\left((0.70)\left(30,000 \ \frac{gal}{day}\right)\left(0.63 \ \frac{lbm}{gal}\right)\right.$$
$$\left. - (1.42)\left(611 \ \frac{lbm}{gal}\right)\right)$$

$$= 69,352 \ ft^3/day$$

The heating value of methane is approximately 960 Btu/ft³.

$$\text{heat produced} = \frac{\left(960 \ \frac{Btu}{ft^3}\right)\left(69,352 \ \frac{ft^3}{day}\right)}{24 \ \frac{hr}{day}}$$

$$= 2.78 \times 10^6 \ Btu/hr$$

$$2.78 \times 10^6 \ Btu/hr > 380,075 \ Btu/hr$$

> Since the heating value of the methane produced is greater than the heat required to raise the sludge temperature, the methane produced is sufficient to provide the energy required.

SOLUTION 3

3.1. The flow is

$$Q = (\text{population})(\text{flow per capita})$$

$$= (300,000 \ \text{capita})\left(250 \ \frac{gal}{\text{capita-day}}\right)$$

$$= 75 \times 10^6 \ gal/day$$

For the existing 24 in (inside diameter) unlined concrete sewer pipe,

$$\text{length} = 4000 \ ft$$

$$\text{vertical drop} = 100 \ ft$$

$$\text{slope} = \frac{100 \ ft}{4000 \ ft} = 0.025 \ ft/ft$$

$$\text{diameter} = \frac{24 \ in}{12 \ \frac{in}{ft}} = 2 \ ft$$

For an unlined concrete pipe, assume $n = 0.013$.

Calculate the capacity of the 2 ft diameter pipe flowing full.

$$A = \frac{\pi D^2}{4} = \frac{\pi(2 \ ft)^2}{4}$$

$$= 3.14 \ ft^2$$

$$R = \frac{A}{P} = \frac{\frac{\pi D^2}{4}}{\pi D}$$

$$= \frac{D}{4}$$

$$= \frac{2 \ ft}{4}$$

$$= 0.5 \ ft$$

Environmental

Use the Chezy-Manning equation.

$$Q = \left(\frac{1.49}{n}\right) A R^{2/3} \sqrt{S}$$

$$= \left(\frac{1.49}{0.013}\right)(3.14 \text{ ft}^2)(0.5 \text{ ft})^{2/3}\sqrt{0.025}$$

$$= 35.85 \text{ ft}^3/\text{sec}$$

Compare this to the flow rate.

$$Q = \left(75 \times 10^6 \ \frac{\text{gal}}{\text{day}}\right)\left(\frac{1.55 \ \frac{\text{ft}^3}{\text{sec}}}{10^6 \ \frac{\text{gal}}{\text{day}}}\right)$$

$$= 116 \text{ ft}^3/\text{sec}$$

A second pipe is required to handle the flow.

$$116 \ \frac{\text{ft}^3}{\text{sec}} - 35.8 \ \frac{\text{ft}^3}{\text{sec}} = 80.2 \text{ ft}^3/\text{sec}$$

Assume that the second pipe is flowing full with the same slope (0.025 ft/ft) and is also an unlined concrete pipe ($n = 0.013$).

$$R = \frac{D}{4}$$

$$A = \frac{\pi D^2}{4}$$

Use the Chezy-Manning equation.

$$80.2 \ \frac{\text{ft}^3}{\text{sec}} = \left(\frac{1.49}{0.013}\right)\left(\frac{\pi D^2}{4}\right)\left(\frac{D}{4}\right)^{2/3}\sqrt{0.025}$$

$$= (14.23 D^2)\left(\frac{D}{4}\right)^{2/3}$$

$$= 5.65 D^{8/3}$$

$$D = \left(\frac{80.2}{5.65}\right)^{3/8}\left(12 \ \frac{\text{in}}{\text{ft}}\right)$$

$$= 32.5 \text{ in}$$

The 32.5 in pipe size is not standard, and sewers are not normally designed to flow full.

> Use a 36 in diameter pipe.

3.2. Two 24 in diameter unlined concrete pipes with a slope of 0.025 ft/ft will flow full and will not handle a peak flow of 75×10^5 gal/day (116 ft^3/sec).

SOLUTION 4

4.1. The typical limits of BOD for the four sources are

wastewater source	expected BOD
sanitary	250 mg/L
hospital	350 mg/L
plating plant	0 mg/L
dairy animal	1200 mg/L

4.2. In addition to organic loading (measured as BOD), the following wastewater characteristics should be considered because these conditions may upset biological treatment processes.

- the presence of heavy metals
- the presence of hexavalent chromium
- the presence of cyanide
- a high or low pH (less than 6 or greater than 9)

The following compounds should also be checked for, although their presence is less likely: phenol, excessive oil and grease, and high levels of hydrogen sulfide.

All these compounds must be removed or neutralized in pretreatment before biological treatment to reduce the organic waste load (BOD).

To help design biological pretreatment, analyze the following constituents.

- dissolved oxygen in incoming waste (COD)
- nutrients such as nitrogen and phosphorus (deficient or rich)
- chlorides and sodium, total dissolved solids (TDS), suspended solids, and the settleability of the waste

4.3. The pretreatment process is as follows.

(a) Use a basin (with aeration if required) to equalize the flow and waste strength.

(b) In the same basin, allow solids to settle out (during quiescent times). A skimmer and/or absorbent boom will be used to remove floating material (especially oil and grease).

(c) Use chemical treatment and precipitation to remove heavy metals such as Cr^{+6}, CN^-, and phenol.

(d) Adjust the pH to 6.0–9.0 (if required).

(e) Add nutrients (N or P) or carbon (if required).

(f) Use a trickling filter or aerated lagoon to provide a moderate amount of biological treatment (i.e., to reduce BOD and NH_3–N or organic nitrogen to levels acceptable to the municipal wastewater treatment facility).

(g) Use final clarification and turbulent discharge to the municipal sewer system to reoxygenate wastewater.

SOLUTION 5

Peak summer wastewater flows are listed in the table.

wastewater source	unit volume	units (gal/unit-day)	flow (gal/day)
summer population	10,000	30	300,000
hospital (beds)	100	165	16,500
laundromat (washers)	20	550	11,000
dining room (meals)	4500[a]	7	31,500
gas station (employees)	4	12	48
		total	359,048

[a]The dining room serves:
  (3 meals/day)(3 servings)(500 seats) = 4500 meals/day

Reference: *Wastewater Engineering*, Metcalf and Eddy

Winter wastewater flows are as follows.

wastewater source	unit volume	units (gal/unit-day)	flow (gal/day)
winter population	300	30	9000
hospital (beds)	10	165	1650
laundromat (washers)	4	550	2200
dining room (meals)	900[a]	7	6300
gas station (employees)	2	12	24
		total	19,174

[a]The dining room serves:
  (3 meals/day)(300 seats) = 900 meals/day

Reference: *Wastewater Engineering*, Metcalf and Eddy

Adjust k for winter and summer temperatures.

$$k_T = k_{20°C}\theta^{T-20°C}$$

In winter,

$$k_{10°C} = \left(0.25\ \frac{1}{\text{day}}\right)(1.06)^{10°C-20°C}$$
$$= 0.14/\text{day}$$

In summer,

$$k_{32°C} = \left(0.25\ \frac{1}{\text{day}}\right)(1.06)^{32°C-20°C}$$
$$= 0.5/\text{day}$$

The detention time in winter is

$$t = \frac{kt}{k_{10}} = \frac{5}{0.14\ \dfrac{1}{\text{day}}} = 35.7\ \text{days}$$

The detention time in summer is

$$t = \frac{5}{0.5\ \dfrac{1}{\text{day}}} = 10\ \text{days}$$

Determine the pond surface area requirements in both summer and winter.

$$A_{\text{summer}} = \frac{Qt}{d}$$

$$= \frac{\left(359,048\ \dfrac{\text{gal}}{\text{day}}\right)(10\ \text{days})}{(2\ \text{ft})\left(7.48\ \dfrac{\text{gal}}{\text{ft}^3}\right)\left(43,560\ \dfrac{\text{ft}^3}{\text{ac}}\right)}$$

$$= 5.5\ \text{ac}$$

$$A_{\text{winter}} = \frac{\left(19,174\ \dfrac{\text{gal}}{\text{day}}\right)(35.7\ \text{days})}{(2\ \text{ft})\left(7.48\ \dfrac{\text{gal}}{\text{ft}^3}\right)\left(43,560\ \dfrac{\text{ft}^3}{\text{ac}}\right)}$$

$$= 1.05\ \text{ac}$$

Check the BOD loading.

$$L_{\text{BOD,summer}} = \frac{\left(250\ \dfrac{\text{mg}}{\text{L}}\right)(0.36\ \text{MGD})\left(8.34\ \dfrac{\text{lbm-L}}{\text{mg-MG}}\right)}{5.5\ \text{ac}}$$

$$= 136\ \text{lbm/ac-day}$$

$$L_{\text{BOD,winter}} = \frac{\left(250\ \dfrac{\text{mg}}{\text{L}}\right)\left(19,174\ \dfrac{\text{gal}}{\text{day}}\right)\left(8.34\ \dfrac{\text{lbm-L}}{\text{mg-MG}}\right)}{\left(1\times10^6\ \dfrac{\text{gal}}{\text{MG}}\right)(1.1\ \text{ac})}$$

$$= 36.3\ \text{lbm/ac-day}$$

For this facility, use a non-aerated stabilization pond with a maximum working depth of 2 ft that has five cells, each 1.1 ac in surface area. The total area is 5.5 ac. Only one cell will be required during the winter months. Additional cells can be brought on-line as the wastewater flows increase in the spring and early summer. All the cells are used during the peak summer season.

SOLUTION 6

- *Landfill:* (a) The principle advantage of landfills is that they can handle all types of waste. (b) The disadvantages include the requirement for large land areas, the slow decomposition of wastes, and the possibility of leachate contaminating nearby ground-water and surface water. (c) The relative cost of landfills is average but increasing. (d) Landfills can serve any size population.

- *Open burning:* (a) The main advantages of open burning are that it can be done on both small and large scales with relatively little cost, and that it reduces the volume of waste, leaving only ash. (b) The disadvantages are that it only works with combustible waste such as paper and wood products. The combustion of the waste may not be complete.

Combustion of large amounts of waste may be difficult to control. No energy is recovered from the combustion of wastes, and the process produces air pollution. (c) The relative cost is low. (d) The population served is usually small.

- *Incineration:* (a) The advantages of incineration include more complete destruction of the wastes and less ash production than in open burning. This process can destroy less-burnable wastes such as plastics and food. Energy from incineration can be captured and used to make steam and electricity. (b) The disadvantages are the high cost and large volumes of waste required. If the waste combustion is not self-sustaining, supplemental fuel may be required to maintain incineration. This process does not work for incombustible materials such as glass and steel. The ash may contain hazardous compounds. This process also produces air pollution. (c) The cost of incineration is high. (d) It best serves a large population.

- *Composting:* (a) The advantages of composting are that it uses the process of natural degradation of wastes and that it can be performed on a small or large scale. (b) The main disadvantage is that this process is only applicable to easily degradable wastes. It is not applicable to plastic, glass, or metal. The addition of moisture and readily degradable material such as yard wastes may be needed to help this process. (c) The relative costs are low to medium. (d) Composting best serves a small to medium population.

- *Recycling:* (a) The main advantages of recycling are that it reuses process materials and saves natural resources. It also reduces the amount of waste that must be disposed by other means. (b) The disadvantage is that it uses more labor than other methods. This method is viable only if there is a company willing to perform the recycling and if there is a market for the recycled materials. The trash must be separated into different waste streams. (c) The relative costs of recycling are medium to low and should decrease as the demand for recycled materials and the cost of other disposal methods increase. (d) The population served is medium to large depending on the market for recycled materials.

SOLUTION 7

7.1. Use the trash disposal in the eighth year as the average over the past 16 years.

The population in year 8 is

$$\frac{500{,}000}{(1.03)^8} = 394{,}705 \text{ people}$$

The solid waste disposed (per capita in year 8) is

$$\frac{5.4 \ \dfrac{\text{lbm}}{\text{day-person}}}{(1.01)^8} = 4.99 \text{ lbm/day-person}$$

The solid waste disposed over the past 16 years is

$$(394{,}705 \text{ people})\left(4.99 \ \frac{\text{lbm}}{\text{day-person}}\right)(16 \text{ yr})$$
$$= 31{,}513{,}247 \text{ lbm-yr/day}$$

Over the next four years, use the average trash disposal as the second year.

The population in two years is

$$(500{,}000)(1.03)^2 = 530{,}450 \text{ people}$$

The solid waste disposed is

$$\left(5.4 \ \frac{\text{lbm}}{\text{day-person}}\right)(1 - 0.03)(1.01)^2$$
$$= 5.34 \text{ lbm/day-person}$$

The solid waste disposed over the next four years is

$$(530{,}450 \text{ people})\left(5.34 \ \frac{\text{lbm}}{\text{day-person}}\right)(4 \text{ yr})$$
$$= 11{,}330{,}412 \text{ lbm-yr/day}$$

The solid waste occupies 75% of the site. The depth of the waste is 40 ft.

The area filled by 16 years of trash is

$$A = \frac{\left(31{,}513{,}247 \ \dfrac{\text{lbm-yr}}{\text{day}}\right)\left(365 \ \dfrac{\text{days}}{\text{yr}}\right)\left(3 \ \dfrac{\text{ft}}{\text{yd}}\right)^3}{\left(1000 \ \dfrac{\text{lbm}}{\text{yd}^3}\right)(40 \text{ ft})\left(43{,}560 \ \dfrac{\text{ft}^2}{\text{ac}}\right)(0.75)}$$
$$= 237.65 \text{ ac}$$

The area filled in the next four years is

$$A = \frac{\left(11{,}330{,}412 \ \dfrac{\text{lbm-yr}}{\text{day}}\right)\left(365 \ \dfrac{\text{days}}{\text{yr}}\right)\left(3 \ \dfrac{\text{ft}}{\text{yd}}\right)^3}{\left(1000 \ \dfrac{\text{lbm}}{\text{yd}^3}\right)(40 \text{ ft})\left(43{,}560 \ \dfrac{\text{ft}^2}{\text{ac}}\right)(0.75)}$$
$$= 85.45 \text{ ac} \quad [135.5 \text{ ac with buffer zone}]$$

Add the areas and include the 250 ft buffer zone.

$$\sqrt{(237.65 \text{ ac} + 85.45 \text{ ac})\left(43{,}560 \ \frac{\text{ft}^2}{\text{ac}}\right)} = 3752 \text{ ft}$$

$$\frac{(3752 \text{ ft} + 250 \text{ ft} + 250 \text{ ft})^2}{43{,}560 \ \dfrac{\text{ft}^2}{\text{ac}}} = \boxed{415 \text{ ac}}$$

7.2. The life of the replacement site is

$$4 \text{ yr} + 5 \text{ yr capacity} = 9 \text{ yr}$$

The average trash disposal at 4.5 years is calculated as follows.

$$\text{population in 4.5 yr} = (500{,}000)(1.03)^{4.5}$$
$$= 571{,}133 \text{ people}$$

$$\text{solid waste disposed} = \left(5.4 \ \frac{\text{lbm}}{\text{person-day}}\right)(1 - 0.03)$$
$$\times (1.01)^{4.5}$$
$$= 5.48 \text{ lbm/person-day}$$

The solid waste disposed over the next nine years is

$$(571{,}133 \text{ people})\left(5.48 \ \frac{\text{lbm}}{\text{person-day}}\right)(9 \text{ yr})$$
$$= 28{,}168{,}280 \text{ lbm-yr/day}$$

The area of the site is

$$A = \frac{\left(28{,}168{,}280 \ \dfrac{\text{lbm-yr}}{\text{day}}\right)\left(365 \ \dfrac{\text{days}}{\text{yr}}\right)\left(3 \ \dfrac{\text{ft}}{\text{yd}}\right)^3}{\left(1000 \ \dfrac{\text{lbm}}{\text{yd}^3}\right)(40 \text{ ft})(0.75)}$$
$$= 9{,}253{,}280 \text{ ft}^2$$

Include the 250 ft buffer zone.

$$\sqrt{9{,}253{,}280 \text{ ft}^2} = 3042 \text{ ft}$$

$$A = \frac{(3042 \text{ ft} + 250 \text{ ft} + 250 \text{ ft})^2}{43{,}560 \ \dfrac{\text{ft}^2}{\text{ac}}}$$
$$= \boxed{288 \text{ ac}}$$

SOLUTION 8

8.1. The spray volume is

$$\frac{\left(2.5 \ \dfrac{\text{in}}{\text{wk}}\right)\left(43{,}560 \ \dfrac{\text{ft}^2}{\text{ac}}\right)\left(7.48 \ \dfrac{\text{gal}}{\text{ft}^3}\right)}{\left(12 \ \dfrac{\text{in}}{\text{ft}}\right)\left(7 \ \dfrac{\text{days}}{\text{wk}}\right)}$$
$$= 9697 \text{ gal/day-ac}$$

The total area required is

$$A_t = \frac{1 \times 10^6 \ \dfrac{\text{gal}}{\text{day}}}{9697 \ \dfrac{\text{gal}}{\text{day-ac}}} = 103.1 \text{ ac}$$

Only 25% of the nozzles are used at one time, so 25% of the total area is used at one time. The area used at any moment is

$$A = (103.1 \text{ ac})(0.25) = \boxed{25.8 \text{ ac}}$$

8.2. The area covered by each nozzle is

$$A_n = \frac{\pi(350 \text{ ft})^2}{(4)\left(43{,}560 \ \dfrac{\text{ft}^2}{\text{ac}}\right)}$$
$$= 2.21 \text{ ac/nozzle}$$

The number of nozzles is

$$n = \frac{103.1 \text{ ac}}{2.21 \ \dfrac{\text{ac}}{\text{nozzle}}} = \boxed{46.7 \text{ nozzles} \quad [\text{use 48}]}$$

8.3. Each nozzle delivers 275 gal/min at 75 lbf/in^2. There are 48 nozzles in use, but only 25% are used at one time.

$$(48 \text{ nozzles})(0.25) = 12 \text{ nozzles at once}$$

The pump flow is

$$Q = \left(275 \ \frac{\text{gal}}{\text{min}}\right)(12 \text{ nozzles})$$
$$= 3300 \text{ gal/min through 12 nozzles}$$
$$= 4.75 \text{ MGD} > 1 \text{ MGD} \quad [\text{OK}]$$

The exit head at the nozzles is

$$h_{\text{exit}} = \left(75 \ \frac{\text{lbf}}{\text{in}^2}\right)\left(2.31 \ \frac{\text{ft}}{\dfrac{\text{lbf}}{\text{in}^2}}\right)$$
$$= 173.25 \text{ ft}$$

The head added by the pumps is

$$h_A = 173.25 \text{ ft} - 45 \text{ ft} + 30 \text{ ft}$$
$$= 158.25 \text{ ft}$$

$$\text{water hp} = \frac{h_A Q(\text{SG of water})}{3956}$$
$$= \frac{(158.25 \text{ ft})\left(3300 \ \dfrac{\text{gal}}{\text{min}}\right)(1.0)}{3956 \ \dfrac{\text{ft-gal}}{\text{hp-min}}}$$
$$= 132.0 \text{ hp}$$

$$\text{motor hp} = \frac{\text{water hp}}{\eta_{\text{pump}}}$$
$$= \frac{132.0 \text{ hp}}{0.82}$$
$$= 161 \text{ hp} \quad (200 \text{ hp})$$

Use one 200 hp pump motor or two 100 hp pump motors. Motors are rated by their output power and motor efficiency is irrelevant.

8.4. The following should be considered when determining the application rate of effluent to spray the fields.

- soil permeability
- soil crop cover (crop uptake of moisture and nutrients)
- average monthly/annual precipitation and evaporation
- organic and nutrient content of effluent to be sprayed
- metal/toxic content of effluent to be sprayed
- depth to groundwater table
- subsurface geology
- site topography and drainage
- distance to nearest surface water

The first five considerations are primarily used in determining the application rate. All the considerations are used in determining the overall site applicability.

SOLUTION 9

Calculate the areas.

$$A_{\text{site}} = 0.25 \text{ ac}$$
$$A_{\text{block}} = (12 \text{ sites})A_{\text{site}}$$
$$= (12 \text{ sites})\left(0.25 \frac{\text{ac}}{\text{site}}\right)$$
$$= 3 \text{ ac}$$
$$\frac{480 \text{ sites}}{12 \frac{\text{sites}}{\text{blocks}}} = 40 \text{ blocks}$$

Assume wastewater flow is 100 gal/capita-day. Wastewater flow from the houses is

$$\left(4.1 \frac{\text{capita}}{\text{site}}\right)\left(100 \frac{\text{gal}}{\text{capita-day}}\right)(480 \text{ sites})$$
$$= 196,800 \text{ gal/day}$$

$$\text{infiltration} + \text{inflow} = 200 \frac{\text{gal}}{\text{in-mi-day}} + 2000 \frac{\text{gal}}{\text{in-mi-day}}$$
$$= 2200 \text{ gal/in-mi-day}$$

The length of the 6 in feeder from the houses is

$$L = (12 \text{ ft of street}) + (50 \text{ ft of lot})$$
$$= 62 \text{ ft}$$
$$L_{\text{total}} = \frac{(62 \text{ ft})(480 \text{ sites})}{5280 \frac{\text{ft}}{\text{mi}}}$$
$$= 5.64 \text{ mi}$$

Infiltration/inflow from the 6 in pipe is

$$\left(2200 \frac{\text{gal}}{\text{in-mi-day}}\right)(6 \text{ in})(5.64 \text{ mi}) = 74,448 \text{ gal/day}$$

Assume the 8 in main is eight blocks long and there are five mains.

$$\frac{((8 \text{ blocks})(540 \text{ ft}) + (8 \text{ blocks})(24 \text{ ft}))(5 \text{ mains})}{5280 \frac{\text{ft}}{\text{mi}}}$$
$$= 4.27 \text{ mi}$$

Infiltration/inflow from the 8 in pipe is

$$\left(2200 \frac{\text{gal}}{\text{in-mi-day}}\right)(8 \text{ in})(4.27 \text{ mi}) = 75,152 \text{ gal/day}$$

Assume that the 10 in main is five blocks long and 200 ft to the pump station.

$$\frac{(5 \text{ blocks})(240 \text{ ft}) + (5 \text{ blocks})(24 \text{ ft}) + 200 \text{ ft}}{5280 \frac{\text{ft}}{\text{mi}}}$$
$$= 0.29 \text{ mi}$$

Infiltration/inflow from the 10 in main is

$$\left(2200 \frac{\text{gal}}{\text{in-mi-day}}\right)(10 \text{ in})(0.29 \text{ mi}) = 6380 \text{ gal/day}$$

The total flow is

$$Q_{\text{total}} = 196,800 \frac{\text{gal}}{\text{day}} + 74,448 \frac{\text{gal}}{\text{day}}$$
$$+ 75,152 \frac{\text{gal}}{\text{day}} + 6380 \frac{\text{gal}}{\text{day}}$$
$$= \boxed{352,780 \text{ gal/day}}$$

SOLUTION 10

10.1. The effluent BOD_5 is the untreated influent soluble BOD_5, S, plus the BOD_5 of the effluent biological solids.

$$20 \ \frac{mg}{L} = S + \left(20 \ \frac{mg}{L}\right)(0.65)(1.42)(0.68)$$

$$= S + 12.6 \ mg/L$$

$$S = 20 \ \frac{mg}{L} - 12.6 \ \frac{mg}{L}$$

$$= 7.4 \ mg/L$$

The biological efficiency, η_b, is

$$\eta_b = \left(\frac{\text{influent } BOD_5 - \text{effluent } BOD_5}{\text{influent } BOD_5}\right) \times 100\%$$

$$= \left(\frac{250 \ \frac{mg}{L} - 7.4 \ \frac{mg}{L}}{250 \ \frac{mg}{L}}\right) \times 100\%$$

$$= \boxed{97\%}$$

10.2. The overall efficiency is

$$\eta = \left(\frac{250 \ \frac{mg}{L} - 20 \ \frac{mg}{L}}{250 \ \frac{mg}{L}}\right) \times 100\% = \boxed{92\%}$$

10.3. Calculate the reactor volume.

$$V_{\text{reactor}} = \theta Q$$

$$X = \frac{\theta_c Y (S_0 - S)}{\theta (1 + R_d \theta_c)}$$

Rearrange for θ and substitute into V_{reactor}.

$$V_{\text{reactor}} = \frac{(5 \ MGD)(10 \ days)\left(0.6 \ \frac{lbm \ VSS}{lbm \ BOD_5}\right)}{\times \left(250 \ \frac{mg}{L} - 7.4 \ \frac{mg}{L}\right)}{\left(3500 \ \frac{mg}{L}\right)\left(1 + \left(0.06 \ \frac{1}{day}\right)(10 \ days)\right)}$$

$$= 1.3 \ MG$$

Sludge production can be calculated from the observed yield.

$$Y_{\text{obs}} = \frac{Y}{1 + k_d \theta_c}$$

$$= \frac{0.6 \ \frac{lbm \ VSS}{lbm \ BOD}}{1 + \left(0.06 \ \frac{1}{day}\right)(10 \ days)}$$

$$= 0.375$$

The biomass production is

$$P_x(\text{MLVSS}) = Y_{\text{obs}}(S_0 - S)Q$$

$$= (0.375)\left(250 \ \frac{mg}{L} - 7.4 \ \frac{mg}{L}\right)$$

$$\times (5 \ MGD)\left(8.34 \ \frac{lbm\text{-}L}{mg\text{-}MGD}\right)$$

$$= \boxed{3794 \ lbm/day}$$

10.4. The increase in MLSS is

$$\frac{\text{MLVSS}}{0.8} = \frac{3794 \ \frac{lbm}{day}}{0.8}$$

$$= \boxed{4743 \ lbm/day}$$

10.5. The sludge to be wasted is the increase in MLSS $-$ MLSS in the effluent.

$$4743 \ \frac{lbm}{day} - \left(\left(20 \ \frac{mg}{L}\right)(5 \ MGD)\left(8.34 \ \frac{lbm\text{-}L}{mg\text{-}MGD}\right)\right)$$

$$= \boxed{3909 \ lbm/day}$$

10.6. If sludge is wasted from the reactor, the sludge wasting rate is Q_w.

$$\theta_c = \frac{V_{\text{reactor}} X}{Q_w X + Q_e X_e}$$

Rearrange this equation, solving for Q_w.

$$Q_w X + Q_e X_e = \frac{V_{\text{reactor}} X}{\theta_c}$$

$$Q_w = \left(\frac{V_{\text{reactor}} X}{\theta_c} - Q_e X_e\right)\left(\frac{1}{X}\right)$$

Assume that $Q_e = Q$, since it is being diminished only by the waste sludge.

$$(\text{VSS})X_e = (0.8)\left(20 \ \frac{mg}{L}\right) = 16 \ mg/L$$

$$Q_w = \left(\frac{(1.3 \ MG)\left(3500 \ \frac{mg}{L}\right)}{10 \ days} - (5 \ MGD)\left(16 \ \frac{mg}{L}\right)\right)$$

$$\times \left(\frac{1}{3500 \ \frac{mg}{L}}\right)$$

$$= \boxed{0.107 \ MGD}$$

10.7. Calculate the recirculation ratio, R, using a mass balance around the reactor.

$$\text{reactor MLVSS} = 3500 \text{ mg/L}$$

$$\text{recycle MLVSS} = 8000 \text{ mg/L}$$

$$\left(3500 \frac{\text{mg}}{\text{L}}\right)(Q + Q_{\text{reactor}}) = \left(8000 \frac{\text{mg}}{\text{L}}\right) Q_{\text{reactor}}$$

$$\frac{Q + Q_{\text{reactor}}}{Q_{\text{reactor}}} = \frac{8000 \frac{\text{mg}}{\text{L}}}{3500 \frac{\text{mg}}{\text{L}}}$$

$$= 2.29$$

$$\frac{Q}{Q_{\text{reactor}}} + 1 = 2.29$$

$$\frac{Q}{Q_{\text{reactor}}} = 2.29 - 1$$

$$= 1.29$$

$$R = \frac{Q_{\text{reactor}}}{Q} = \frac{1}{1.29}$$

$$= \boxed{0.78}$$

10.8. The hydraulic retention time is

$$\theta = \frac{V}{Q} = \frac{1.3 \text{ MG}}{5 \text{ MGD}} = \boxed{0.26 \text{ day}}$$

10.9. Check the specific substrate utilization rate.

$$U = \frac{S_0 - S}{\theta X}$$

$$= \frac{250 \frac{\text{mg}}{\text{L}} - 7.4 \frac{\text{mg}}{\text{L}}}{(0.26 \text{ day})\left(3500 \frac{\text{mg}}{\text{L}}\right)}$$

$$= \boxed{0.27 \text{ mg BOD}_5 \text{ utilized/mg MLVSS-day}}$$

10.10. Check the food-to-microorganism ratio.

$$\text{F:M} = \frac{S_0}{\theta X}$$

$$= \frac{250 \frac{\text{mg}}{\text{L}}}{(0.26 \text{ day})\left(3500 \frac{\text{mg}}{\text{L}}\right)}$$

$$= \boxed{0.275 \text{ mg BOD/mg MLVSS-day}}$$

10.11. Check the volumetric loading rate (VLR).

$$\text{VLR} = \frac{S_0 Q}{V}$$

$$= \frac{\left(250 \frac{\text{mg}}{\text{L}}\right)(5 \text{ MGD})\left(8.34 \frac{\text{lbm-L}}{\text{mg-MG}}\right)}{\frac{1.3 \times 10^6 \text{ gal}}{7.48 \frac{\text{gal}}{\text{ft}^3}}} \times \left(1000 \frac{\text{ft}^3}{1000 \text{ ft}^3}\right)$$

$$= 60 \text{ lbm/day-BOD/1000 ft}^3$$

Calculate the oxygen required.

$$m_{O_2} = \frac{Q(S_0 - S)}{f} - 1.42 P_x$$

$$f = \frac{\text{BOD}_5}{\text{BOD}_L} = 0.68$$

$$m_{O_2} = \frac{(5 \text{ MGD})\left(250 \frac{\text{mg}}{\text{L}} - 7.4 \frac{\text{mg}}{\text{L}}\right)\left(8.34 \frac{\text{lbm-L}}{\text{mg-MG}}\right)}{0.68}$$
$$- (1.42)\left(3794 \frac{\text{lbm}}{\text{day}}\right)$$

$$= 9490 \text{ lbm/day}$$

Air is 23.2% O_2 by weight. Since the transfer efficiency is 8%, the normal airflow is

$$\text{airflow} = \frac{9490 \frac{\text{lbm}}{\text{day}}}{\left(0.075 \frac{\text{lbm}}{\text{ft}^3}\right)(0.232)(0.08)\left(1440 \frac{\text{min}}{\text{day}}\right)}$$

$$= 4734 \text{ ft}^3/\text{min}$$

With the 100% excess capacity, the blower capacity is

$$(2)\left(4734 \frac{\text{ft}^3}{\text{min}}\right) = \boxed{9468 \text{ ft}^3/\text{min}}$$

SOLUTION 11

11.1. Reuse as much flume/wash water as possible after sedimentation and chlorination to reduce the wastewater flow.

The sedimentation tank volume is

$$V = Qt = \frac{\left(2.2 \times 10^6 \frac{\text{gal}}{\text{day}}\right)(2 \text{ hr})}{24 \frac{\text{hr}}{\text{day}}}$$

$$= 1.83 \times 10^5 \text{ gal}$$

$$A = \frac{V}{D} = \frac{1.83 \times 10^5 \text{ gal}}{\left(7.48 \frac{\text{gal}}{\text{ft}^3}\right)(10 \text{ ft})}$$

$$= 2451 \text{ ft}^2$$

Use a length-to-width ratio of 4:1.

$$LW = L\left(\frac{L}{4}\right) = 2451 \text{ ft}^2$$

$$L = \sqrt{(4)(2451 \text{ ft}^2)} = 99 \text{ ft}$$

$$W = \frac{L}{4} = \frac{99 \text{ ft}}{4} = 24.75 \text{ ft}$$

The surface overflow rate is

$$\frac{2.2 \times 10^6 \frac{\text{gal}}{\text{day}}}{2451 \text{ ft}^2} = 898 \text{ gal/day-ft}^2$$

The required weir length is

$$L_{\text{weir}} = \frac{2.2 \times 10^6 \frac{\text{gal}}{\text{day}}}{15,000 \frac{\text{gal}}{\text{day-ft}}} = 147 \text{ ft}$$

The design summary of the sedimentation basin for the flume/wash water is

detention time, t	2 hr
volume, V	1.83×10^5 gal
dimensions, $L \times W \times D$	99 ft × 24.75 ft × 10 ft
weir length, L_{weir}	147 ft
overflow rate	900 gal/day-ft^2

11.2. Design an aerobic lagoon to treat all the wastewater.

The composition of the combined wastes is

$$Q = 2.2 \times 10^6 \frac{\text{gal}}{\text{day}} + 6.6 \times 10^5 \frac{\text{gal}}{\text{day}}$$

$$+ 7.5 \times 10^4 \frac{\text{gal}}{\text{day}}$$

$$= 2.935 \times 10^6 \text{ gal/day}$$

$$\text{BOD} = \frac{\left(2.2 \times 10^6 \frac{\text{gal}}{\text{day}}\right)\left(200 \frac{\text{mg}}{\text{L}}\right) + \left(6.6 \times 10^5 \frac{\text{gal}}{\text{day}}\right) \times \left(1230 \frac{\text{mg}}{\text{L}}\right) + \left(7.5 \times 10^4 \frac{\text{gal}}{\text{day}}\right)\left(1420 \frac{\text{mg}}{\text{L}}\right)}{2.935 \times 10^6 \frac{\text{gal}}{\text{day}}}$$

$$= 463 \text{ mg/L}$$

The BOD organic loading is

$$(2.935 \text{ MGD})\left(463 \frac{\text{mg}}{\text{L}}\right)\left(8.34 \frac{\text{lbm-L}}{\text{mg-MG}}\right)$$

$$= 11,333 \text{ lbm/day}$$

Size the aerobic lagoon for a 2 day detention time at 1.5 ft depth.

$$A = \frac{Qt}{D} = \frac{\left(2.935 \times 10^6 \frac{\text{gal}}{\text{day}}\right)(2 \text{ days})}{(1.5 \text{ ft})\left(7.48 \frac{\text{gal}}{\text{ft}^3}\right)\left(43,560 \frac{\text{ft}^2}{\text{ac}}\right)}$$

$$= 12.0 \text{ ac}$$

Check the organic loading.

$$\frac{11,333 \frac{\text{lbm}}{\text{day}}}{12.0 \text{ ac}} = 944.4 \text{ lbm/ac-day}$$

Check the lagoon size based on an organic loading of 86 lbm/ac-day and 1.5 ft depth.

$$A = \frac{11,333 \frac{\text{lbm}}{\text{day}}}{86 \frac{\text{lbm}}{\text{ac-day}}} = 131.8 \text{ ac}$$

$$V = (131.8 \text{ ac})\left(43,560 \frac{\text{ft}^2}{\text{ac}}\right)(1.5 \text{ ft})\left(7.48 \frac{\text{gal}}{\text{ft}^3}\right)$$

$$= 64.4 \times 10^6 \text{ gal}$$

$$t = \frac{V}{Q} = \frac{64.4 \times 10^6 \text{ gal}}{2.935 \times 10^6 \frac{\text{gal}}{\text{day}}} = 21.9 \text{ days}$$

A lagoon that is 64.4×10^6 gal in volume with an area of 131.8 ac is too large to be practical or economical for this facility.

Design a lagoon for a composite of process water and lime drainage only.

$$Q = 6.6 \times 10^5 \frac{\text{gal}}{\text{day}} + 7.5 \times 10^4 \frac{\text{gal}}{\text{day}}$$

$$= 7.35 \times 10^5 \text{ gal/day}$$

$$\text{BOD} = \frac{\left(6.6 \times 10^5 \frac{\text{gal}}{\text{day}}\right)\left(1230 \frac{\text{mg}}{\text{L}}\right) + \left(7.5 \times 10^4 \frac{\text{gal}}{\text{day}}\right)\left(1420 \frac{\text{mg}}{\text{L}}\right)}{7.35 \times 10^5 \frac{\text{gal}}{\text{day}}}$$

$$= 1249 \text{ mg/L}$$

The BOD organic loading is

$$(0.735 \text{ MGD})\left(1249 \ \frac{\text{mg}}{\text{L}}\right)\left(8.34 \ \frac{\text{lbm-L}}{\text{mg-MG}}\right)$$
$$= 7658 \ \text{lbm/day}$$

Design a lagoon for a 2 day detention time at 1.5 ft depth.

$$A = \frac{\left(7.35 \times 10^5 \ \dfrac{\text{gal}}{\text{day}}\right)(2 \ \text{days})}{(1.5 \ \text{ft})\left(7.48 \ \dfrac{\text{gal}}{\text{ft}^3}\right)\left(43{,}560 \ \dfrac{\text{ft}^2}{\text{ac}}\right)}$$
$$= 3.0 \ \text{ac}$$

$$\text{organic loading} = \frac{7658 \ \dfrac{\text{lbm}}{\text{day}}}{3.0 \ \text{ac}}$$
$$= 2553 \ \text{lbm/ac-day}$$

A lagoon that is only 3.0 ac has a very high organic loading and will not effectively treat the waste.

Use an aerobic lagoon, built in four cells, to treat all the waste. Each cell is 3.0 ac to accommodate the flow of the process water and lime drainage only.

The design summary for an aerobic lagoon to treat all the wastewater is

total area, A	12 ac
number of cells and size	four 3.0 ac cells
detention time, t	2 days
working depth, D	1.5 ft
organic loading	944.4 lbm/ac-day

The treated effluent will be disposed by land application.

11.3. The following alternatives may also be considered for treating the wastewater from this facility.

(a) Since the beet sugar plant operates only three months of the year, the wastewater could be stored and treated slowly over the entire year. More storage capacity would be required. This option would reduce the organic loading on the aerobic lagoon.

(b) Treat the composite flow of process water and lime drainage by a high-rate-activated sludge system.

SOLUTION 12

12.1. The treatment process is shown in the following schematic diagram.

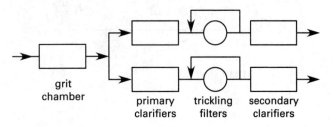

grit chamber   primary clarifiers   trickling filters   secondary clarifiers

12.2. According to the *Ten States' Standards* Sec. 61.1, this facility requires two primary and secondary clarifiers, each designed for the AADF. Since two sets of clarifiers are required, two trickling filters should also be provided. Each set of units can handle the full flow, and either may be taken out of service for maintenance.

From Table 29.10, assume a surface loading rate (SLR) of 1000 gal/day-ft^2.

$$A = \frac{Q}{\text{SLR}} = \frac{2 \times 10^6 \ \dfrac{\text{gal}}{\text{day}}}{1000 \ \dfrac{\text{gal}}{\text{day-ft}^2}}$$
$$= 2000 \ \text{ft}^2$$

Assume a detention time of two hours.

$$V = tQ = \frac{(2 \ \text{hr})\left(2 \times 10^6 \ \dfrac{\text{gal}}{\text{day}}\right)}{\left(24 \ \dfrac{\text{hr}}{\text{day}}\right)\left(7.48 \ \dfrac{\text{gal}}{\text{ft}^3}\right)}$$
$$= 22{,}282 \ \text{ft}^3$$

$$D = \frac{V}{A} = \frac{22{,}282 \ \text{ft}^3}{2000 \ \text{ft}^2} = 11.1 \ \text{ft}$$

For each clarifier, the length-to-width ratio is 4:1.

$$LW = L\left(\frac{L}{4}\right) = 2000 \ \text{ft}^2$$

$$L = \sqrt{(4)(2000 \ \text{ft}^2)} = 89.4 \ \text{ft}$$

$$W = \frac{L}{4} = \frac{89.4 \ \text{ft}}{4} = 22.4 \ \text{ft}$$

Assume a weir loading rate of 20,000 gal-day/ft.

$$L_{\text{weir}} = \frac{2 \times 10^6 \ \dfrac{\text{gal}}{\text{day}}}{20{,}000 \ \dfrac{\text{gal-day}}{\text{ft}}} = 100 \ \text{ft}$$

The design summary of the clarifiers is

dimensions, $L \times W \times D$	89.4 ft × 22.4 ft × 11.1 ft
detention time, t	2 hr
volume, V	22,282 ft^3
weir length, L_{weir}	100 ft

In the primary clarifiers, assume the BOD$_5$ reduction is 30% and the TSS reduction is 60%.

12.3. For a trickling filter using plastic media, the NRC formulas are not applicable.

The influent BOD$_5$ to the trickling filter is

$$S_i = \left(250 \ \frac{\text{mg}}{\text{L}}\right)(1 - 0.3)$$
$$= 175 \ \text{mg/L}$$

Assume the effluent BOD_5 from the trickling filter: $S_e \leq$ 30 mg/L. The treatment efficiency is

$$\left(\frac{S_i - S_e}{S_i}\right) \times 100\% = \left(\frac{175 \frac{mg}{L} - 30 \frac{mg}{L}}{175 \frac{mg}{L}}\right) \times 100\%$$

$$= 82.9\%$$

Assume the depth of the trickling filter is 10 ft. The area of the trickling filter can be calculated by

$$\frac{S_e}{S_i} = e^{-k_{20}D(Q/A)^{-n}}$$

Rearranging,

$$A = Q \left(\frac{-\ln \frac{S_e}{S_i}}{k_{20}D}\right)^{1/n}$$

Assume average values (modified for depth) of $k_{20} =$ 0.085 gal/min and $n = 0.5$. (Corrections for variations with depth and/or temperature would normally be required. However, no supporting information was given, so already corrected values are assumed.)

$$Q = \frac{2 \times 10^6 \frac{gal}{day}}{1440 \frac{min}{day}} = 1389 \text{ gal/min}$$

$$A = \left(1389 \frac{gal}{min}\right) \left(\frac{-\ln \frac{30 \text{ mg}}{175 \text{ L}}}{\left(0.085 \frac{gal}{min}\right)(10 \text{ ft})}\right)^{1/0.5}$$

$$= 5979 \text{ ft}^2$$

$$D = \sqrt{\frac{4A}{\pi}} = \sqrt{\frac{(4)(5979 \text{ ft}^2)}{\pi}}$$

$$= 87.25 \text{ ft}$$

The trickling filter dimensions are 87.25 ft diameter $\times$ 10 ft deep.

$$V = (5979 \text{ ft}^2)(10 \text{ ft}) = 59{,}790 \text{ ft}^3$$

The organic loading is

$$\frac{(2 \text{ MGD})\left(175 \frac{mg}{L}\right)\left(8.34 \frac{lbm\text{-}L}{mg\text{-}MG}\right)}{59{,}790 \text{ ft}^3}$$

$$= \boxed{48.8 \text{ lbm/day-1000 ft}^3}$$

The hydraulic loading is

$$\frac{1389 \frac{gal}{min}}{5979 \text{ ft}^2} = \boxed{0.23 \text{ gal/min-ft}^2}$$

These values fit the criteria for a high-rate trickling filter.

SOLUTION 13

13.1. The design procedure is as follows. Calculate the cell volume from the flow and detention time. (Assume all three ponds have the same volume.)

$$V = \boxed{Qt}$$

Calculate the cell surface area.

$$A = \boxed{\frac{V}{D}}$$

13.2. The lagoon operating temperatures are

$$T_{winter} = \boxed{\frac{A_{ft^2}f T_{aw,°F} + Q_{MGD} T_{ww,°F}}{A_{ft^2}f + Q_{MGD}}}$$

$$T_{summer} = \boxed{\frac{A_{ft^2}f T_{as,°F} + Q_{MGD} T_{ww,°F}}{A_{ft^2}f + Q_{MGD}}}$$

13.3. To correct k for another operating temperature,

$$\theta = \text{temperature coefficient}$$

$$k_2 = \boxed{k_1 \theta^{T_2 - T_1}}$$

13.4. Assume there is no algal growth in any of the three cells.

$$S_1 = \boxed{\frac{1 + k_d \theta_c}{Y_T k \theta_c}}$$

13.5. The concentration is

$$X_{1,VSS} = \boxed{\frac{(S_0 - S_1)Q}{S_1 k V}}$$

13.6. Assume all the soluble BOD_u entering the cell is removed.

$$X_{2,VSS} = \boxed{\frac{X_1 + Y_T S_1}{\frac{k_d}{t} + 1}}$$

13.7. Assume the soluble BOD_u entering the cell is 0 mg/L.

$$X_3 = \boxed{\frac{X_2}{\frac{k_d}{t} + 1}}$$

13.8. $\boxed{\mathrm{BOD}_{5,\mathrm{eff}} = 0.54X_3(\mathrm{VSS}_{\mathrm{eff}}) + S_e}$

13.9. The correction factor for pH is

$$\mathrm{pH}_{\mathrm{factor}} = \frac{1}{1 + (0.04)(10^{\mathrm{pH}_{\mathrm{opt}}-\mathrm{pH}} - 1)}$$

The correction for temperature is

$$T_{\mathrm{factor}} = 10^{(0.033)(T-20°\mathrm{C})}$$

$$\mu_{\mathrm{max,corrected}} = \mu_{\mathrm{max,assumed}}(\mathrm{pH}_{\mathrm{factor}})\,T_{\mathrm{factor}}$$

Calculate the retention time above which nitrification will occur.

$$\theta_c = \boxed{\frac{1}{\mu_{\mathrm{max}}}}$$

13.10. To find the oxygen requirements,

$$\mathrm{NOD} = 3.18 Q N_{\mathrm{available\,mg/L}}$$

$$\Delta \mathrm{O}_2 = \boxed{\begin{array}{c} 8.34 Q(1 - 1.424T)(S_0 - S_e) \\ + (8.34)(1.42)k_d V + \mathrm{NOD} \end{array}}$$

13.11. The motor horsepower required is

$$\mathrm{hp} = \frac{1\ \mathrm{lbm\ O_2\ required}}{\left(1.8\ \dfrac{\mathrm{lbm\ O_2}}{\mathrm{hp\text{-}hr}}\right)(24\ \mathrm{hr})} = \boxed{0.023\ \mathrm{hp}}$$

13.12. Assume the mixing power is approximately 30 horsepower per million gallons.

$$P = \left(30\ \frac{\mathrm{hp}}{\mathrm{MG}}\right) V_{\mathrm{MG}}$$

Use the greater value of this and the power requirement from Sol. 13.11 in each cell. Determine the number of aerators needed to achieve the total horsepower required.

13.13. Use a length-to-width ratio of 3:1 minimum and determine the position of the aerators.

13.14. The first cell controls the nutrient requirements.

$$\Delta X = 8.34 Q X_1$$

$$N_{\mathrm{requirement}} = 0.122 \Delta X$$

$$P_{\mathrm{requirement}} = 0.023 \Delta X$$

$$N_{\mathrm{available}} = 8.34 Q N_{\mathrm{influent,mg/L}}$$

$$P_{\mathrm{available}} = 8.34 Q P_{\mathrm{influent,mg/L}}$$

If the amount of N or P available is less than the amount required, add the difference of N or P.

SOLUTION 14

14.1. For the pipe,

$$A = \frac{\pi D^2}{4} = \frac{\pi \left(\dfrac{8\ \mathrm{in}}{12\ \dfrac{\mathrm{in}}{\mathrm{ft}}}\right)^2}{4}$$

$$= 0.349\ \mathrm{ft}^2$$

$$Q = \mathrm{v}A = \left(0.6\ \frac{\mathrm{ft}}{\mathrm{sec}}\right)(0.349\ \mathrm{ft}^2)$$

$$= \left(0.21\ \frac{\mathrm{ft}^3}{\mathrm{sec}}\right)\left(7.48\ \frac{\mathrm{gal}}{\mathrm{ft}^3}\right)\left(60\ \frac{\mathrm{sec}}{\mathrm{min}}\right)$$

$$= \boxed{94\ \mathrm{gal/min}}$$

14.2. The diluted characteristics can be calculated from

$$C = \frac{Q_1 C_1 + Q_2 C_2}{Q_1 + Q_2}$$

$$\mathrm{BOD}_5 = \frac{\left(150\ \dfrac{\mathrm{mg}}{\mathrm{L}}\right)\left(0.21\ \dfrac{\mathrm{ft}^3}{\mathrm{sec}}\right) + \left(4\ \dfrac{\mathrm{mg}}{\mathrm{L}}\right)\left(2\ \dfrac{\mathrm{ft}^3}{\mathrm{sec}}\right)}{0.21\ \dfrac{\mathrm{ft}^3}{\mathrm{sec}} + 2\ \dfrac{\mathrm{ft}^3}{\mathrm{sec}}}$$

$$= \boxed{17.9\ \mathrm{mg/L}}$$

14.3. The dissolved oxygen is

$$\mathrm{DO} = \frac{\left(1\ \dfrac{\mathrm{mg}}{\mathrm{L}}\right)\left(0.21\ \dfrac{\mathrm{ft}^3}{\mathrm{sec}}\right) + \left(10\ \dfrac{\mathrm{mg}}{\mathrm{L}}\right)\left(2\ \dfrac{\mathrm{ft}^3}{\mathrm{sec}}\right)}{0.21\ \dfrac{\mathrm{ft}^3}{\mathrm{sec}} + 2\ \dfrac{\mathrm{ft}^3}{\mathrm{sec}}}$$

$$= \boxed{9.1\ \mathrm{mg/L}}$$

14.4. The summer temperature is

$$T_{\mathrm{summer}} = \frac{(22°\mathrm{C})\left(0.21\ \dfrac{\mathrm{ft}^3}{\mathrm{sec}}\right) + (11.1°\mathrm{C})\left(2\ \dfrac{\mathrm{ft}^3}{\mathrm{sec}}\right)}{0.21\ \dfrac{\mathrm{ft}^3}{\mathrm{sec}} + 2\ \dfrac{\mathrm{ft}^3}{\mathrm{sec}}}$$

$$= \boxed{12.14°\mathrm{C}}$$

14.5. The winter temperature is

$$T_{\mathrm{winter}} = \frac{(22°\mathrm{C})\left(0.21\ \dfrac{\mathrm{ft}^3}{\mathrm{sec}}\right) + (7.2°\mathrm{C})\left(2\ \dfrac{\mathrm{ft}^3}{\mathrm{sec}}\right)}{0.21\ \dfrac{\mathrm{ft}^3}{\mathrm{sec}} + 2\ \dfrac{\mathrm{ft}^3}{\mathrm{sec}}}$$

$$= \boxed{8.61°\mathrm{C}}$$

Environmental

14.6. Calculate the dissolved oxygen concentration 20 mi downstream using the summer temperature and the following equation.

$$K_{D,T} = 1.047^{T-20°C} K_{D,20°C}$$

$$K_{D,12.1°C} = \left((1.047)^{12.1°C-20°C}\right)\left(0.1 \, \frac{1}{\text{day}}\right)$$

$$= 0.07 \, 1/\text{day}$$

$$K_{R,T} = (1.016)^{T-20°C} K_{R,20°C}$$

$$K_{R,12.1°C} = \left((1.016)^{12.1°C-20°C}\right)\left(0.25 \, \frac{1}{\text{day}}\right)$$

$$= 0.22 \, 1/\text{day}$$

$$t = \frac{x}{v} = \frac{(20 \text{ mi})\left(5280 \, \frac{\text{ft}}{\text{mi}}\right)}{\left(0.2 \, \frac{\text{ft}}{\text{sec}}\right)\left(3600 \, \frac{\text{sec}}{\text{hr}}\right)\left(24 \, \frac{\text{hr}}{\text{day}}\right)}$$

$$= 6.1 \text{ days}$$

$$\text{BOD}_u = \frac{\text{BOD}_t}{1 - 10^{-K_D t}}$$

$$= \frac{17.9 \, \frac{\text{mg}}{\text{L}}}{1 - 10^{-(0.07 \, 1/\text{day})(5 \text{ days})}}$$

$$= 32.35 \text{ mg/L}$$

Use the Streeter-Phelps equation to calculate the dissolved oxygen concentration 20 mi downstream.

$$D_t = \left(\frac{K_D \text{BOD}_u}{K_R - K_D}\right)\left(10^{-K_D t} - 10^{-K_R t}\right) + D_o 10^{-K_R t}$$

D_t is the oxygen deficit at time t and D_o is the oxygen deficit after initial mixing.

DO_{sat} at $12.1°C = 10.8$ mg/L

$$D_o = 10.8 \, \frac{\text{mg}}{\text{L}} - 9.1 \, \frac{\text{mg}}{\text{L}}$$

$$= 1.7 \text{ mg/L}$$

$$D_t = \left(\frac{\left(0.07 \, \frac{1}{\text{day}}\right)\left(32.35 \, \frac{\text{mg}}{\text{L}}\right)}{0.22 \, \frac{1}{\text{day}} - 0.07 \, \frac{1}{\text{day}}}\right)$$

$$\times \left(10^{-(0.07 \, 1/\text{day})(6.1 \text{ days})}\right.$$

$$\left. - 10^{-(0.22 \, 1/\text{day})(6.1 \text{ days})}\right)$$

$$+ \left(1.7 \, \frac{\text{mg}}{\text{L}}\right)\left(10^{-(0.22 \, 1/\text{day})(6.1 \text{ days})}\right)$$

$$= 5.04 \text{ mg/L}$$

The dissolved oxygen at 20 mi is

$$\text{DO}_{\text{sat}} - D_t = 10.8 \, \frac{\text{mg}}{\text{L}} - 5.04 \, \frac{\text{mg}}{\text{L}}$$

$$= \boxed{5.76 \text{ mg/L}}$$

SOLUTION 15

15.1. Assume the wastewater flow is 125 gal/capita-day. Add 10% for infiltration.

$$Q = \left(125 \, \frac{\text{gal}}{\text{capita-day}}\right)(2000 \text{ capita})(1 + 0.10)$$

$$= 275,000 \text{ gal/day}$$

Design a pipe to flow half full at a rate of 275,000 gal/day. (A peak flow factor could also be used.) Use the Chezy-Manning equation.

$$Q = \left(\frac{1.49}{n}\right) A R^{2/3} \sqrt{S}$$

Assume $n = 0.013$.

$$S = \frac{0.25 \text{ in}}{12 \text{ in}} = 0.021 \text{ ft/ft}$$

The flow area is

$$A_{\text{flow}} = \frac{\pi D^2}{(4)(2)} = \frac{\pi D^2}{8}$$

$$R = \frac{A}{P} = \frac{\frac{\pi D^2}{8}}{\frac{\pi D}{2}} = \frac{D}{4}$$

$$Q_1 = \frac{275,000 \, \frac{\text{gal}}{\text{day}}}{\left(7.48 \, \frac{\text{gal}}{\text{ft}^3}\right)\left(24 \, \frac{\text{hr}}{\text{day}}\right)\left(3600 \, \frac{\text{sec}}{\text{hr}}\right)}$$

$$= 0.426 \text{ ft}^3/\text{sec}$$

$$0.426 \, \frac{\text{ft}^3}{\text{sec}} = \left(\frac{1.49}{0.013}\right)\left(\frac{\pi D^2}{8}\right)\left(\frac{D}{4}\right)^{2/3} \sqrt{0.021}$$

$$D^2 \left(\frac{D}{4}\right)^{2/3} = 0.0653$$

Simplifying,

$$D_{\text{ft}}^{8/3} = 0.1646$$

$$D_{\text{ft}} = 0.51 \text{ ft}$$

$$D_{\text{in}} = (0.51 \text{ ft})\left(12 \, \frac{\text{in}}{\text{ft}}\right)$$

$$= 6.1 \text{ in}$$

Use an 8 in pipe (the recommended minimum size for a sewer main).

Environmental

15.2. Calculate the full pipe flow, Q_2, for an 8 in pipe at a slope $S = 0.021$ and $n = 0.013$.

$$A = \frac{\pi D^2}{4}$$

$$R = \frac{D}{4}$$

$$Q_2 = \left(\frac{1.49}{0.013}\right)\left(\frac{\pi\left(\dfrac{8\text{ in}}{12\,\dfrac{\text{in}}{\text{ft}}}\right)^2}{4}\right)\left(\frac{\dfrac{8\text{ in}}{12\,\dfrac{\text{in}}{\text{ft}}}}{4}\right)^{2/3}\sqrt{0.021}$$

$$= 1.76\text{ ft}^3/\text{sec}$$

$$\frac{Q_1}{Q_2} = \frac{0.426\,\dfrac{\text{ft}^3}{\text{sec}}}{1.76\,\dfrac{\text{ft}^3}{\text{sec}}} = 0.242$$

From a graph or table of circular channel ratios,

$$\frac{d}{D} = 0.38$$

The depth of flow is

$$d = (0.38)(8\text{ in}) = \boxed{3.0\text{ in}}$$

From a table of properties of partially filled circular pipes, for $d/D = 0.38$, $A/D^2 = 0.2739$.

$$A = (0.2739)\left(\frac{8\text{ in}}{12\,\dfrac{\text{in}}{\text{ft}}}\right)^2 = 0.122\text{ ft}^2$$

$$\text{v} = \frac{Q}{A} = \frac{0.426\,\dfrac{\text{ft}^3}{\text{sec}}}{0.122\text{ ft}^2} = \boxed{3.49\text{ ft/sec}}$$

SOLUTION 16

The following steps should be taken to neutralize and clean up the site.

(a) Secure the site by a fence and/or other barrier to prevent unauthorized entry, possible injury, and damage to the confinement dike.

(b) Implement the required health and safety plan, measures, and procedures. These include protective clothing and equipment, monitors, an exclusion zone, and a decontamination area.

(c) Remove the liquid chemical from the diked area by pumping it into a tanker truck for transport off-site to be treated and disposed of or recycled.

(d) Visually assess the area of contamination. Perform soil borings or excavation of pits or trenches to characterize the soil, determine the groundwater elevation, locate confining layers, and define the depth and area of contamination. Sample the contaminated soil and water to determine the concentrations of the toxic chemical around the site. This will help define the area that must be treated.

(e) Install groundwater-monitoring wells to determine if the groundwater becomes contaminated and to monitor its depth and movement. This step assumes that the groundwater is fairly close to the surface.

(f) Cover the contaminated area and install hay bales or silt fences around it to minimize the spread of contamination by rainwater and the formation of leachate. Perform the same measures for any excavated piles of contaminated soil during the site cleanup process. This step is based on the assumption that the area of contamination is small.

(g) Determine the best methods to treat the contaminated soil, and begin treatment. These methods may include on-site treatment, in situ treatment, excavation and on-site treatment, or excavation and off-site treatment.

(h) Begin treating any contaminated groundwater and surface water by on-site pumping, if necessary.

(i) Once the contaminated soil is removed and/or treated, replace the treated soil on-site to fill any excavations (if permitted by the regulatory agencies), or fill the excavations with clean soil. Finish cleaning up the site and return it to its original condition.

(j) Regulatory agencies (U.S. EPA as well as state and local) must be notified of the spill and the cleanup activities. Regulatory approval may be required before site cleanup can begin. If the spill is in a heavily populated area, a public relations/information campaign may be warranted to gain the support of the residents.

SOLUTION 17

A lagoon is typically designed for at least several days' detention time of wastewater flow. Detention times of lagoons range from one day to more than 100 days. This large volume of wastewater can readily absorb peak flow rates of several times the average flow rate, as long as the capacity of the lagoon is not exceeded. The wastewater in the lagoon can be discharged at lower flow rates to other treatment processes. A lagoon serves as a good equalization basin to dampen peak flow rates to lower average flow rates.

SOLUTION 18

18.1. The self-cleansing velocity is used in sewer design. It is the minimum velocity required to pick up and carry solid particles of a typical size with the flow. This velocity is usually between 2.0 ft/sec and 2.5 ft/sec.

The scour velocity is used in the design of aerated grit chambers. It is the velocity required to scour organic particles off the grit. This velocity is usually between 1.5 ft/sec and 2.0 ft/sec.

18.2. For a spherical particle of a 0.5 in diameter, the minimum velocity should be the self-cleansing velocity, which can be calculated with the following equation.

$$v_{minimum} = C\sqrt{\frac{k(\gamma_s - \gamma)}{\gamma}d}$$

γ = specific weight of water = 62.4 lbf/ft^3

γ_s = specific weight of the particle = 100 lbf/ft^3

C = Chezy coefficient = 100

k = 0.04 (initiation of scour) to 0.8

(effective cleansing)

d = particle diameter

Use $k = 0.8$.

$$v_{minimum} = 100 \sqrt{\frac{(0.8)\left(100\ \frac{lbf}{ft^3} - 62.4\ \frac{lbf}{ft^3}\right)}{62.4\ \frac{lbf}{ft^3}}}$$
$$\times \left(\frac{0.5\ in}{12\ \frac{in}{ft}}\right)$$
$$= \boxed{2.9\ ft/sec}$$

SOLUTION 19

19.1. The state regulatory agencies and the U.S. EPA control sanitary landfill design.

19.2. The average density of shredded waste is approximately $\boxed{450\ lbm/yd^3}$ before compaction in the landfill.

19.3. The average density of baled and compacted waste is $\boxed{1600-1800\ lbm/yd^3}$.

19.4. Daily soil cover is typically $\boxed{6\ in\ deep.}$

19.5. Intermediate soil cover is typically $\boxed{2\ ft\ deep.}$

19.6. Final soil cover is typically $\boxed{2\ ft\ deep.}$

19.7. Intermediate soil cover can be used for $\boxed{up\ to\ 30\ days.}$

19.8. The clay liner serves to separate the waste from the underlying soil and groundwater. It reduces the chance of landfill leachate seeping into the nearby groundwater and surface water.

19.9. The limitation of clay liners is that they are somewhat permeable and their properties can change considerably over time due to interactions with materials in the landfill and surrounding forces. When clay dries out (becomes desiccated) it is much more permeable. Chemicals in the landfill may break down and permeate the clay. The clay liner may fracture or rupture due to the weight of the landfill, shifting of the underlying soil or landfill material, or other subsurface disturbances.

19.10. NIMBY stands for "not in my backyard." This is the prevailing attitude of people regarding the siting of landfills (and other waste-treatment and disposal facilities). People agree that these facilities are necessary, but they do not want to live near them. The NIMBY attitude makes landfill siting difficult.

Environmental

Topic IV: Structural

Chapter

Structural

9 Concrete

PROBLEM 1

1.1. What is an admixture?

1.2. Give six examples of admixtures and explain the purpose of each.

PROBLEM 2

A 1:1.6:2.6 (by weight) mixture of cement, sand, and coarse aggregate is produced with the following specifications.

cement	specific gravity = 3.15
	94 lbf per sack
sand	specific gravity = 2.62
	SSD
coarse aggregate	specific gravity = 2.65
	SSD
water	5.8 gal per sack cement
entrained air	3%

2.1. How much cement, sand, coarse aggregate, and water are required to produce 1 yd^3 of concrete? Express all answers in pounds-force.

2.2. If the sand absorbs 1.6% moisture and the coarse aggregate has 3.2% excess moisture (based on saturated surface dry conditions), what weight (in pounds-force) of water is needed?

2.3. Discuss four techniques that can be used to ensure proper curing of concrete pavement.

2.4. Discuss two methods of installing transverse joints in concrete pavement.

PROBLEM 3

A concrete mixture uses the following materials.

cement	34 sacks
	specific gravity = 3.15
	94 lbf per sack
fine aggregate	6500 lbf (dry basis)
	specific gravity = 2.67
	−2% moisture absorption from SSD
gravel	11,500 lbf (dry basis)
	specific gravity = 2.64
	+1.5% moisture excess from SSD
water	142 gal (as delivered)
entrained air	4%

3.1. What is the concrete yield (in cubic yards)?

3.2. What is the water-cement ratio in gallons per sack of cement?

3.3. What is the cement factor in sacks per cubic yard?

PROBLEM 4

A 16 in (gross dimension) square, tied column must carry 220 kip dead and 250 kip live loads. The dead load includes the column self-weight. The column is not exposed to any moments. Sidesway is prevented at the top, and slenderness effects are to be disregarded. The concrete compressive strength is 4000 lbf/in^2; the steel tensile yield strength is 60,000 lbf/in^2.

4.1. Design the column. Specify longitudinal steel size and tie spacing.

4.2. Sketch the column cross section.

PROBLEM 5

A simply supported exterior concrete beam has a length of 22 ft. The beam carries a live load of 3.5 kips/ft and a dead load of 1.5 kips/ft. The dead load does not include the beam weight. The beam width is 16 in. The concrete compressive strength is 3000 lbf/in^2; the steel yield strength is 36,000 lbf/in^2. Design the beam depth and steel reinforcement. No compressive reinforcement is to be used. Do not design shear reinforcement.

PROBLEM 6

Determine the deflection of the concrete beam designed in the previous problem. Assume a depth to reinforcement, d, of 27.3 in.

PROBLEM 7

A cantilever beam extends 8 ft from a rigid support and carries 100 lbf/ft dead load and 350 lbf/ft live load. The dead load does not include the concrete weight. The beam width is 10 in. 4000 lbf/in^2 concrete and 60 kips/in^2 steel will be used. No. 3 bar is available for stirrup and miscellaneous use. Design the beam.

7.1. Specify the beam depth.

7.2. Give the amount, type, and location of the longitudinal steel and shear reinforcement.

7.3. Draw a cross section of the beam.

PROBLEM 8

A combined footing supports two 18 in columns separated by a distance of 18 ft (center-to-center). The right column (column A) is located near a property line and cannot extend more than 2 ft to the right. There is no limitation on how far the footing can extend to the left of column B, nor is there any limitation on the footing width. The footing thickness is limited to 2 ft overall. Overburden loading and punching shear are to be neglected in the design. Use the following data.

column A	80 kips dead load
	55 kips live load
column B	110 kips dead load
	105 kips live load
concrete compressive strength	3000 lbf/in^2
steel yield strength	60,000 lbf/in^2
allowable soil bearing pressure	3500 lbf/ft^2

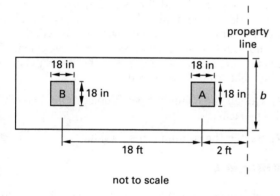

not to scale

8.1. Size the footing for a uniform pressure distribution.

8.2. Draw the shear and moment diagrams for the footing.

PROBLEM 9

A 10 in × 10 in column carries a service-level axial load of 10,000 lbf and a moment of 5000 ft-lbf, and it rests on a square footing. The allowable soil pressure is 1500 lbf/ft^2. Ultimate or factored loads may be taken as 1.6 times service loads. 3000 lbf/in^2 concrete and 60 kips/in^2 steel are used.

9.1. Determine the size of the footing required to keep the soil pressure below the allowable value.

9.2. Check for shear.

9.3. Design the amount, spacing, and location of the steel reinforcement.

PROBLEM 10

A water tank weighs 40,000 lbf when full. It is supported by four square legs that, in turn, are supported by two isolated, 2 ft wide footings with the dimensions illustrated. The footings rest on a 6 in thick concrete slab with an effective depth of 3 in. All concrete has a compressive strength of 3000 lbf/in^2. The allowable bearing pressure of the soil is 1300 lbf/ft^2.

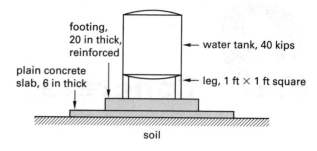

elevation view

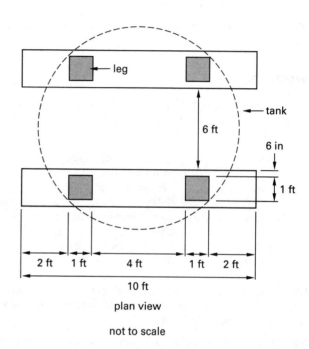

plan view

not to scale

10.1. Determine if the two footings are of sufficient size.

10.2. Determine if the slab shear strength is sufficient.

PROBLEM 11

A 32 ft diameter tank is used to hold 55,000 gal of a water-based solution from a plating operation. The specific gravity of the solution is 1.0. The tank itself weighs 200,000 lbf. The tank is supported by a ringwall (circular) foundation, also 32 ft in diameter, that supports all of the tank and content weight. The soil in which the ringwall is located has a frost-line depth of 18 in and an allowable soil pressure of 1500 lbf/ft^2. Use the following material properties: $f'_c = 3000$ lbf/in^2, $f_y = 60$ kips/in^2. Do not design for wind or seismic loading.

11.1. Determine the required depth of the ringwall.

11.2. Design the steel reinforcement. State the assumptions or the basis for the design.

11.3. Draw a cross section of the wall and label all dimensions and steel bars.

PROBLEM 12

A 20 ft long rib-slab system has the dimensions shown. Each beam uses three no. 9 longitudinal bars. The concrete has a compressive strength of 3000 lbf/in². The steel has a tensile yield strength of 60,000 lbf/in². Determine the ultimate moment capacity of this design per repeating unit.

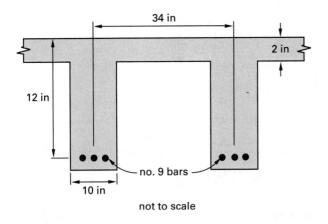

not to scale

PROBLEM 13

A 4 ft wide, 30 ft long simply supported beam is part of a roof section and is constructed as a double-tee, prestressed section as shown. The total superimposed dead load is 45 lbf/ft²; the live load is 40 lbf/ft². The initial concrete compressive strength is 3500 lbf/in²; the ultimate concrete compressive strength is 5000 lbf/in². 250 kips/in² tendons are used, and each web contains two tendons.

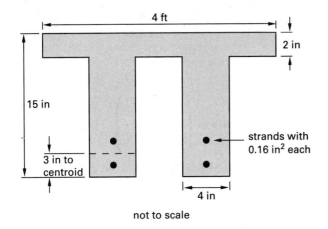

not to scale

13.1. Check all permissible concrete stresses immediately after prestress transfer.

13.2. Check all permissible concrete stresses with service loading.

13.3. What is the section's nominal moment strength?

PROBLEM 14

A reinforced concrete beam has a trough in the compression region as shown. Concrete is normalweight with specified compressive strength of 4000 lbf/in², and grade 60, no. 11 rebars are specified. Given that the steel yields when flexural failure occurs, what is the strain in the tension reinforcement when the compression strain in concrete reaches the ultimate value of 0.003?

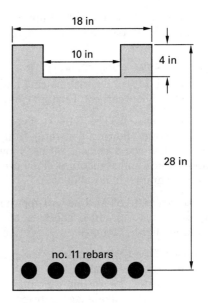

PROBLEM 15

A two-span post-tensioned concrete beam has an effective strand prestress of 280 kips in the idealized strand sag profile shown. The rigidity (product of modulus of elasticity and moment of inertia, EI) of the member is 250×10^6 kips-in². What is the secondary downward reaction at support B induced by the prestress?

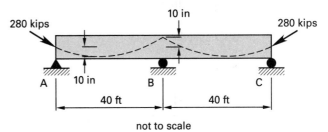

not to scale

SOLUTION 1

1.1. Ingredients added to concrete immediately before or during mixing (other than portland cement, aggregates, and water) are known as *admixtures*.

1.2. Six types of admixtures are

(a) *Accelerators* (ASTM C494, Type C): Accelerate setting and enhance early strength. Example: calcium chloride (ASTM D98).

(b) *Air entraining* (ASTM C260): Improve durability and workability. Example: salts of wood resins (vinsol resins).

(c) *Retarders* (ASTM C494, Type B): Retard the setting time to avoid difficulties with placing and finishing (typically used in hot weather). Example: lignins.

(d) *Superplasticizers* (ASTM C1017, Type 1): Make high-slump concrete (flowing concrete) from concrete with normal to low water-cement ratios, allow for easy placing, and reduce and sometimes eliminate the need for vibration. Example: lignosulfonates.

(e) *Water reducers* (ASTM C494, Type A): Reduce water requirement to produce concrete of a certain slump. Example: lignosulfonates.

(f) *Pozzolans* (ASTM C618): Improve the properties of concrete by changing the properties of the various types of cement; substituted for certain amounts of cement; reduce temperature rise, alkali-aggregate expansion, and harmful effects of tricalcium aluminate. Examples: fly ash, blast furnace slag, ground pumice.

SOLUTION 2

2.1. Calculate the volume for one sack of cement.

$$V = \frac{\left(94 \ \frac{\text{lbf}}{\text{sack}}\right)(\text{proportion})}{\left(62.4 \ \frac{\text{lbf}}{\text{ft}^3}\right)(\text{SG})}$$

$$\text{SG} = \text{SSD specific gravity}$$

For cement, the volume is

$$V = \frac{\left(94 \ \frac{\text{lbf}}{\text{sack}}\right)(1)}{\left(62.4 \ \frac{\text{lbf}}{\text{ft}^3}\right)(3.15)} = 0.48 \ \text{ft}^3/\text{sack}$$

For fine aggregate, the volume is

$$V = \frac{\left(94 \ \frac{\text{lbf}}{\text{sack}}\right)(1.6)}{\left(62.4 \ \frac{\text{lbf}}{\text{ft}^3}\right)(2.62)} = 0.92 \ \text{ft}^3/\text{sack}$$

For coarse aggregate, the volume is

$$V = \frac{\left(94 \ \frac{\text{lbf}}{\text{sack}}\right)(2.6)}{\left(62.4 \ \frac{\text{lbf}}{\text{ft}^3}\right)(2.65)} = 1.48 \ \text{ft}^3/\text{sack}$$

For water, the volume is

$$V = \frac{5.8 \ \text{gal}}{7.48 \ \frac{\text{gal}}{\text{ft}^3}} = 0.78 \ \text{ft}^3/\text{sack}$$

The yield volume is

$$\frac{\begin{array}{c} 0.48 \ \text{ft}^3 + 0.92 \ \text{ft}^3 \\ + 1.48 \ \text{ft}^3 + 0.78 \ \text{ft}^3 \end{array}}{(1 - 0.03)\left(3 \ \frac{\text{ft}}{\text{yd}}\right)^3} = 0.140 \ \text{yd}^3/\text{sack}$$

For 1 yd³,

$$\text{cement} = \left(\frac{1 \ \text{yd}^3}{0.140 \ \frac{\text{yd}^3}{\text{sack}}}\right)\left(94 \ \frac{\text{lbf}}{\text{sack}}\right)$$

$$= \boxed{671 \ \text{lbf}}$$

$$\text{fine aggregate} = \left(\frac{1 \ \text{yd}^3}{0.140 \ \frac{\text{yd}^3}{\text{sack}}}\right)(1.6)\left(94 \ \frac{\text{lbf}}{\text{sack}}\right)$$

$$= \boxed{1074 \ \text{lbf}}$$

$$\text{coarse aggregate} = \left(\frac{1 \ \text{yd}^3}{0.140 \ \frac{\text{yd}^3}{\text{sack}}}\right)(2.6)\left(94 \ \frac{\text{lbf}}{\text{sack}}\right)$$

$$= \boxed{1746 \ \text{lbf}}$$

$$\text{water} = \left(\frac{1 \ \text{yd}^3}{0.140 \ \frac{\text{yd}^3}{\text{sack}}}\right)\left(0.78 \ \frac{\text{ft}^3}{\text{sack}}\right)\left(62.4 \ \frac{\text{lbf}}{\text{ft}^3}\right)$$

$$= \boxed{348 \ \text{lbf}}$$

2.2. The ordered material weights are based on SSD density. The water absorbed by sand is

$$(-0.016)(1074 \text{ lbf}) = -17.18 \text{ lbf}$$

The excess water in coarse aggregate is

$$(0.032)(1746 \text{ lbf}) = 55.87 \text{ lbf}$$

The adjustment to required water is

$$-55.87 \text{ lbf} + 17.18 \text{ lbf} = -38.69 \quad (-39 \text{ lbf})$$

The weight of the water needed is

$$348 \text{ lbf} - 39 \text{ lbf} = \boxed{309 \text{ lbf}}$$

2.3. Four techniques that can be used to obtain proper curing are

- ponding, which uses earth or sand dikes around the perimeter to retain a pond of water. It requires considerable labor and is generally restricted to small jobs.

- covering with plastic sheets, which provides an effective moisture barrier and is easily applied

- spraying or fogging, which is effective and can be carried out using ordinary lawn sprinklers. This method requires an ample water supply and may be costly.

- applying commercial curing compounds to the surface

2.4. One way to install transverse joints is to saw a continuous straight slot in the top of the slab. Another method is to place strips of wood, plastic, or metal in the fresh concrete. Dimensions vary greatly depending on conditions.

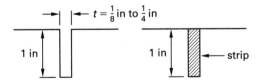

SOLUTION 3

3.1. Calculate the volumes based on SSD density.

For cement,

$$\text{weight} = (34 \text{ sacks})\left(94 \frac{\text{lbf}}{\text{sack}}\right) = 3196 \text{ lbf}$$

$$V = \frac{3196 \text{ lbf}}{(3.15)\left(62.4 \frac{\text{lbf}}{\text{ft}^3}\right)} = 16.3 \text{ ft}^3$$

For fine aggregate,

$$V = \frac{6500 \text{ lbf}}{(2.67)\left(62.4 \frac{\text{lbf}}{\text{ft}^3}\right)} = 39.0 \text{ ft}^3$$

For the water deficit of fine aggregate,

$$(-0.02)(6500 \text{ lbf}) = -130.0 \text{ lbf}$$

For coarse aggregate,

$$V = \frac{11,500 \text{ lbf}}{(2.64)\left(62.4 \frac{\text{lbf}}{\text{ft}^3}\right)} = 69.8 \text{ ft}^3$$

The excess water in coarse aggregate is

$$(0.015)(11,500 \text{ lbf}) = 172.5 \text{ lbf}$$

For water,

$$V = \frac{142 \text{ gal}}{7.48 \frac{\text{gal}}{\text{ft}^3}} = 19.0 \text{ ft}^3$$

The correction from aggregate deviations from SSD is

$$\frac{-130 \text{ lbf} + 172.5 \text{ lbf}}{62.4 \frac{\text{lbf}}{\text{ft}^3}} = 0.7 \text{ ft}^3$$

The total volume of water is

$$V = 19.0 \text{ ft}^3 + 0.7 \text{ ft}^3 = 19.7 \text{ ft}^3$$

$$\text{yield} = \left(\frac{16.3 \text{ ft}^3 + 39.0 \text{ ft}^3 + 69.8 \text{ ft}^3 + 19.7 \text{ ft}^3}{1 - 0.04}\right)$$

$$\times \left(\frac{1}{\left(3 \frac{\text{ft}}{\text{yd}}\right)^3}\right)$$

$$= \boxed{5.6 \text{ yd}^3}$$

3.2. The water-cement ratio is

$$\text{water-cement ratio} = (19.7 \text{ ft}^3)\left(7.48 \frac{\text{gal}}{\text{ft}^3}\right)$$

$$\times \left(\frac{1}{34 \text{ sacks}}\right)$$

$$= \boxed{4.3 \text{ gal/sack}}$$

3.3. The cement factor is

$$\text{cement factor} = \frac{34 \text{ sacks}}{5.6 \text{ yd}^3}$$

$$= \boxed{6.1 \text{ sacks/yd}^3}$$

SOLUTION 4

4.1. Determine the design load, P_u. From ACI 318 Eq. 9-2,

$$P_u = 1.2 P_{\text{dead}} + 1.6 P_{\text{live}}$$
$$= (1.2)(220 \text{ kips}) + (1.6)(250 \text{ kips})$$
$$= 664 \text{ kips}$$

Determine the column capacity, $\phi P_{n,\text{max}}$. From ACI 318 Eq. 10-2,

$$\phi P_{n,\text{max}} = 0.80\phi\left(0.85f'_c(A_g - A_{\text{st}}) + f_y A_{\text{st}}\right)$$
$$A_g = (16 \text{ in})(16 \text{ in}) = 256 \text{ in}^2$$

For a tied column, $\phi = 0.65$. Substituting numerical values, the capacity is

$$\phi P_{n,\text{max}} = (0.80)(0.65)\left((0.85)\left(4 \frac{\text{kips}}{\text{in}^2}\right)(256 \text{ in}^2 - A_{\text{st}})\right.$$
$$\left. + \left(60 \frac{\text{kips}}{\text{in}^2}\right) A_{\text{st}}\right)$$
$$= 452.6 \text{ kips} + \left(29.4 \frac{\text{kips}}{\text{in}^2}\right) A_{\text{st}}$$

The design criterion is

$$\phi P_{n,\text{max}} \geq P_u$$
$$664 \text{ kips} = 452.6 \text{ kips} + \left(29.4 \frac{\text{kips}}{\text{in}^2}\right) A_{\text{st}}$$
$$A_{\text{st}} = 7.19 \text{ in}^2$$

Check the limits.

$$\rho_{\min} < \rho < \rho_{\max}$$
$$\rho = \frac{A_{\text{st}}}{A_g}$$

From ACI 318 Sec. 10.9.1,

$$\rho_{\min} = 0.01$$
$$\rho_{\max} = 0.08$$
$$\rho = \frac{7.19 \text{ in}^2}{256 \text{ in}^2} = 0.028 \quad \text{[within limits—OK]}$$

Select the longitudinal reinforcement. Various combinations are possible—for example, using no. 9 bars.

For eight no. 9 bars,

$$A_{s,\text{provided}} = 8 \text{ in}^2$$

Select the ties. The tie spacing is governed by ACI 318 Sec. 7.10.5.

$$16 d_b = (16)(1.125 \text{ in}) = 18 \text{ in}$$
$$48 d_t = (48)(0.375 \text{ in}) = 18 \text{ in} \quad \text{[for no. 3 tie]}$$

least column dimension $= 16 \text{ in}$ [controls]

4.2.

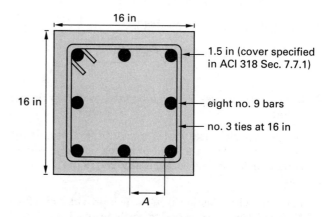

In accordance with ACI 318 Sec. 7.10.5.3, if clear distance dimension A in the illustration is larger than 6 in, supplementary ties would be needed. Calculate A.

$$A = \frac{16 \text{ in} - (2)(1.5 \text{ in} + 0.375 \text{ in}) - (3)(1.125 \text{ in})}{2}$$
$$= 4.44 \text{ in} < 6 \text{ in}$$

Supplementary ties are not needed.

SOLUTION 5

To control deflection, select a depth that satisfies the requirement of ACI 318 Table 9.5(a).

$$h \geq \left(\frac{l}{16}\right)\left(0.4 + \frac{f_y}{100}\right)$$
$$\geq \left(\frac{(22 \text{ ft})\left(12 \frac{\text{in}}{\text{ft}}\right)}{16}\right)\left(0.4 + \frac{36 \frac{\text{kips}}{\text{in}^2}}{100}\right)$$
$$\geq 12.54 \text{ in}$$

The nominal moment of capacity for a singly reinforced section is given by

$$M_n = f'_c b d^2 \omega(1 - 0.59\omega)$$
$$\omega = \frac{\rho f_y}{f'_c}$$

Large ω values will result in small depths and heavy reinforcement, while the opposite occurs when ω is small. Use the common intermediate value of $\omega = 0.18$.

$$M_n = \left(3\ \frac{kips}{in^2}\right)(16\ in)\,d^2(0.18)\big(1-(0.59)(0.18)\big)$$
$$= 7.72d^2$$

Obtain a first estimate of the applied moment by neglecting self-weight.

$$w_u = 1.2w_{dead} + 1.6w_{live}$$
$$= (1.2)\left(1.5\ \frac{kips}{ft}\right) + (1.6)\left(3.5\ \frac{kips}{ft}\right)$$
$$= 7.4\ kips/ft$$
$$M_u = \tfrac{1}{8}w_u l^2$$
$$= \left(\tfrac{1}{8}\right)\left(7.4\ \frac{kips}{ft}\right)(22\ ft)^2$$
$$= 447.7\ ft\text{-}kips$$

The design requirement is

$$\phi M_n \ge M_u$$
$$(0.9)(7.72d^2) = (447.7\ ft\text{-}kips)\left(12\ \frac{in}{ft}\right)$$
$$d = 27.8\ in$$

On the basis of the previous calculations, select a depth, d, of 30 in.

Calculate the actual value of M_u (including self-weight).

$$w_{self} = \left(0.15\ \frac{kips}{ft^3}\right)(16\ in)(30\ in)\left(\frac{1}{\left(12\ \frac{in}{ft}\right)^2}\right)$$
$$= 0.50\ kips/ft$$
$$w_u = (1.2)\left(1.5\ \frac{kips}{ft} + 0.5\ \frac{kips}{ft}\right)$$
$$\quad + (1.6)\left(3.5\ \frac{kips}{ft}\right)$$
$$= 8.0\ kips/ft$$
$$M_u = \left(\tfrac{1}{8}\right)\left(8.0\ \frac{kips}{ft}\right)(22\ ft)^2$$
$$= 484\ ft\text{-}kips$$

Using the actual moment and a depth, d, of 27.5 in, solve for ω.

$$M_u = \phi M_n$$
$$(484\ ft\text{-}kips)\left(12\ \frac{in}{ft}\right) = (0.9)\left(3\ \frac{kips}{in^2}\right)(16\ in)$$
$$\times\ (27.5\ in)^2\omega(1-0.59\omega)$$
$$\omega(1-0.59\omega) = 0.178$$
$$\omega = 0.203$$

From this value of ω, the area of steel is

$$A_s = \omega\left(\frac{f'_c}{f_y}\right)bd$$
$$= (0.203)\left(\frac{3\ \frac{kips}{in^2}}{36\ \frac{kips}{in^2}}\right)(16\ in)(27.5\ in)$$
$$= 7.44\ in^2$$

Various arrangements and bar selections are possible. Using no. 11 bars, the number of bars is

$$n = \frac{A_s}{A_{one\ bar}} = \frac{7.44\ in^2}{1.56\ in^2}$$
$$= 4.77\quad [\text{use five no. 11 bars}]$$

In order to fit these bars in the 16 in width available and satisfy the spacing requirements, it is necessary to use bundles as shown.

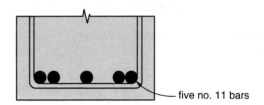

five no. 11 bars

beam size $= \boxed{16\ in \times 30\ in}$

SOLUTION 6

Determine the cracked moment of inertia, I_{cr}. From ACI 318 Sec. 8.5.1,

$$E_c = 57{,}000\sqrt{f'_c} = \frac{57{,}000\sqrt{3000\ \frac{lbf}{in^2}}}{1000\ \frac{lbf}{kip}}$$
$$= 3122\ kips/in^2$$

$$n = \frac{E_s}{E_c} = \frac{29{,}000 \; \frac{\text{kips}}{\text{in}^2}}{3122 \; \frac{\text{kips}}{\text{in}^2}}$$

$$= 9.3$$

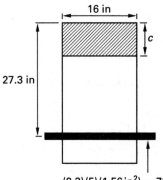

$$(9.3)(5)(1.56 \, \text{in}^2) = 72.54 \, \text{in}^2$$

$$\frac{16c^2}{2} = (72.54)(27.3 - c)$$

$$c^2 + 9.1c - 247.5 = 0$$

$$c = 11.83 \, \text{in}$$

$$I_{cr} = \left(\tfrac{1}{3}\right)(16 \, \text{in})(11.83 \, \text{in})^3 + (72.54 \, \text{in}^2)(15.47 \, \text{in})^2$$

$$= 26{,}190 \, \text{in}^4$$

Calculate the gross moment of inertia, I_g.

$$I_g = \tfrac{1}{12}bh^3 = \left(\tfrac{1}{12}\right)(16 \, \text{in})(30 \, \text{in})^3$$

$$= 36{,}000 \, \text{in}^4$$

Determine the cracking moment, M_{cr}. From ACI 318 Eq. 9-10,

$$f_r = 7.5\sqrt{f_c'} = 7.5\sqrt{3000 \; \frac{\text{lbf}}{\text{in}^2}}$$

$$= 410.8 \, \text{lbf/in}^2$$

The cracking moment is given by ACI 318 Eq. 9-9.

$$M_{cr} = \frac{f_r I_g}{y_t}$$

$$= \frac{\left(410.8 \; \frac{\text{lbf}}{\text{in}^2}\right)(36{,}000 \, \text{in}^4)}{\left(\frac{30 \, \text{in}}{2}\right)\left(1000 \; \frac{\text{lbf}}{\text{kip}}\right)\left(12 \; \frac{\text{in}}{\text{ft}}\right)}$$

$$= 82.2 \, \text{ft-kips}$$

Obtain the maximum service load moment, M_a.

$$w_{\text{service}} = 3.5 \; \frac{\text{kips}}{\text{ft}} + 1.5 \; \frac{\text{kips}}{\text{ft}} + 0.5 \; \frac{\text{kip}}{\text{ft}}$$

$$= 5.5 \, \text{kips/ft}$$

$$M_a = \tfrac{1}{8}w_{\text{service}}l^2$$

$$= \left(\tfrac{1}{8}\right)\left(5.5 \; \frac{\text{kips}}{\text{ft}}\right)(22 \, \text{ft})^2$$

$$= 332.8 \, \text{ft-kips}$$

Calculate the effective moment of inertia using ACI 318 Eq. 9-8.

$$I_e = \left(\frac{M_{cr}}{M_a}\right)^3 I_g + \left(1 - \left(\frac{M_{cr}}{M_a}\right)^3\right) I_{cr}$$

$$= \left(\frac{82.2 \, \text{ft-kips}}{332.8 \, \text{ft-kips}}\right)^3 (36{,}000 \, \text{in}^4)$$

$$+ \left(1 - \left(\frac{82.2 \, \text{ft-kips}}{332.8 \, \text{ft-kips}}\right)^3\right)(26{,}190 \, \text{in}^4)$$

$$= 26{,}338 \, \text{in}^4$$

Calculate the immediate deflections.

$$\Delta_{\text{immediate}} = \frac{5w_{\text{service}}l^4}{384 E I_e}$$

$$= \frac{(5)\left(5.5 \; \frac{\text{kips}}{\text{ft}}\right)(264 \, \text{in})^4}{(384)\left(3122 \; \frac{\text{kips}}{\text{in}^2}\right)(26{,}338 \, \text{in}^4)\left(12 \; \frac{\text{in}}{\text{ft}}\right)}$$

$$= 0.35 \, \text{in}$$

Find the long-term additional deflection using ACI 318 Eq. 9-11.

$$\Delta(\text{creep} + \text{shrinkage}) = \lambda \Delta_{\text{immediate}}$$

$$\lambda = \frac{\xi}{1 + 50\rho'}$$

$$= \frac{2}{1 + (50)(0)}$$

$$= 2$$

$$\Delta(\text{creep} + \text{shrinkage}) = (2)(0.35 \, \text{in}) = 0.70 \, \text{in}$$

$$\text{total deflection} = 0.35 \, \text{in} + 0.70 \, \text{in}$$

$$= \boxed{1.05 \, \text{in} \quad (1 \, \text{in})}$$

SOLUTION 7

7.1.

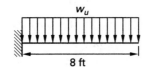

Select the preliminary beam depth based on deflection. From ACI 318 Table 9.5(a),

$$h_{\min} = \frac{l}{8} = \frac{(8)(12 \text{ in})}{8} = 12 \text{ in}$$

Estimate the beam weight as that of a 10 in × 12 in beam.

$$w_d = (1.2)\left(0.1 \ \frac{\text{kip}}{\text{ft}}\right) + \frac{(1.2)(120 \text{ in}^2)\left(0.15 \ \frac{\text{kip}}{\text{ft}^3}\right)}{\left(12 \ \frac{\text{in}}{\text{ft}}\right)^2}$$

$$= 0.12 \ \frac{\text{kip}}{\text{ft}} + 0.15 \ \frac{\text{kip}}{\text{ft}}$$

$$= 0.27 \text{ kip/ft}$$

$$w_u = 0.27 \ \frac{\text{kip}}{\text{ft}} + (1.6)\left(0.35 \ \frac{\text{kip}}{\text{ft}}\right)$$

$$= 0.83 \text{ kip/ft}$$

$$M_u = \frac{w_u l^2}{2}$$

$$= \left(\tfrac{1}{2}\right)\left(0.83 \ \frac{\text{kip}}{\text{ft}}\right)(8 \text{ ft})^2\left(12 \ \frac{\text{in}}{\text{ft}}\right)$$

$$= 318.7 \text{ in-kips}$$

Calculate the required bd^2.

$$M_u = \phi\left(f_c' bd^2 \omega(1 - 0.59\omega)\right)$$

$$\omega = \frac{\rho f_y}{f_c'}$$

Use $\omega = 0.18$ to define the section size.

$$\rho = \omega\left(\frac{f_c'}{f_y}\right) = (0.18)\left(\frac{4 \ \frac{\text{kips}}{\text{in}^2}}{60 \ \frac{\text{kips}}{\text{in}^2}}\right)$$

$$= 0.012$$

$$bd^2 = \frac{M_u}{\phi f_c' \omega(1 - 0.59\omega)}$$

$$= \frac{318.7 \text{ in-kips}}{(0.9)\left(4 \ \frac{\text{kips}}{\text{in}^2}\right)(0.18)(1 - (0.59)(0.18))}$$

$$= 550.3 \text{ in}^3$$

$$d = \sqrt{\frac{bd^2}{b}} = \sqrt{\frac{550.3 \text{ in}^3}{10 \text{ in}}}$$

$$= 7.42 \text{ in}$$

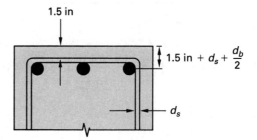

7.2. Assuming no. 6 bars and no. 3 stirrups,

$$h \geq 7.42 \text{ in} + 1.5 \text{ in} + \frac{3}{8} \text{ in} + \frac{3}{8} \text{ in} = 9.67 \text{ in}$$

Use $h = 10$ in.

Refine the value of ω by using the exact selected dimensions.

$$d = 10 \text{ in} - 1.5 \text{ in} - \frac{3}{8} \text{ in} - \frac{3}{8} \text{ in} = 7.75 \text{ in}$$

$$bd^2 = (10 \text{ in})(7.75 \text{ in})^2$$

$$= 600.6 \text{ in}^3$$

$$(1 - 0.59\omega)\omega = \frac{M_u}{\phi f_c' bd^2}$$

$$= \frac{318.7 \text{ in-kips}}{(0.9)\left(4 \ \frac{\text{kips}}{\text{in}^2}\right)(600.6 \text{ in}^3)}$$

$$= 0.147$$

$$\omega = 0.163$$

$$\rho = \omega\left(\frac{f_c'}{f_y}\right)$$

$$= (0.163)\left(\frac{4 \ \frac{\text{kips}}{\text{in}^2}}{60 \ \frac{\text{kips}}{\text{in}^2}}\right)$$

$$= 0.011$$

$$A_s = \rho bd = (0.011 \text{ in})(10 \text{ in})(7.75 \text{ in})$$

$$= 0.85 \text{ in}^2$$

Two no. 6 bars provide 0.88 in^2.

Use two no. 6 bars.

The shear at the critical section is

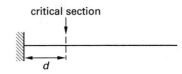

critical section

$$V_u = \left(0.83 \; \frac{\text{kip}}{\text{ft}}\right)\left(8 \text{ ft} - \frac{7.75 \text{ in}}{12 \; \frac{\text{in}}{\text{ft}}}\right) = 6.11 \text{ kips}$$

From ACI 318 Eq. 11-3,

$$V_c = 2\sqrt{f_c'}b_w d$$

$$= 2\sqrt{4000 \; \frac{\text{lbf}}{\text{in}^2}}(10 \text{ in})(7.75 \text{ in})$$

$$= 9803 \text{ lbf} \quad (9.80 \text{ kips})$$

$$\phi V_c = (0.75)(9.80 \text{ kips})$$

$$= 7.35 \text{ kips}$$

Even though $V_u > \phi V_c/2$, the beam depth is 10 in, so no shear reinforcement is needed (ACI 318 Sec. 11.4.6.1).

7.3.

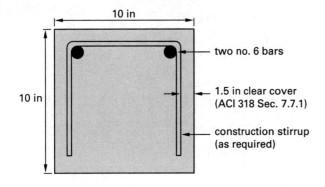

10 in

10 in

two no. 6 bars

1.5 in clear cover
(ACI 318 Sec. 7.7.1)

construction stirrup
(as required)

SOLUTION 8

8.1. Size the footing to attain a uniform pressure distribution under service loads. Locate the resultant of applied service loads.

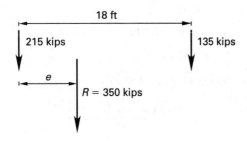

18 ft

215 kips

135 kips

e

$R = 350$ kips

$$(350 \text{ kips})e = (135 \text{ kips})(18 \text{ ft})$$

$$e = 6.94 \text{ ft}$$

$$\text{footing length} = \big((18 \text{ ft} - e) + 2 \text{ ft}\big)(2)$$

$$= \big((18 \text{ ft} - 6.94 \text{ ft}) + 2 \text{ ft}\big)(2)$$

$$= 26.1 \text{ ft}$$

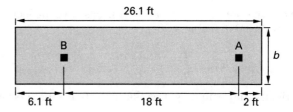

26.1 ft

B

A

b

6.1 ft

18 ft

2 ft

The footing thickness is likely to be governed by punching shear around the columns. Because this problem does not require checking punching shear, assume the footing thickness is the maximum allowed (i.e., 2 ft).

Determine the dimension b. Equating the total load to the soil capacity times the bearing area gives

$$350 \text{ kips} + \left(0.15 \; \frac{\text{kip}}{\text{ft}^3}\right)(26.1 \text{ ft})(2 \text{ ft})b$$

$$= \left(3.5 \; \frac{\text{kips}}{\text{ft}^2}\right)(26.1 \text{ ft})b$$

$$b = \boxed{4.2 \text{ ft}}$$

8.2. Calculate the shear and moment diagrams. The design of the footing uses factored loads.

$$\text{column A} = (1.2)(80 \text{ kips}) + (1.6)(55 \text{ kips})$$

$$= 184 \text{ kips}$$

$$\text{column B} = (1.2)(110 \text{ kips}) + (1.6)(105 \text{ kips})$$

$$= 300 \text{ kips}$$

$$\text{total} = 184 \text{ kips} + 300 \text{ kips}$$

$$= 484 \text{ kips}$$

(The footing weight does not induce shear or moment, so these are not included.)

The resultant of the factored load is not exactly in the same position as the resultant for the service load, but this shift is typically neglected and the pressure is assumed uniform. For this reason, the reactions on the inverted footing are not exactly equal to the factored column loads.

$$q_u = \frac{484 \text{ kips}}{26.1 \text{ ft}} = 18.5 \text{ kips/ft}$$

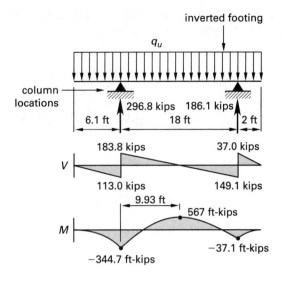

SOLUTION 9

9.1. The problem can be illustrated as shown.

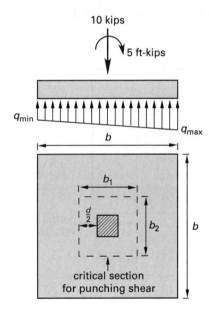

$$I = \tfrac{1}{12}b^4$$

$$q_{max} = \frac{10 \text{ kips}}{b^2} + \frac{(5 \text{ ft-kips})(0.5b)}{I}$$

$$= \frac{10}{b^2} + \frac{30}{b^3}$$

To determine the size of the footing, b, solve by trial and error. To allow for a 1 ft thick footing, limit q_{max} to $1.5 \text{ kips/ft}^2 - 0.15 \text{ kip/ft}^2 = 1.35 \text{ kips/ft}^2$.

b (ft)	q_{max} (kips/ft^2)
2	6.25
3	2.22
3.5	1.52
3.6	1.42
3.7	1.32

$$q_{min} = 0.14 \text{ kip/ft}^2$$

$$\boxed{b = 3.7 \text{ ft (square)}}$$

9.2. Check the 1 ft depth for punching shear. Find the fraction of the moment transferred by shear, γ_v, from ACI 318 Eq. 11-37 and Eq. 13-1. Since $b_1 = b_2$,

$$\gamma_v = 1 - \frac{1}{1 + \tfrac{2}{3}\sqrt{\dfrac{b_1}{b_2}}}$$

$$= 1 - \frac{1}{1 + \left(\tfrac{2}{3}\right)(1)}$$

$$= 0.4$$

Find the factored moment affecting the punching shear.

$$\gamma_v M_u = (0.4)(1.6)(5 \text{ ft-kips}) = 3.2 \text{ ft-kips}$$

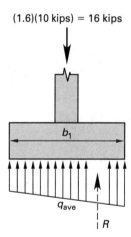

Assuming a single layer of no. 6 bars and a clear cover of 3 in,

$$d = 12 \text{ in} - 3 \text{ in} - 0.75 \text{ in} = 8.25 \text{ in}$$

$$R = q_{ave} b_1 b_2$$

$$b_1 = b_2 = 10 \text{ in} + d$$

$$= 10 \text{ in} + 8.25 \text{ in}$$

$$= 18.25 \text{ in}$$

$$q_{ave} = \frac{P}{A} = \frac{16 \text{ kips}}{(3.7 \text{ ft})^2}$$

$$= 1.17 \text{ kips/ft}^2$$

$$V_u = 16 \text{ kips} - R$$

$$= 16 \text{ kips} - \left(1.17 \, \frac{\text{kips}}{\text{ft}^2}\right)\left(\frac{18.25 \text{ in}}{12 \, \frac{\text{in}}{\text{ft}}}\right)^2$$

$$= 13.29 \text{ kips}$$

From ACI 318 Sec. R11.11-7,

$$J_c = \frac{db_1^3}{6} + \frac{d^3 b_1}{6} + \frac{db_2 b_1^2}{2}$$

$$= \frac{(8.25 \text{ in})(18.25 \text{ in})^3}{6} + \frac{(8.25 \text{ in})^3(18.25 \text{ in})}{6}$$

$$+ \frac{(8.25 \text{ in})(18.25 \text{ in})(18.25 \text{ in})^2}{2}$$

$$= 35{,}139 \text{ in}^4$$

$$A_c = 2d(b_1 + b_2) = (2)(8.25 \text{ in})(18.25 \text{ in} + 18.25 \text{ in})$$

$$= 602.25 \text{ in}^4$$

Calculate the maximum punching shear stress using ACI 318 Sec. R11.11.7.2.

$$v_u = \frac{V_u}{A_c} + \frac{\gamma_v M_u \left(\frac{b_1}{2}\right)}{J_c}$$

$$= \left(\frac{13.29 \text{ kips}}{602.25 \text{ in}^2} + \frac{(3.2 \text{ ft-kips})\left(\frac{18.25 \text{ in}}{2}\right)\left(12 \, \frac{\text{in}}{\text{ft}}\right)}{35{,}139 \text{ in}^4}\right)$$

$$\times \left(1000 \, \frac{\text{lbf}}{\text{kip}}\right)$$

$$= 32.0 \text{ lbf/in}^2$$

Calculate the allowable stress using ACI 318 Eq. 11-38.

$$\phi v_n = \frac{\phi V_c}{b_o d}$$

Using ACI 318 Eq. 11-31,

$$V_c = \left(2 + \frac{4}{\beta_c}\right)\sqrt{f_c'} b_o d$$

Using ACI 318 Sec. R11.11.2.1,

$$\beta_c = \frac{\text{long column dimension}}{\text{short column dimension}} = 1$$

$$b_o = 2(b_1 + b_2)$$

$$= (2)(18.25 \text{ in} + 18.25 \text{ in})$$

$$= 73 \text{ in}$$

From ACI 318 Eq. 11-32 (with $\alpha = 40$) and Eq. 11-33, V_c should not exceed the smallest of

$$\left(\frac{40d}{b_o} + 2\right)\sqrt{f_c'} b_o d = \left(\frac{(40)(8.25 \text{ in})}{73 \text{ in}} + 2\right)$$

$$\times \sqrt{3000 \, \frac{\text{lbf}}{\text{in}^2}}(73 \text{ in})(8.25 \text{ in})$$

$$= 215{,}091 \text{ lbf}$$

$$4\sqrt{f_c'} b_o d = 4\sqrt{3000 \, \frac{\text{lbf}}{\text{in}^2}}(73 \text{ in})(8.25 \text{ in})$$

$$= 131{,}946 \text{ lbf} \quad \text{[controls]}$$

$$\phi v_n = \phi 4\sqrt{f_c'} = (0.75)(4)\sqrt{3000 \, \frac{\text{lbf}}{\text{in}^2}}$$

$$= 164 \text{ lbf/in}^2$$

Since $\phi v_n > v_u$, the punching shear capacity is adequate. Check the one-way shear. The capacity is ϕV_c.

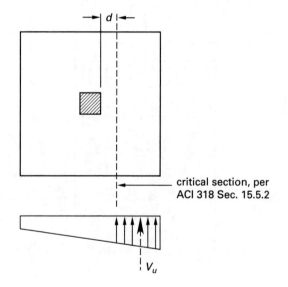

critical section, per ACI 318 Sec. 15.5.2

$$\phi V_c = \phi 2\sqrt{f_c'} bd$$

$$= \frac{(0.75)(2)\sqrt{3000 \, \frac{\text{lbf}}{\text{in}^2}}(3.7 \text{ ft})\left(12 \, \frac{\text{in}}{\text{ft}}\right)(8.25 \text{ in})}{1000 \, \frac{\text{lbf}}{\text{kip}}}$$

$$= 30.0 \text{ kips}$$

ϕV_c is greater than the total factored load, so there is no need to obtain V_u. The one-way shear is adequate.

9.3. Calculate the maximum moment.

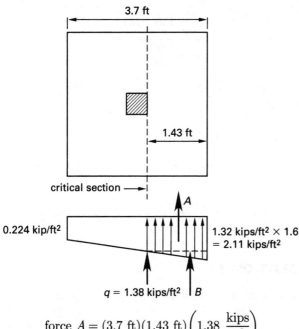

force $A = (3.7 \text{ ft})(1.43 \text{ ft})\left(1.38 \, \dfrac{\text{kips}}{\text{ft}^2}\right)$

$= 7.30 \text{ kips}$

Using values calculated in the illustration,

force $B = \left(2.11 \, \dfrac{\text{kips}}{\text{ft}^2} - 1.38 \, \dfrac{\text{kips}}{\text{ft}^2}\right)\left(\dfrac{1.43 \text{ ft}}{2}\right)(3.7 \text{ ft})$

$= 1.93 \text{ kips}$

$M_u = (7.30 \text{ kips})\left(\dfrac{1.43 \text{ ft}}{2}\right)$

$\qquad + (1.93 \text{ kips})\left(\dfrac{2}{3}\right)(1.43 \text{ ft})$

$= 7.06 \text{ ft-kips}$

Use trial and error to compute the required steel area.

$$A_s = \frac{M_u}{0.9 f_y j d}$$

Assume $j = 0.95$.

$$A_s = \frac{(7.06 \text{ ft-kips})\left(12 \, \dfrac{\text{in}}{\text{ft}}\right)}{(0.9)\left(60 \, \dfrac{\text{kips}}{\text{in}^2}\right)(0.95)(8.25 \text{ in})}$$

$= 0.20 \text{ in}^2 \quad$ [clearly less than minimum]

Using ACI 318 Sec. 7.12.2.1,

$A_{s,\text{min}} = \rho_{\text{min}} A$

$\qquad = (0.0018)(44.4 \text{ in})(12 \text{ in})$

$\qquad = 0.96 \text{ in}^2$

Use five no. 4 bars in each direction.

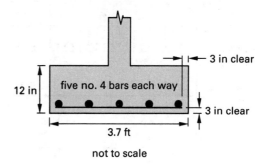

not to scale

SOLUTION 10

Since the plan dimension of the plain concrete slab is not specified, calculate the minimum required.

$$\sigma_{\text{soil net}} = 1300 \, \frac{\text{lbf}}{\text{ft}^2} - \left(150 \, \frac{\text{lbf}}{\text{ft}^3}\right)\left(\frac{6 \text{ in}}{12 \, \frac{\text{in}}{\text{ft}}}\right)$$

$$= 1225 \text{ lbf/ft}^2$$

$$A = \frac{40{,}000 \text{ lbf}}{1225 \, \dfrac{\text{lbf}}{\text{ft}^2}}$$

$$= 32.65 \text{ ft}^2$$

Given that the clear distance between the footings is 6 ft and that the length is 10 ft, the minimum area exceeds 60 ft^2.

10.1. Neglect the slab. Assuming all the load is dead,

$$w_u = \frac{(2)(10 \text{ kips})(1.2)}{10 \text{ ft}}$$

$$= 2.4 \text{ kips/ft}$$

The shear at the tank leg is

$$V_u = w_u x = \left(2.4 \, \frac{\text{kips}}{\text{ft}}\right)(2.5 \text{ ft})$$

$$= 6 \text{ kips}$$

The moment at the tank leg is

$$M_u = \tfrac{1}{2} w_u x^2 = \left(\tfrac{1}{2}\right)\left(2.4 \, \frac{\text{kips}}{\text{ft}}\right)(2.5 \text{ ft})^2$$

$$= 7.5 \text{ ft-kips}$$

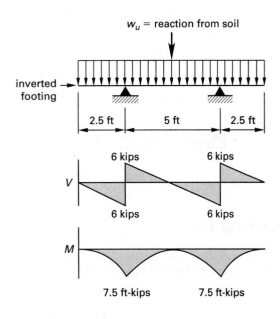

Taking $d \approx (0.8)(20 \text{ in}) = 16 \text{ in}$,

$$V_c = 2\sqrt{f'_c}bd$$

$$= \frac{2\sqrt{3000 \frac{\text{lbf}}{\text{in}^2}}(24 \text{ in})(16 \text{ in})}{1000 \frac{\text{lbf}}{\text{kip}}}$$

$$= 42.07 \text{ kips}$$

$$V_u < 7 \text{ kips}$$

$$\phi V_c = (0.75)(42.07 \text{ kips})$$

$$= 31.5 \text{ kips} > V_u$$

The footing size is sufficient for shear. (The moment of 8.75 ft-kips is small and can be easily accommodated by the footing.)

10.2. The critical section for shear in the plain slab is at a distance, d, from the face of the footings (see illustration).

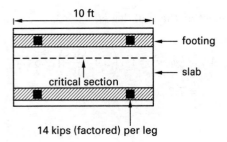

$$V_c = 2\sqrt{f'_c}bd \quad [\text{normalweight concrete}]$$

$$= \frac{2\sqrt{3000 \frac{\text{lbf}}{\text{in}^2}}(120 \text{ in})(3 \text{ in})}{1000 \frac{\text{lbf}}{\text{kip}}}$$

$$= 39.44 \text{ kips}$$

$$V_u \le 28 \text{ kips}$$

$$\phi V_c = (0.75)(39.44 \text{ kips})$$

$$= 29.6 \text{ kips} > V_u \quad [\text{conservatively estimated}]$$

The plain slab shear strength is sufficient.

SOLUTION 11

11.1. Calculate the load per foot of ringwall.

$$\text{weight of liquid} = (55{,}000 \text{ gal})\left(0.13368 \frac{\text{ft}^3}{\text{gal}}\right)$$

$$\times \left(62.4 \frac{\text{lbf}}{\text{ft}^3}\right)$$

$$= 458{,}789 \text{ lbf} \quad (458.8 \text{ kips})$$

$$\text{weight of tank} = 200{,}000 \text{ lbf} \quad (200 \text{ kips})$$

$$\text{total weight} = 458.8 \text{ kips} + 200 \text{ kips}$$

$$= 658.8 \text{ kips}$$

$$\begin{array}{c}\text{perimeter of}\\ \text{centerline of ringwall}\end{array} = \pi(32 \text{ ft})$$

$$= 100.53 \text{ ft}$$

The service load per foot of ringwall is

$$q = \frac{658.8 \text{ kips}}{100.53 \text{ ft}} = 6.55 \text{ kips/ft}$$

Neglecting the weight of the ringwall, the required width to keep the soil pressure within allowances is

$$b = \frac{6.55 \frac{\text{kips}}{\text{ft}}}{1.5 \frac{\text{kips}}{\text{ft}^2}} = 4.37 \text{ ft}$$

This dimension is large and requires the use of a footing under the ringwall. Because of the 18 in deep frost line, transfer the load at a depth of 24 in. Assuming access to the underside of the tank is unnecessary, the projection of the ringwall above the ground surface is arbitrary. Select 6 in.

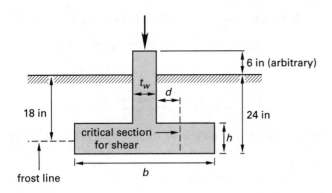

Assuming the full loading is dead load, the factored load per foot is

$$p_u = 1.2q$$

Using ACI 318 Sec. 14.5.3.2,

$$t_w \geq 7.5 \text{ in} \quad (8 \text{ in})$$

Check the 8 in value for strength.

Estimate h. The shear force at the critical section, V_u, is

$$V_u = P_u \left(\frac{b - (t_w + 2d)}{2b} \right)$$

$$= P_u \left(\frac{1}{2} - \frac{t_w}{2b} - \frac{d}{b} \right)$$

The shear capacity of a 1 ft long section is ϕV_c.

$$\phi V_c = (0.75)(2)\sqrt{f'_c}\, d$$

$$\frac{(0.75)(2)\sqrt{3000 \dfrac{\text{lbf}}{\text{in}^2}}(1 \text{ ft})d\left(12 \dfrac{\text{in}}{\text{ft}}\right)}{1000 \dfrac{\text{lbf}}{\text{kip}}} = 0.986d \text{ kips}$$

Taking tentative values of b and t_w as 60 in and 8 in, respectively, and working on a per foot of perimeter basis,

$$P_u = \frac{1.2(\text{total weight})}{\text{centerline perimeter}}$$

$$= \frac{(1.2)(658.8 \text{ kips})}{100.53 \text{ ft}}$$

$$= 7.86 \text{ kips/ft}$$

$$V_u = \frac{P_u}{b}\left(\frac{b}{2} - \frac{t_w}{2} - d \right)$$

$$= \left(\frac{7.86 \dfrac{\text{kips}}{\text{ft}}}{60 \text{ in}} \right)\left(\frac{60 \text{ in}}{2} - \frac{8 \text{ in}}{2} - d \right)$$

$$= 3.41 - 0.131d$$

Equating V_u to ϕV_c,

$$3.41 - 0.131d = 0.986d$$

$$d = 3.05 \text{ in}$$

However, ACI 318 Sec. 15.7, prescribes a minimum effective depth, d, of 6 in. With a 3 in clear cover, the total depth is

$$h = d + 3 + d_b \approx d + 4 \text{ in}$$

Use $d = 6$ in, and try $h = 10$ in. Check the flexure before finalizing h.

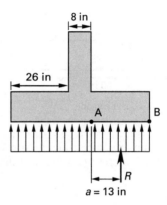

Again taking $b = 60$ in, the load between points A and B in the illustration is

$$R = \frac{P_u}{b}\left(\frac{b}{2} - \frac{t_w}{2} \right)$$

$$= \left(\frac{7.86 \dfrac{\text{kips}}{\text{ft}}}{60 \text{ in}} \right)\left(\frac{60 \text{ in}}{2} - \frac{8 \text{ in}}{2} \right)$$

$$= 3.41 \text{ kips/ft}$$

$$M_u = Ra = \left(3.41 \dfrac{\text{kips}}{\text{ft}} \right)(13 \text{ in})$$

$$= 44.3 \text{ in-kips/ft of ringwall footing}$$

Assume $j = 0.95$.

$$A_{s,\text{req}} = \frac{M_u}{\phi f_y j d}$$

$$= \frac{44.3 \dfrac{\text{in-kips}}{\text{ft}}}{(0.9)\left(60 \dfrac{\text{kips}}{\text{in}^2} \right)(0.95)(6 \text{ in})}$$

$$= 0.144 \text{ in}^2/\text{ft}$$

The reinforcement area per circumferential foot (12 in) of ringwall is

$$A_{s,min} = \rho_{min} bh = (0.0018)(12 \text{ in})(10 \text{ in})$$
$$= 0.216 \text{ in}^2/\text{ft} \quad [\text{controls}]$$

Use no. 3 bars at 6 in on center spacing (provides $0.22 \text{ in}^2/\text{ft}$).

$\boxed{h = 10 \text{ in can be used.}}$

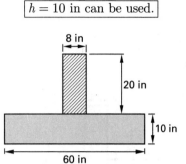

not to scale

Check the bearing pressure, taking into consideration the weight of the concrete (150 lbf/ft^3) in the wall and footing.

$$\begin{aligned}\text{added weight} \atop \text{per foot of wall} &= \left(\frac{(8 \text{ in})(20 \text{ in}) + (10 \text{ in})(60 \text{ in})}{\left(12 \frac{\text{in}}{\text{ft}}\right)^2} \right) \\ &\quad \times \left(0.15 \frac{\text{kip}}{\text{ft}^3} \right) \\ &= 0.79 \text{ kip/ft}\end{aligned}$$

$$\begin{aligned}\text{actual bearing} \atop \text{pressure} &= \frac{6.55 \frac{\text{kips}}{\text{ft}} + 0.79 \frac{\text{kip}}{\text{ft}}}{5 \text{ ft}} \\ &= 1.47 \text{ kips/ft}^2 < 1.5 \text{ kips/ft}^2 \\ &\quad\quad\quad\quad\quad\quad\quad\quad\quad [\text{allowable}]\end{aligned}$$

11.2. Select the steel for the ringwall. The load per foot is small, so minimum requirements will govern.

For vertical steel, from ACI 318 Sec. 14.3.2, $\rho = 0.0012$. The reinforcement area per circumferential foot (12 in) of ringwall is

$$A_{s,v} = \rho_v t_w b = (0.0012)(8 \text{ in})(12 \text{ in})$$
$$= 0.12 \text{ in}^2/\text{ft}$$

$\boxed{\text{Use no. 3 bars at 11 in.}}$

For horizontal steel, from ACI 318 Sec. 14.3.3, $\rho = 0.002$.

$$A_{s,h} = \rho_h t_w h = (0.002)(8 \text{ in})(20 \text{ in})$$
$$= 0.32 \text{ in}^2/\text{ft} \quad [\text{in shaded area}]$$

$\boxed{\text{Use three no. 3 or two no. 4 bars.}}$

This solution assumes that the bottom of the tank is adequate to carry all of the content's weight. This will require a very stiff tank bottom. If the bottom is flexible, there will be a radial force on the ringwall, translating into a hoop stress in the ringwall steel. The minimum requirements for horizontal steel in the ringwall would not be adequate in that case.

11.3. A summary of the final design is shown.

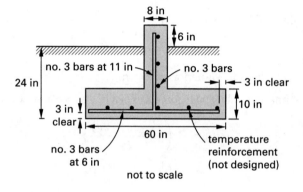

not to scale

SOLUTION 12

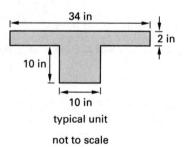

typical unit

not to scale

Use ACI 318 Sec. 8.12.2 to check the width of the flange.

$$b_{eff,max} = (2)(12 \text{ in}) + 10 \text{ in}$$
$$= 34 \text{ in}$$

The full width is available.

Calculate the area of concrete in compression at ultimate tension.

$$T = f_y A_s = \left(60 \ \frac{\text{kips}}{\text{in}^2} \right)(3 \ \text{in}^2)$$

$$= 180 \ \text{kips}$$

$$A_f = b_{\text{eff}} h = (34 \ \text{in})(2 \ \text{in}) = 68 \ \text{in}^2$$

$$A_c = \frac{T}{0.85 f_c'} = \frac{180 \ \text{kips}}{(0.85)\left(3 \ \frac{\text{kips}}{\text{in}^2} \right)}$$

$$= 70.59 \ \text{in}^2 > 68 \ \text{in}^2 \quad [\text{OK}]$$

$$A_w = A_c - A_f$$

$$= 70.59 \ \text{in}^2 - (2 \ \text{in})(34 \ \text{in})$$

$$= 2.59 \ \text{in}^2 \quad \begin{bmatrix} \text{required} \\ \text{from web} \end{bmatrix}$$

$$a_w = \frac{A_w}{b_w} = \frac{2.59 \ \text{in}^2}{10 \ \text{in}}$$

$$= 0.259 \ \text{in}^2 \quad \begin{bmatrix} \text{from underside of flange} \\ \text{to compression depth} \end{bmatrix}$$

Compute the moment capacity, ϕM_n.

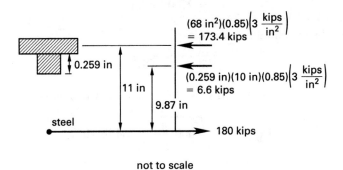

not to scale

$$M_n = \frac{(173.4 \ \text{kips})(11 \ \text{in}) + (6.6 \ \text{kips})(9.87 \ \text{in})}{12 \ \frac{\text{in}}{\text{ft}}}$$

$$= 164.4 \ \text{ft-kips}$$

$$\phi M_n = (0.9)(164.4 \ \text{ft-kips})$$

$$= \boxed{148.0 \ \text{ft-kips}}$$

SOLUTION 13

13.1. Locate the centroid and calculate the moment of inertia.

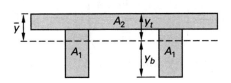

	area (in^2)	y (in)	(area)y (in^3)	(area)$(y - \bar{y})^2$ (in^4)	I_o (in^4)
$2A_1$	104	8.5	884	1347.8	1464.7
A_2	96	1.0	96	1460.2	32.0
totals	200		980	2808.0	1496.7

$$\bar{y} = \frac{980 \ \text{in}^3}{200 \ \text{in}^2} = 4.9 \ \text{in}$$

$$I_x = 2808 \ \text{in}^4 + 1496.7 \ \text{in}^4$$

$$= 4304.7 \ \text{in}^4$$

$$y_t = 4.9 \ \text{in}$$

$$y_b = 10.1 \ \text{in}$$

$$e = 7.1 \ \text{in}$$

Check stresses at transfer. Assume tendon stress after transfer, f_{pi}, is the maximum permitted. From ACI 318 Sec. 18.5.1,

$$f_{pi} = 0.7 f_{pu} = (0.7)\left(250 \ \frac{\text{kips}}{\text{in}^2} \right)$$

$$= 175 \ \text{kips/in}^2$$

$$\text{area of tendons} = (4)(0.16 \ \text{in}^2)$$

$$= 0.64 \ \text{in}^2$$

The prestressing force at transfer is

$$P_i = (0.64 \ \text{in}^2)\left(175 \ \frac{\text{kips}}{\text{in}^2} \right)$$

$$= 112 \ \text{kips}$$

The stresses should be checked at 50 diameters of the strand from the ends and at the centerline. $(d_s = 0.375 \ \text{in}.)$

Calculate the bending moments from self-weight at sections 1 and 2.

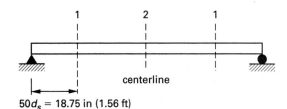

centerline

$50d_s = 18.75 \ \text{in} \ (1.56 \ \text{ft})$

$$w_{\text{self}} = A\gamma = \frac{(200 \ \text{in}^2)\left(0.15 \ \frac{\text{kip}}{\text{ft}^3} \right)}{\left(12 \ \frac{\text{in}}{\text{ft}} \right)^2}$$

$$= 0.208 \ \text{kips/ft}$$

At section 1,

$$M_1 = \left(0.208 \ \frac{\text{kip}}{\text{ft}}\right)(15 \ \text{ft})(1.56 \ \text{ft})$$
$$- \left(\tfrac{1}{2}\right)\left(0.208 \ \frac{\text{kip}}{\text{ft}}\right)(1.56 \ \text{ft})^2$$
$$= 4.62 \ \text{ft-kips}$$

At section 2,

$$M_2 = \tfrac{1}{8}wL^2 = \left(\tfrac{1}{8}\right)\left(0.208 \ \frac{\text{kip}}{\text{ft}}\right)(30 \ \text{ft})^2$$
$$= 23.4 \ \text{ft-kips}$$

Calculate the stresses at section 1.

$$\sigma_t = \frac{P}{A} + \frac{Pec}{I}$$
$$= -\frac{112 \ \text{kips}}{200 \ \text{in}^2} + \frac{(112 \ \text{kips})(7.1 \ \text{in})(4.9 \ \text{in})}{4304.7 \ \text{in}^4}$$
$$- \frac{(4.62 \ \text{ft-kips})(4.9 \ \text{in})\left(12 \ \frac{\text{in}}{\text{ft}}\right)}{4304.7 \ \text{in}^4}$$
$$= \left(-0.560 \ \frac{\text{kip}}{\text{in}^2} + 0.905 \ \frac{\text{kip}}{\text{in}^2} - 0.063 \ \frac{\text{kip}}{\text{in}^2}\right)\left(1000 \ \frac{\text{lbf}}{\text{kip}}\right)$$
$$= \boxed{282 \ \text{lbf/in}^2}$$

$$\sigma_b = -0.560 \ \frac{\text{kip}}{\text{in}^2} - \left(0.905 \ \frac{\text{kip}}{\text{in}^2}\right)\left(\frac{10.1 \ \text{in}}{4.9 \ \text{in}}\right)$$
$$+ \left(0.063 \ \frac{\text{kip}}{\text{in}^2}\right)\left(\frac{10.1 \ \text{in}}{4.9 \ \text{in}}\right)$$
$$= \left(-0.560 \ \frac{\text{kip}}{\text{in}^2} - 1.865 \ \frac{\text{kips}}{\text{in}^2} + 0.130 \ \frac{\text{kip}}{\text{in}^2}\right)$$
$$\times \left(1000 \ \frac{\text{lbf}}{\text{kip}}\right)$$
$$= \boxed{-2295 \ \text{lbf/in}^2}$$

At section 2,

$$\sigma_t = -0.560 \ \frac{\text{kip}}{\text{in}^2} + 0.905 \ \frac{\text{kip}}{\text{in}^2}$$
$$- \frac{(23.4 \ \text{ft-kips})(4.7 \ \text{in})\left(12 \ \frac{\text{in}}{\text{ft}}\right)}{4304.7 \ \text{in}^4}$$
$$= \left(-0.560 \ \frac{\text{kip}}{\text{in}^2} + 0.905 \ \frac{\text{kip}}{\text{in}^2} - 0.31 \ \frac{\text{kip}}{\text{in}^2}\right)\left(1000 \ \frac{\text{lbf}}{\text{kip}}\right)$$
$$= \boxed{35 \ \text{lbf/in}^2}$$

$$\sigma_b = -0.560 \ \frac{\text{kip}}{\text{in}^2} - 1.865 \ \frac{\text{kips}}{\text{in}^2}$$
$$+ \left(0.32 \ \frac{\text{kip}}{\text{in}^2}\right)\left(\frac{10.1 \ \text{in}}{4.9 \ \text{in}}\right)$$
$$= \left(-0.560 \ \frac{\text{kip}}{\text{in}^2} - 1.865 \ \frac{\text{kips}}{\text{in}^2} + 0.660 \ \frac{\text{kip}}{\text{in}^2}\right)$$
$$\times \left(1000 \ \frac{\text{lbf}}{\text{kip}}\right)$$
$$= \boxed{-1765 \ \text{lbf/in}^2}$$

In summary, at transfer the stresses are as follows. The maximum compression is

$$\boxed{2295 \ \text{lbf/in}^2 \quad [\text{not OK}]}$$

According to ACI 318 Sec. 18.4.1, the allowable compression is

$$0.6f'_{ci} = (0.6)\left(3500 \ \frac{\text{lbf}}{\text{in}^2}\right)$$
$$= 2100 \ \text{lbf/in}^2$$

The maximum tension is

$$\boxed{282 \ \text{psi} \quad [\text{OK}]}$$

According to ACI 318 Sec. 18.4.1, the allowable tension is

$$6\sqrt{f'_{ci}} = 6\sqrt{3500 \ \frac{\text{lbf}}{\text{in}^2}} = 354 \ \text{lbf/in}^2$$

13.2. Check stresses at the service load level. Assume total losses = 20% of f_{pi}.

$$f_{pf} = 0.8f_{pf} = (0.8)\left(175 \ \frac{\text{kips}}{\text{in}^2}\right)$$
$$= 140 \ \text{kips/in}^2$$

The final tendon force is

$$P_f = \left(140 \ \frac{\text{kips}}{\text{in}^2}\right)(0.64 \ \text{in}^2) = 89.6 \ \text{kips}$$

The service load is

$$w_{\text{service}} = \left(\frac{45 \ \frac{\text{lbf}}{\text{ft}^2} + 40 \ \frac{\text{lbf}}{\text{ft}^2}}{1000 \ \frac{\text{lbf}}{\text{kip}}}\right)(4 \ \text{ft}) + 0.208 \ \frac{\text{kip}}{\text{ft}}$$
$$= 0.548 \ \text{kip/ft}$$

The moments at sections 1 and 2 are

$$M_1 = (4.62 \text{ ft-kips}) \left(\dfrac{0.548 \frac{\text{kip}}{\text{ft}}}{0.208 \frac{\text{kip}}{\text{ft}}} \right)$$

$$= 12.17 \text{ ft-kips}$$

$$M_2 = (23.4 \text{ ft-kips}) \left(\dfrac{0.548 \frac{\text{kip}}{\text{ft}}}{0.208 \frac{\text{kip}}{\text{ft}}} \right)$$

$$= 61.65 \text{ ft-kips}$$

Calculate the stresses at section 1.

$$\sigma_t = \frac{P}{A} + \frac{Pec}{I} = -\frac{89.6 \text{ kips}}{200 \text{ in}^2} + \frac{(89.6 \text{ kips})(7.1 \text{ in})(4.9 \text{ in})}{4304.7 \text{ in}^4}$$

$$- \frac{(12.17 \text{ ft-kips})(12 \text{ in})(4.9 \text{ in})}{4304.7 \text{ in}^4}$$

$$= \left(-0.448 \frac{\text{kip}}{\text{in}^2} + 0.724 \frac{\text{kip}}{\text{in}^2} - 0.166 \frac{\text{kip}}{\text{in}^2} \right) \left(1000 \frac{\text{lbf}}{\text{kip}} \right)$$

$$= \boxed{110 \text{ lbf/in}^2}$$

$$\sigma_b = -0.448 \frac{\text{kip}}{\text{in}^2} - \left(0.724 \frac{\text{kip}}{\text{in}^2} \right) \left(\frac{10.1 \text{ in}}{4.9 \text{ in}} \right)$$

$$+ \left(0.166 \frac{\text{kip}}{\text{in}^2} \right) \left(\frac{10.1 \text{ in}}{4.9 \text{ in}} \right)$$

$$= \left(-0.448 \frac{\text{kip}}{\text{in}^2} - 1.492 \frac{\text{kips}}{\text{in}^2} + 0.342 \frac{\text{kip}}{\text{in}^2} \right)$$

$$\times \left(1000 \frac{\text{lbf}}{\text{kip}} \right)$$

$$= \boxed{-1598 \text{ lbf/in}^2}$$

At section 2,

$$\sigma_t = -0.448 \frac{\text{kip}}{\text{in}^2} + 0.724 \frac{\text{kip}}{\text{in}^2}$$

$$- \frac{(61.65 \text{ ft-kips})(12 \text{ in})(4.9 \text{ in})}{4304.7 \text{ in}^4}$$

$$= \left(-0.448 \frac{\text{kip}}{\text{in}^2} + 0.724 \frac{\text{kip}}{\text{in}^2} - 0.842 \frac{\text{kip}}{\text{in}^2} \right) \left(1000 \frac{\text{lbf}}{\text{kip}} \right)$$

$$= \boxed{-566 \text{ lbf/in}^2}$$

$$\sigma_b = -0.448 \frac{\text{kip}}{\text{in}^2} - 1.492 \frac{\text{kips}}{\text{in}^2}$$

$$+ \left(0.842 \frac{\text{kip}}{\text{in}^2} \right) \left(\frac{10.1 \text{ in}}{4.9 \text{ in}} \right)$$

$$= \left(-0.448 \frac{\text{kip}}{\text{in}^2} - 1.492 \frac{\text{kips}}{\text{in}^2} + 1.736 \frac{\text{kips}}{\text{in}^2} \right)$$

$$\times \left(1000 \frac{\text{lbf}}{\text{kip}} \right)$$

$$= \boxed{-204 \text{ lbf/in}^2}$$

In summary, at service the stresses are as follows. The maximum compression is

$$\boxed{1598 \text{ lbf/in}^2 \quad [\text{OK}]}$$

According to ACI 318 Sec. 18.4.2, the allowable compression is

$$0.45 f_c' = (0.45) \left(5000 \frac{\text{lbf}}{\text{in}^2} \right)$$

$$= 2250 \text{ lbf/in}^2$$

The maximum tension is

$$\boxed{110 \text{ lbf/in}^2 \quad [\text{OK}]}$$

This is less than $7.5 \sqrt{f_c'}$ permitted by ACI 318 Sec. 18.3.3.

13.3. Find the section's nominal moment strength. Calculate the stress in the tendon at ultimate, f_{ps}. Using ACI 318 Eq. 18.1 and omitting the terms that account for compression reinforcement,

$$f_{ps} = f_{pu} \left(1 - \left(\frac{\gamma_p}{\beta_1} \right) \rho_p \left(\frac{f_{pu}}{f_c'} \right) \right)$$

f_{pu} is the specified tensile strength of the tendons. Using ACI 318 Chap. 18 notation,

$$\frac{f_{py}}{f_{pu}} = 0.85$$

$$\gamma_p = 0.4$$

$$f_c' = 5000 \text{ lbf/in}^2$$

$$\beta_1 = 0.80$$

$$\rho_p = \frac{A_{ps}}{bd_p}$$

$$= \frac{0.64 \text{ in}^2}{(4 \text{ ft})(12 \text{ in})\left(12 \frac{\text{in}}{\text{ft}}\right)}$$

$$= 0.0011$$

$$f_{ps} = \left(250 \frac{\text{kips}}{\text{in}^2}\right)\left(1 - \left(\frac{0.4}{0.80}\right)(0.0011)\left(\frac{250}{5}\right)\right)$$

$$= 243.1 \text{ kips/in}^2$$

The total tension at ultimate is

$$T = A_{ps}f_{ps}$$

$$= \left(243.1 \frac{\text{kips}}{\text{in}^2}\right)(0.64 \text{ in}^2)$$

$$= 155.58 \text{ kips}$$

The depth of the compression block is

$$a = \frac{T}{0.85f_c'b}$$

$$= \frac{155.58 \text{ kips}}{(0.85)\left(5 \frac{\text{kips}}{\text{in}^2}\right)(48 \text{ in})}$$

$$= 0.76 \text{ in}$$

The flexural capacity is

$$\phi M_n = \phi A_{ps}f_{ps}\left(d_p - \frac{a}{2}\right)$$

$$= \frac{(0.9)(155.58 \text{ kips})\left(12 \text{ in} - \frac{0.76 \text{ in}}{2}\right)}{12 \frac{\text{in}}{\text{ft}}}$$

$$= \boxed{135.6 \text{ ft-kips}}$$

SOLUTION 14

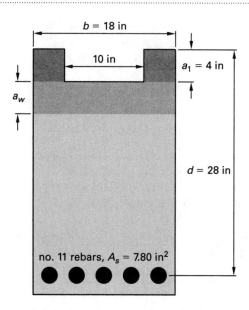

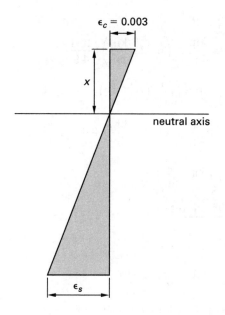

Structural

The area of concrete in compression when the section reaches its moment capacity is

$$A_c = \frac{f_y A_s}{0.85 f'_c} = \frac{\left(60{,}000 \ \frac{\text{lbf}}{\text{in}^2}\right)(7.80 \ \text{in}^2)}{(0.85)\left(4000 \ \frac{\text{lbf}}{\text{in}^2}\right)}$$

$$= 137.6 \ \text{in}^2$$

This is greater than the area to the sides of the trough.

$$A_{\text{top}} = (4 \ \text{in})(18 \ \text{in} - 10 \ \text{in}) = 32 \ \text{in}^2$$

Therefore, the compression area extends below the trough to a depth of

$$a_w = \frac{A_c - A_{\text{top}}}{b}$$

$$= \frac{137.6 \ \text{in}^2 - 32 \ \text{in}^2}{18 \ \text{in}}$$

$$= 5.87 \ \text{in}$$

The depth of the neutral axis is

$$x = \frac{a}{\beta_1} = \frac{A_{\text{top}} + a_w}{\beta_1}$$

$$= \frac{4.00 \ \text{in} + 5.87 \ \text{in}}{0.85}$$

$$= 11.6 \ \text{in}$$

From similar triangles,

$$\frac{\epsilon_s}{d - x} = \frac{0.003}{x}$$

$$\epsilon_s = (28 \ \text{in} - 11.6 \ \text{in})\left(\frac{0.003}{11.6 \ \text{in}}\right)$$

$$= \boxed{0.004}$$

SOLUTION 15

Applying the usual assumptions for the analysis of pre-stressed beams, the tendon profile over each 40 ft segment can be represented by superposition of the chord, which is inclined upward 10 in, and by a parabolic strand that has an equivalent sag of

$$s = 10 \ \text{in} + (0.5)(10 \ \text{in})$$

$$= 15 \ \text{in}$$

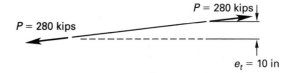

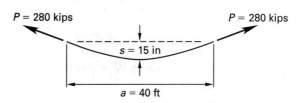

The equivalent prestress forces are those of the reactions of the stressed strand against the rigid concrete.

$$w_e = \frac{8Ps}{a^2} = \frac{(8)(280 \ \text{kips})\left(\dfrac{15 \ \text{in}}{12 \ \frac{\text{in}}{\text{ft}}}\right)}{(40 \ \text{ft})^2}$$

$$= 1.75 \ \text{kips/ft} \uparrow$$

$$F = w_e a + \frac{2P e_t}{a}$$

$$= \left(1.75 \ \frac{\text{kips}}{\text{ft}}\right)(40 \ \text{ft}) + \frac{(2)(280 \ \text{kips})(10 \ \text{in})}{(40 \ \text{ft})\left(12 \ \frac{\text{in}}{\text{ft}}\right)}$$

$$= 81.67 \ \text{kips} \downarrow$$

Structural

The structure is statically indeterminate to one degree. Let the reaction at point B be the unknown force, and release it.

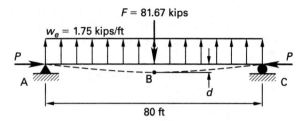

$F = 81.67$ kips

$w_e = 1.75$ kips/ft

P

A   B   C

d

80 ft

$$EI = 250 \times 10^6 \text{ kips-in}^2$$

$$d = \frac{5w_e L^4}{384EI} - \frac{FL^3}{48EI}$$

$$= \frac{(5)\left(1.75 \ \dfrac{\text{kips}}{\text{ft}}\right)(80 \ \text{ft})^4 \left(12 \ \dfrac{\text{in}}{\text{ft}}\right)^3}{(384)(250 \times 10^6 \ \text{kips-in}^2)}$$

$$- \frac{(81.67 \ \text{kips})(80 \ \text{ft})^3 \left(12 \ \dfrac{\text{in}}{\text{ft}}\right)^3}{(48)(250 \times 10^6 \ \text{kips-in}^2)}$$

$$= 0.430 \text{ in} \uparrow$$

The flexibility coefficient is obtained by applying a unit force upward at the released point and computing the deflection caused by that force.

$$f = \frac{L^3}{48EI} = \frac{(80 \ \text{ft})^3 \left(12 \ \dfrac{\text{in}}{\text{ft}}\right)^3}{(48)(250 \times 10^6 \ \text{kips-in}^2)}$$

$$= 0.0737 \text{ in/kip} \uparrow$$

For consistent displacement,

$$d + R_B f = 0$$

$$0.430 \text{ in} + R_B \left(0.0737 \ \frac{\text{in}}{\text{kip}}\right) = 0$$

$$R_B = \boxed{5.8 \text{ kips} \downarrow}$$

Structural

10 Steel

PROBLEM 1

A moment-resisting connection between a W18 × 46 beam and a W10 × 45 column is designed to resist a moment of 140 ft-kips and a 70 kip vertical load. All steel is A36. Use $^7/_8$ in A325 bolts.

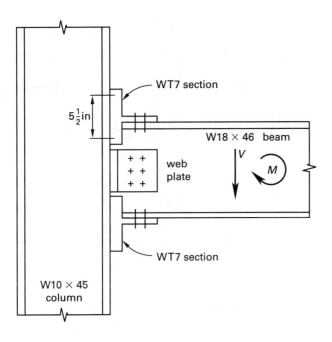

1.1. How many bolts are needed to attach the WT7 section to the column flange?

1.2. Specify the required flange thickness of the WT7 sections. Prying force is to be reduced to an insignificant amount.

1.3. If four $^5/_8$ in diameter A325 bolts are used in each WT section instead of $^7/_8$ in bolts, what is the minimum beam size (depth)?

PROBLEM 2

A 3.4 kips sign is supported by three equally spaced cables from a W8 beam, the exact type to be determined. The beam, in turn, is supported by two side wires and a threaded tie rod. Assume all beam, wire, and rod connections are pinned.

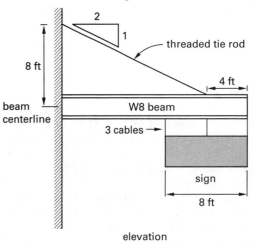

elevation

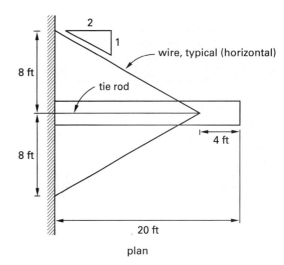

plan

2.1. What is the lightest W8 beam that can be used to support the sign? Use A992 steel.

2.2. What are the reactions in the wires and rods?

PROBLEM 3

Design the interior columns of the frame shown in the figure. The columns are braced continuously in the plane perpendicular to the frame. The girders are W12 × 96. Use A992 steel. Left-to-right sidesway is uninhibited. Connections of columns to their supports are rigid.

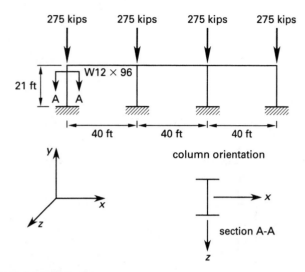

column orientation

section A-A

PROBLEM 4

A framing bent consists of W-shaped vertical steel columns with a truss girder constructed of double-angle tension and compression members. All steel is A36. Truss members are welded to the mounting plate (WT shapes) that are, in turn, bolted to the columns. (The centerlines of the truss members intersect the centroids of the bolt groups.) Bents are spaced every 22 ft and carry a uniformly distributed roof load of 70 lbf/ft^2, which includes an allowance for the truss girder itself.

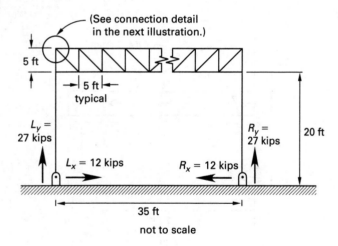

not to scale

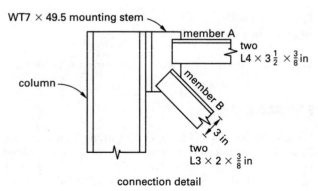

connection detail

4.1. Use A325 bolts in a bearing-type connection with threads excluded from shear planes to design the bolted connections between the mounting stem and the column. Disregard prying action.

4.2. Design the weld connection for member B. Assume E70 electrodes will be used.

PROBLEM 5

A 22 ft long W24 × 94 beam crosses at right angles and sits on an 18 ft long W27 × 146 beam. The W27 × 146 beam is, in turn, supported on unstiffened L8 × 4 × 1 seat angles that are welded to the flanges of the columns. A 180 kip concentrated load is applied at midlength along the W24 × 94 beam. All beams are simply supported on seats that offer no moment resistance.

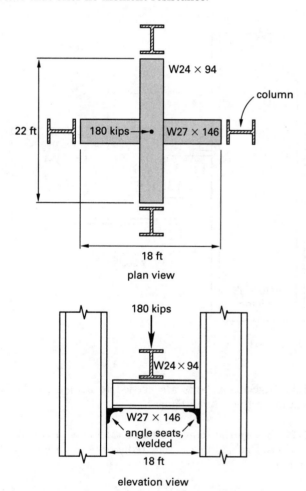

plan view

elevation view

5.1. What fraction of the 180 kip load is carried by each beam?

5.2. What are the maximum bending stresses in each beam?

5.3. Two parallel vertical welds connect each seat to the column. E70 electrodes are used. Evaluate the adequacy of the seat angles that support the W27 × 146 beam and design the welds. (The 4 in leg is horizontal.) Assume reasonable values for clearances and underruns. Beams are A992 steel, and seat angles are A36.

PROBLEM 6

A horizontal roof is constructed with 36 ft girders supporting 32 ft joists. The girders are supported by 32 ft high columns. All members are A992 steel. All horizontal members can be assumed to be simply supported. The columns can be assumed to be pin-ended at both ends. The roof carries a live load of 0.02 kips/ft² and a dead load of 0.06 kips/ft². Wind and earthquake forces are carried by an external bracing system and should not be considered in this design. The roof decking provides continuous bracing for the joist.

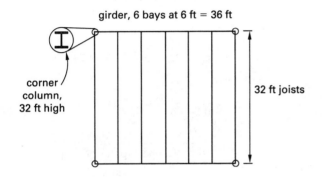

6.1. Select the lightest W-section for the joists such that the maximum deflection does not exceed 1/360th of the length.

6.2. Select the lightest W-section for the girders.

6.3. Select the lightest available W14 column.

PROBLEM 7

An exterior two-story, 24 ft high steel column supports an exterior sign, a distributed roof load, and story loads. The column is pinned at the base. Refer to the illustration for details. Select a wide-flange column according to AISC procedures.

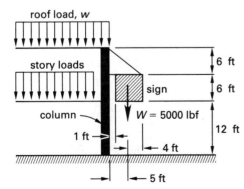

vertical reaction from roof loads: R = 20,000 lbf
vertical reaction from story loads: R = 30,000 lbf

PROBLEM 8

A single plate bracket is welded to the face of a column as shown. The column flanges have a thickness in excess of 1 in. Determine the size of the fillet weld required. Use E70 electrodes and A36 steel.

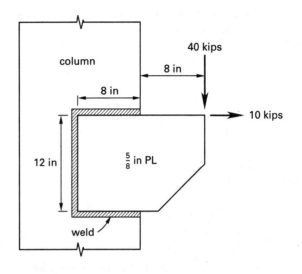

PROBLEM 9

Design an A992 wide-flange column to support the loads indicated.

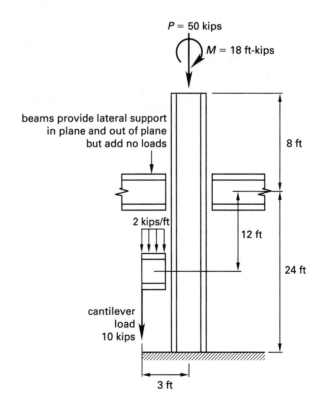

SOLUTION 1

1.1. The shear of 70 kips is assumed to be carried entirely by the web plate. This problem is not concerned with the web plate, so the 70 kips shear is disregarded. Only the 140 ft-kip moment is carried by the WT7 sections.

The depth of the W18 × 46 beam is 18 in. The horizontal tensile (and compressive) forces resisted by all of the bolts in the column-to-T-section connection is

$$H = \frac{M}{h} = \frac{(140 \text{ ft-kips})\left(12 \frac{\text{in}}{\text{ft}}\right)}{18 \text{ in}}$$
$$= 93.33 \text{ kips}$$

From *AISC Manual* Table 7-2, the tensile capacity of $7/8$ in A325 bolts is $B = 27.1$ kips/bolt. The number of bolts required per T-section is

$$n = \frac{H}{B} = \frac{93.33 \text{ kips}}{27.1 \frac{\text{kips}}{\text{bolt}}}$$
$$= \boxed{3.44 \text{ bolts}} \quad \text{[use four bolts per T-section]}$$

1.2. The base of the T-section is bolted to the flange of the column. Although the beam-column connection is a moment connection, the top T-section is loaded in pure tension. The procedure in Part 9 of the *AISC Manual* can be used.

step 1: From Sol. 1.1, four bolts are needed. The tensile force per bolt is

$$T = \frac{H}{n} = \frac{93.33 \text{ kips}}{4 \text{ bolts}} = 23.33 \text{ kips/bolt}$$

step 2: Assume that the thickness of the T-section's web will be approximately 1 in. Since the bolt-to-bolt spacing is 5.5 in, the free distance between the T-section web and the bolt hole is

$$b = \frac{5.5 \text{ in} - 1 \text{ in}}{2} = 2.25 \text{ in}$$

The flange width, b_f, of the W18 × 46 is 6.06 in. The flange width of the W10 × 45 beam is 8 in, which permits a WT 8 in long. Take the tributary length per bolt pair, p, as half of the WT length.

$$p = \frac{L}{2} = \frac{8 \text{ in}}{2} = 4 \text{ in}$$

For a trial WT, assume a stem thickness, t_w, of 0.5 in. This gives

$$b' = b - 0.5t_w = 2.25 \text{ in} - (0.5)(0.5 \text{ in})$$
$$= 2.0 \text{ in}$$

For A36 steel, $F_u = 58$ kips/in². The required flange thickness to eliminate prying action is

$$t_{min} = \sqrt{\frac{6.66 T b'}{p F_u}}$$
$$= \sqrt{\frac{(6.66)(23.33 \text{ kips})(2.0 \text{ in})}{(4 \text{ in})\left(58 \frac{\text{kips}}{\text{in}^2}\right)}}$$
$$= 1.16 \text{ in}$$

Try a WT7 × 79.5.

$$t_f = 1.19 \text{ in}$$
$$t_w = 0.745 \text{ in}$$

The revised values are

$$b' = b - 0.5t_w = 2.25 \text{ in} - (0.5)(0.745 \text{ in})$$
$$= 1.88 \text{ in}$$

$$t_{min} = \sqrt{\frac{6.66 T b'}{p F_u}}$$
$$= \sqrt{\frac{(6.66)(23.33 \text{ kips})(1.88 \text{ in})}{(4 \text{ in})\left(58 \frac{\text{kips}}{\text{in}^2}\right)}}$$
$$= \boxed{1.12 \text{ in}}$$

The WT7 × 79.5 is satisfactory. Use a WT7 × 79.5 × 8 in with four $7/8$ in A325 bolts.

1.3. From *AISC Manual* Table 7-2, the capacity of $5/8$ in A325 bolts in tension is

$$B = 13.8 \text{ kips/bolt}$$

The total capacity of four bolts is

$$(4 \text{ bolts})\left(13.8 \frac{\text{kips}}{\text{bolt}}\right) = 55.2 \text{ kips}$$
$$d_{min} = \frac{M}{F}$$
$$= \frac{(140 \text{ ft-kips})\left(12 \frac{\text{in}}{\text{ft}}\right)}{55.2 \text{ kips}}$$
$$= 30.4 \text{ in}$$

$$\boxed{\text{A W33 beam size is required.}}$$

SOLUTION 2

2.1.

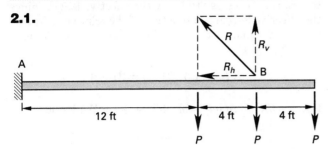

From $\sum M_A = 0$, $R_v = 3P$. From the slope of the tie rod, $R_h = 2R_v = 6P$.

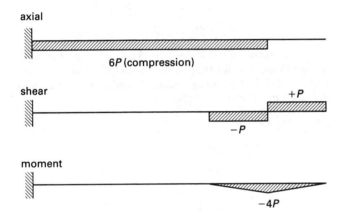

Point B is braced by the combined effect of the lateral wires and the restoring twist provided by the applied loads and the tie rod reaction.

$$P = \frac{3.4 \text{ kips}}{3} = 1.13 \text{ kips} \quad \text{[neglecting self-weight]}$$

Segment AB will control the selection of the lightest W8 section. Find the self-weight contributions. w is in lbf/ft.

The contribution to the moment is

$$\frac{w(4 \text{ ft})(2)}{1000 \ \frac{\text{lbf}}{\text{kip}}} = 0.008w$$

The contribution to the axial loading is

$$\left(\frac{w(20 \text{ ft})(10 \text{ ft})}{\left(1000 \ \frac{\text{lbf}}{\text{kip}} \right)(16 \text{ ft})} \right)(2) = 0.025w$$

Try a W8 × 18 ($KL/r_y < 200$).

$$A = 5.26 \text{ in}^2$$
$$r_y = 1.23 \text{ in}$$
$$r_x = 3.43 \text{ in}$$
$$w = 18 \text{ lbf/ft}$$

$$M = (4 \text{ ft})\left(\frac{3.4 \text{ kips}}{3} \right) + \left(0.008 \ \frac{\text{kip-ft}^2}{\text{lbf}} \right)\left(18 \ \frac{\text{lbf}}{\text{ft}} \right)$$
$$= 4.677 \text{ ft-kips}$$

$$F = (6)\left(\frac{3.4 \text{ kips}}{3} \right) + \left(18 \ \frac{\text{lbf}}{\text{ft}} \right)\left(0.025 \ \frac{\text{ft-kip}}{\text{lbf}} \right)$$
$$= 7.25 \text{ kips}$$

Check the member for combined axial compression and bending using ASD. Compression strength is controlled by the y-axis.

$$\frac{KL}{r_y} = \frac{(16 \text{ ft})\left(12 \ \frac{\text{in}}{\text{ft}} \right)}{1.23 \text{ in}} = 156$$

From *AISC Manual* Table 4-22,

$$\frac{F_{cr}}{\Omega} = 6.18 \text{ kips/in}^2$$

$$\frac{P_n}{\Omega} = \left(\frac{F_{cr}}{\Omega} \right)A = \left(6.18 \ \frac{\text{kips}}{\text{in}^2} \right)(5.26 \text{ in}^2)$$
$$= 32.5 \text{ kips}$$

For bending about the x-axis, use $L_b = 16$ ft and $C_b = 1.0$ (conservative). From *AISC Manual* Table 3-10,

$$\frac{M_n}{\Omega} = 21 \text{ ft-kips}$$

For combined bending,

$$P_r = F = 7.25 \text{ kips}$$
$$P_c = \frac{P_n}{\Omega} = 32.5 \text{ kips}$$
$$\frac{P_r}{P_c} = \frac{7.25 \text{ kips}}{32.5 \text{ kips}}$$
$$= 0.22 > 0.2$$

AISC Specification Eq. H1-1a controls.

$$M_r = M = 4.677 \text{ ft-kips}$$

$$M_c = \frac{M_n}{\Omega} = 21 \text{ ft-kips}$$

$$\frac{P_r}{P_c} + \left(\frac{8}{9}\right)\left(\frac{M_r}{M_c}\right) = \frac{7.25 \text{ kips}}{32.5 \text{ kips}} + \left(\frac{8}{9}\right)\left(\frac{4.677 \text{ ft-kips}}{21 \text{ ft-kips}}\right)$$

$$= 0.42 < 1.0$$

The W8 × 18 is satisfactory.

The next smallest W8 is a W8 × 15, which is likely to be adequate for stress but should not be used because KL/r_y exceeds 200 (the *AISC Specification* preferred maximum slenderness ratio listed in Sec. B7).

$$\boxed{\text{Use a W8} \times 18.}$$

2.2. The vertical reaction in each wire is 0. The ratio of the resultant force in the tie rod to the vertical component is

$$\frac{\sqrt{(2)^2 + (1)^2}}{1} = \sqrt{5}$$

The beam is simply supported at the wall and tie rod. The reaction in the tie rod is

$$R = \sqrt{5}\,(3.4 \text{ kips}) + \frac{\left(18 \, \frac{\text{lbf}}{\text{ft}}\right)(20 \text{ ft})(10 \text{ ft})\sqrt{5}}{(16 \text{ ft})\left(1000 \, \frac{\text{lbf}}{\text{kip}}\right)}$$

$$= \boxed{8.11 \text{ kips}}$$

SOLUTION 3

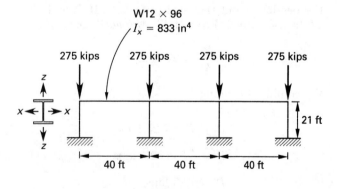

In order to select a section from the *AISC Manual* tables, it is necessary to know the effective length. Since the effective length is a function of the relative rigidities, assume a value of the effective length factor. Assume $K = 1.5$.

$$KL = (1.5)(21 \text{ ft}) = 31.5 \text{ ft}$$

From *AISC Manual* Table 4-1, select a W12 × 96.

$$I_y = 270 \text{ in}^4$$

Calculate the effective length.

$$G_t = \frac{\left(\frac{I}{L}\right)_{\text{column}}}{\left(\frac{I}{L}\right)_{\text{girder}}} = \frac{\dfrac{270 \text{ in}^4}{21 \text{ ft}}}{\dfrac{(2)(833 \text{ in}^4)}{40 \text{ ft}}}$$

$$= 0.309$$

$G_b = 1$ is recommended by the *AISC Manual* for rigid connections to footings. From *AISC Commentary* Fig. C-A-7.2 (sidesway uninhibited), $K \approx 1.2$ and $KL = 25$ ft.

Try a W12 × 72.

$$I_y = 195 \text{ in}^4$$

$$G_t = \frac{\left(\frac{I}{L}\right)_{\text{column}}}{\left(\frac{I}{L}\right)_{\text{girder}}} = \frac{\dfrac{195 \text{ in}^4}{21 \text{ ft}}}{\dfrac{(2)(833 \text{ in}^4)}{40 \text{ ft}}}$$

$$= 0.223$$

$$G_b = 1$$

$$K \approx 1.19$$

$$KL = 25 \text{ ft} \quad [\text{OK}]$$

Interpolating between 24 ft and 26 ft, the capacity from *AISC Manual* Table 4-1 is

$$\frac{P_n}{\Omega_c} = \frac{328 \text{ kips} + 292 \text{ kips}}{2.0} = 310 \text{ kips}$$

The actual loading is

$$P = 275 \text{ kips} + \frac{\left(72 \, \frac{\text{lbf}}{\text{ft}}\right)(21 \text{ ft})}{1000 \, \frac{\text{lbf}}{\text{kip}}}$$

$$= 276.5 \text{ kips} < \frac{P_n}{\Omega_c} \quad [\text{OK}]$$

$$\boxed{\text{Use a W12} \times 72.}$$

SOLUTION 4

4.1. Determine the design forces.

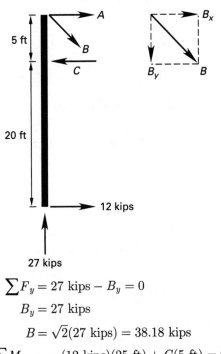

$$\sum F_y = 27 \text{ kips} - B_y = 0$$
$$B_y = 27 \text{ kips}$$
$$B = \sqrt{2}(27 \text{ kips}) = 38.18 \text{ kips}$$
$$\sum M_{\text{top}} = -(12 \text{ kips})(25 \text{ ft}) + C(5 \text{ ft}) = 0$$
$$C = 60 \text{ kips}$$
$$\sum F_x = A + B_x - C + 12 \text{ kips} = 0$$
$$A = -12 \text{ kips} + 60 \text{ kips} - 27 \text{ kips}$$
$$= 21 \text{ kips}$$

The centroids of members A and B intersect at the centerline of the bolted connections to the columns.

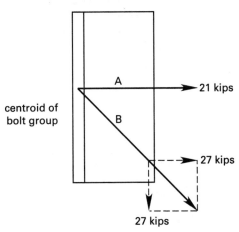

tension, $T = 21 \text{ kips} + 27 \text{ kips} = 48 \text{ kips}$

shear, $V = 27 \text{ kips}$

Find the allowable bolt stresses for A325 (threads not excluded from shear plane) from *AISC Specification* Table J3.2.

$$F_{nt} = 90 \text{ kips/in}^2$$
$$F_{nv} = 68 \text{ kips/in}^2$$

From *AISC Specification* Sec. J3.6, for ASD ($\Omega = 2.0$), the nominal tensile stress for combined tension and shear is given by *AISC Specification* Eq. J3-3b.

$$F'_{nt} = 1.3F_{nt} - \frac{\Omega F_{nt}}{F_{nv}}f_v \leq F_{nt}$$

f_v is the required shear stress in kips/in². Try six $^5/_8$ in bolts. ($A_b = 0.307 \text{ in}^2$.)

$$f_v = \frac{V}{nA_b} = \frac{27 \text{ kips}}{(6)(0.307 \text{ in}^2)} = 14.7 \text{ kips/in}^2$$

$$F'_{nt} = 1.3F_{nt} - \frac{\Omega F_{nt}}{F_{nv}}f_v \leq F_{nt}$$

$$= (1.3)\left(90 \frac{\text{kips}}{\text{in}^2}\right) - \left(\frac{(2.0)\left(90 \frac{\text{kips}}{\text{in}^2}\right)}{68 \frac{\text{kips}}{\text{in}^2}}\right)$$

$$\times \left(14.7 \frac{\text{kips}}{\text{in}^2}\right)$$

$$= 78.2 \text{ kips/in}^2 \leq 90 \text{ kips/in}^2 \quad [\text{OK}]$$

The available tensile strength of the six bolts is

$$\frac{nR_n}{\Omega} = \frac{nF'_{nt}A_b}{\Omega} = \frac{(6)\left(78.2 \frac{\text{kips}}{\text{in}^2}\right)(0.307 \text{ in}^2)}{2.0}$$

$$= 72.0 \text{ kips} > T$$

$$\boxed{\text{Use six } ^5/_8 \text{ in diameter A325-X bolts.}}$$

4.2.

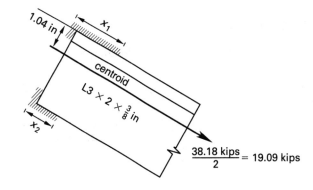

From *AISC Specification* Tables J2.4 and J2.5,

$$\text{minimum weld size} = \frac{3}{16} \text{ in}$$

$$\text{maximum weld size} = \frac{3}{8} \text{ in} - \frac{1}{16} \text{ in}$$

$$= \frac{5}{16} \text{ in}$$

$$F_{v,\text{fillet}} = \frac{0.6F_{\text{Exx}}}{\Omega} = \frac{(0.6)\left(70 \ \dfrac{\text{kips}}{\text{in}^2}\right)}{2.0}$$

$$= 21 \text{ kips/in}^2$$

Select a $^3/_{16}$ in fillet. The allowable strength, R_n/Ω, is

$$(0.923)(3) = 2.78 \text{ kips/in}$$

The length of fillet required is

$$\frac{19.09 \text{ kips}}{2.78 \ \dfrac{\text{kips}}{\text{in}}} = 6.87 \text{ in}$$

Neglecting end returns, the centroid of the weld coincides with the centroid of the member when

$$1.04x_1 = (3 - 1.04)x_2$$

$$x_1 = 1.885x_2$$

$$x_1 + x_2 = 6.87 \text{ in}$$

$$x_2 = 2.38 \text{ in}$$

$$x_1 = 4.49 \text{ in}$$

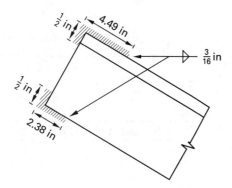

SOLUTION 5

5.1. For each of the beams, the loading is the same as for a simple beam.

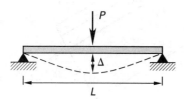

The stiffness is

$$k = \frac{P}{\Delta} = \frac{48EI}{L^3}$$

For the W27 × 146 beam,

$$I = 5630 \text{ in}^4$$

$$L = (18 \text{ ft})\left(12 \ \frac{\text{in}}{\text{ft}}\right) = 216 \text{ in}$$

$$k = \frac{(48E)(5630 \text{ in}^4)}{(216 \text{ in})^3} = 0.0268E$$

For the W24 × 94 beam,

$$I = 2700 \text{ in}^4$$

$$L = (22 \text{ ft})\left(12 \ \frac{\text{in}}{\text{ft}}\right) = 264 \text{ in}$$

$$k = \frac{(48E)(2700 \text{ in}^4)}{(264 \text{ in})^3} = 0.00704E$$

The load on the W27 × 146 beam is

$$F = \left(\frac{0.0268E}{0.0268E + 0.00704E}\right)(180 \text{ kips})$$

$$= \boxed{142.6 \text{ kips} \quad (79\%)}$$

The load on the W24 × 94 beam is

$$F = \left(\frac{0.00704E}{0.0268E + 0.00704E}\right)(180 \text{ kips})$$

$$= \boxed{37.4 \text{ kips} \quad (21\%)}$$

5.2. For the W27 × 146 beam,

$$M = \frac{PL}{4} + \frac{wL^2}{8}$$

$$= \frac{(142.6 \text{ kips})(18 \text{ ft})}{4} + \frac{\left(0.146 \ \dfrac{\text{kips}}{\text{ft}}\right)(18 \text{ ft})^2}{8}$$

$$= 647.6 \text{ ft-kips}$$

$$f_b = \frac{M}{S} = \frac{(647.6 \text{ ft-kips})\left(12 \ \dfrac{\text{in}}{\text{ft}}\right)}{411 \text{ in}^3}$$

$$= \boxed{18.91 \text{ kips/in}^2}$$

For the W24 × 94 beam,

$$M = \frac{PL}{4} + \frac{wL^2}{8}$$

$$= \frac{(37.4 \text{ kips})(22 \text{ ft})}{4} + \frac{\left(0.094 \frac{\text{kips}}{\text{ft}}\right)(22 \text{ ft})^2}{8}$$

$$= 211.4 \text{ ft-kips}$$

$$f_b = \frac{(211.4 \text{ ft-kips})\left(12 \frac{\text{in}}{\text{ft}}\right)}{222 \text{ in}^3} = \boxed{11.43 \text{ kips/in}^2}$$

5.3. The reaction is

$$R_a = \frac{142.6 \text{ kips}}{2} + \frac{\left(0.146 \frac{\text{kips}}{\text{ft}}\right)(18 \text{ ft})}{2}$$

$$= 72.61 \text{ kips}$$

For the W27 × 146 beam,

$$b_f = 13.965 \text{ in}$$
$$t_w = 0.605 \text{ in}$$
$$k = 1.6875 \text{ in}$$

The minimum bearing length, N, is the larger of the two values for local web yielding and web crippling. Use *AISC Manual* Table 9-4 to find the values of R_1/Ω, R_2/Ω, R_3/Ω, and R_4/Ω.

For local web yielding,

$$N = \frac{R_a - \frac{R_1}{\Omega}}{\frac{R_2}{\Omega}} \geq k$$

$$= \frac{72.61 \text{ kips} - 88.7 \text{ kips}}{20.2 \frac{\text{kips}}{\text{in}}} \geq 1.6875 \text{ in}$$

$$= -0.80 \text{ in} < 1.6875 \text{ in}$$

Since N cannot be less than k, $N = 1.6875$ in.

For web crippling,

$$\left(\frac{N}{d}\right)_{\text{max}} = \frac{1.6875 \text{ in}}{27.4 \text{ in}}$$
$$= 0.062 < 0.20$$

Since N/d is less than 0.20, use the following equation.

$$N = \frac{R_a - \frac{R_3}{\Omega}}{\frac{R_4}{\Omega}} \geq k$$

$$= \frac{72.61 \text{ kips} - 112 \text{ kips}}{5.99 \frac{\text{kips}}{\text{in}}} \geq 1.6875 \text{ in}$$

$$= -6.58 \text{ in} < 1.6875 \text{ in}$$

Since N cannot be less than k, $N = 1.6875$ in.

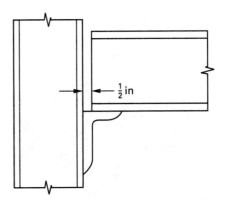

Check shear yielding and flexural yielding of the angle. Check local yielding and crippling of the beam web. From *AISC Manual* Table 10-6, for an L8 × 4 × 1 with $N_{\text{req}} = 2$ in,

$$\frac{R_n}{\Omega} = 115 \text{ kips} > 72.61 \text{ kips} \quad [\text{OK}]$$

Design the fillet weld connection for the 8 in vertical leg. From *AISC Manual* Table 10-6, for a $^9/_{16}$ fillet,

$$\frac{R_n}{\Omega} = 80.1 \text{ kips} > 72.61 \text{ kips} \quad [\text{OK}]$$

Bolt the beam to the seat angle, and provide a top angle connection per *AISC Specification* Sec. J2.2b.

$$\boxed{\text{The L8} \times 4 \times 1 \text{ is adequate.}}$$

The final bracket design is

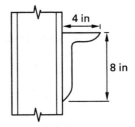

SOLUTION 6

6.1. Select a W-section for the joists.

$$l = (32 \text{ ft})\left(12 \frac{\text{in}}{\text{ft}}\right) = 384 \text{ in}$$

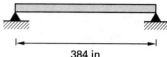

384 in

$$w_D = \left(0.06 \frac{\text{kips}}{\text{ft}^2}\right)(6 \text{ ft}) + w_{\text{self}}$$

$$= 0.36 \text{ kips/ft} + w_{\text{self}}$$

$$w_L = \left(0.02 \frac{\text{kips}}{\text{ft}^2}\right)(6 \text{ ft})$$

$$= 0.12 \text{ kips/ft}$$

For uniform loading,

$$\Delta = \frac{5wl^4}{384EI}$$

The deflection criterion is

$$\frac{\Delta}{l} \le \frac{1}{360}$$

$$\frac{5wl^3}{384EI} \le \frac{1}{360}$$

The moment of inertia required to meet the deflection criterion is

$$I \ge \frac{4.69wl^3}{E}$$

Neglecting self-weight,

$$I > \frac{(4.69)\left(0.48 \frac{\text{kips}}{\text{ft}}\right)(384 \text{ in})^3}{\left(29,000 \frac{\text{kips}}{\text{in}^2}\right)\left(12 \frac{\text{in}}{\text{ft}}\right)}$$

$$\ge 366 \text{ in}^4$$

Try a W16 × 31. (From *AISC Manual* Table 3-3, $I = 375 \text{ in}^4$.) Adjust the load to account for the self-weight.

$$w = 0.48 \frac{\text{kips}}{\text{ft}} + 0.031 \frac{\text{kips}}{\text{ft}} = 0.51 \text{ kips/ft}$$

Check the expression for the required moment of inertia.

$$I > (366 \text{ in}^4)\left(\frac{0.51 \frac{\text{kips}}{\text{ft}}}{0.48 \frac{\text{kips}}{\text{ft}}}\right)$$

$$= 389 \text{ in}^4 > 375 \text{ in}^4 \quad \text{[not acceptable]}$$

Next, try a W18 × 35. (From *AISC Manual* Table 3-3, $I = 510 \text{ in}^4$, and from Table 3-6, $Z_x = 66.5 \text{ in}^3$.) The deflection requirement is met with this section. Check the capacity.

The applied moment is

$$w = 0.48 \frac{\text{kips}}{\text{ft}} + 0.035 \frac{\text{kips}}{\text{ft}}$$

$$= 0.515 \text{ kips/ft}$$

$$M = \left(\tfrac{1}{8}\right)\left(0.515 \frac{\text{kips}}{\text{ft}}\right)(32 \text{ ft})^2$$

$$= 65.92 \text{ ft-kips}$$

The capacity is

$$\frac{M_n}{\Omega_b} = \frac{F_y Z_x}{\Omega_b} = \frac{\left(50 \frac{\text{kips}}{\text{in}^2}\right)(66.5 \text{ in}^3)}{(1.67)\left(12 \frac{\text{in}}{\text{ft}}\right)}$$

$$= 165.9 \text{ ft-kips} > M \quad \text{[OK]}$$

Use a W18 × 35 for the joists.

6.2. Select a W-shape for girders. Treat the reaction from the joists as an equivalent uniform load (which is justifiable, given the total number of joists). Assume bracing at each joist location.

The deflection criterion is

$$\frac{\Delta}{l} \le \frac{1}{360}$$

The required moment of inertia is

$$I \ge \frac{4.69wl^3}{E}$$

Calculate the load, neglecting self-weight for a first approximation.

$$w = \left(0.06 \frac{\text{kips}}{\text{ft}^2}\right)(16 \text{ ft}) + \left(0.02 \frac{\text{kips}}{\text{ft}^2}\right)(16 \text{ ft})$$

$$= 1.28 \text{ kips/ft}$$

$$I = \frac{(4.69)\left(1.28 \frac{\text{kips}}{\text{ft}}\right)(432 \text{ in})^3}{\left(29,000 \frac{\text{kips}}{\text{in}^2}\right)\left(12 \frac{\text{in}}{\text{ft}}\right)}$$

$$= 1391 \text{ in}^4 \quad \text{[required]}$$

Structural

Try a W24 × 62 section.

$$I = 1550 \text{ in}^4$$

$$Z = 144 \text{ in}^3$$

$$b_f = 7.04 \text{ in}$$

Check the moment capacity.

$$w = 1.28 \, \frac{\text{kips}}{\text{ft}} + 0.062 \, \frac{\text{kips}}{\text{ft}}$$

$$= 1.34 \text{ kips/ft}$$

$$M = \tfrac{1}{8}wl^2 = \left(\tfrac{1}{8}\right)\left(1.34 \, \frac{\text{kips}}{\text{ft}}\right)(36 \text{ ft})^2$$

$$= 217.1 \text{ ft-kips}$$

From *AISC Manual* Table 3-2, $L_p = 4.87$ ft is less than the actual unbraced length, $L_b = 6$ ft. Therefore, the moment strength is controlled by the fully plastic capacity of the section. From *AISC Manual* Table 3-2, $M_{px}/\Omega_b = 382$ kips-ft $> M$.

The W24 × 62 is adequate in flexure.

> Use a W24 × 62 for the girders.

6.3. The load on the column is

$$P = \frac{\left(1.34 \, \frac{\text{kips}}{\text{ft}}\right)(36 \text{ ft})}{2} = 24.12 \text{ kips}$$

Because of the extreme height, the KL/r limit of 200 is likely to control.

$$\frac{(32 \text{ ft})\left(12 \, \frac{\text{in}}{\text{ft}}\right)}{r_y} \leq 200$$

$$r_y > 1.92 \text{ in}$$

From *AISC Manual* Table 4-22, at $KL/r = 200$, $F_{cr}/\Omega = 3.76$ kips/in². A W14 × 53 has $r_y = 1.92$ in and $A = 15.6$ in², so the allowable load is

$$P_{\text{allowable}} = \left(3.76 \, \frac{\text{kips}}{\text{in}^2}\right)(15.6 \text{ in}^2) = 58.7 \text{ kips}$$

This is much larger than the design load.

> Use a W14 × 53 for the columns.

SOLUTION 7

Calculate the horizontal reaction, H.

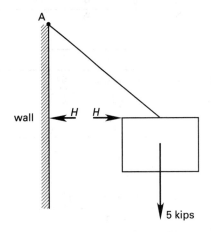

Taking clockwise moments as positive,

$$\sum M_A = -H(6 \text{ ft}) + (5 \text{ kips})(5 \text{ ft})$$

$$= 0$$

$$H = 4.17 \text{ kips}$$

The loads acting on the column are as illustrated.

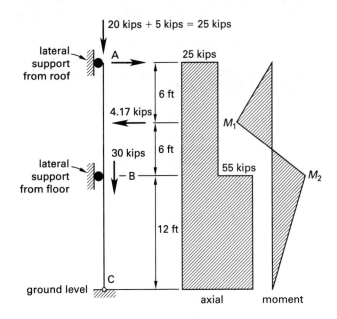

Structural

Calculate the moments using moment distribution.

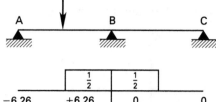

		$\frac{1}{2}$	$\frac{1}{2}$	
FEM	−6.26	+6.26	0	0
BAL	+6.26	−3.13	−3.13	
COM		+3.13	0	
BAL		−1.56	−1.56	
	0	+4.69	−4.69	0

(all units in ft-kips)

$$\text{FEM} = \frac{Pl}{8} = \frac{(4.17 \text{ kips})(12 \text{ ft})}{8} = 6.26 \text{ ft-kips}$$

$$M_2 = 4.69 \text{ ft-kips}$$

$$M_1 = \frac{(4.17 \text{ kips})(12 \text{ ft})}{4} - \frac{4.69 \text{ ft-kips}}{2}$$
$$= 10.17 \text{ ft-kips}$$

To simplify connections, the *AISC Manual* recommends that W-shape columns have a nominal depth of 8 in or greater. Select a trial segment based on the flexural strength for an unbraced length of compression flange of 12 ft in the upper segment. *AISC Manual* Table 3-10 shows that a W8 × 21 has a moment strength approximately triple the required value of 10.17 ft-kips, based on a conservative value of $C_b = 1.0$. *AISC Manual* Table 3-2 shows that the section is compact for $F_y = 50 \text{ kips/in}^2$. Try a W8 × 21.

$$A = 6.16 \text{ in}^2$$
$$r_x = 3.49 \text{ in}^2$$
$$r_y = 1.26 \text{ in}^2 \quad \text{[controls]}$$

For segment AB,

$$P_r = 25 \text{ kips}$$
$$M_r = M_1 = 10.17 \text{ ft-kips}$$
$$\frac{KL}{r_y} = \frac{(12 \text{ ft})\left(12 \frac{\text{in}}{\text{ft}}\right)}{1.26 \text{ in}}$$
$$= 114$$

From *AISC Manual* Table 4-22,

$$\frac{F_{cr}}{\Omega} = 11.6 \text{ kips/in}^2$$
$$P_c = \frac{P_n}{\Omega} = \left(\frac{F_{cr}}{\Omega}\right)A$$
$$= \left(11.6 \frac{\text{kips}}{\text{in}^2}\right)(6.16 \text{ in}^2)$$
$$= 71.5 \text{ kips}$$

From *AISC Manual* Table 3-10, for the unbraced length of 12 ft,

$$M_c = 37 \text{ ft-kips}$$

Check combined axial compression plus bending.

$$\frac{P_r}{P_c} = \frac{25 \text{ kips}}{71.5 \text{ kips}}$$
$$= 0.35 > 0.2$$

Use *AISC Specification* Eq. H1-1a with loading in only one direction.

$$\frac{P_r}{P_c} + \left(\frac{8}{9}\right)\left(\frac{M_r}{M_c}\right) \leq 1.0$$
$$= 0.35 + \left(\frac{8}{9}\right)\left(\frac{10.17 \text{ ft-kips}}{37.0 \text{ ft-kips}}\right) \leq 1.0$$
$$0.59 \leq 1.0 \quad \text{[OK]}$$

The W8 × 21 is OK for segment AB. Next, check segment BC. For segment BC,

$$P_r = 55 \text{ kips}$$
$$M_r = M_2 = 4.69 \text{ ft-kips}$$
$$\frac{P_r}{P_c} = \frac{55 \text{ kips}}{71.5 \text{ kips}}$$
$$= 0.77 > 0.2$$

Use *AISC Specification* Eq. H1-1a.

$$\frac{P_r}{P_c} + \left(\frac{8}{9}\right)\left(\frac{M_r}{M_c}\right) \leq 1.0$$
$$= 0.77 + \left(\frac{8}{9}\right)\left(\frac{4.69 \text{ ft-kips}}{37 \text{ ft-kips}}\right) \leq 1.0$$
$$0.88 \leq 1.0 \quad \text{[OK]}$$

The W8 × 21 is adequate.

SOLUTION 8

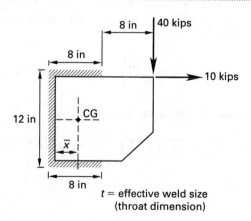

t = effective weld size (throat dimension)

Determine the location of the centroid of the welds.

$$\bar{x} = \frac{(8t \text{ in}^2)(4 \text{ in})(2)}{((8 \text{ in})(2) + 12 \text{ in})t \text{ in}} = 2.286 \text{ in}$$

The forces and moment on the weld group are

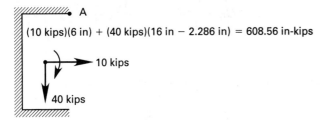

(10 kips)(6 in) + (40 kips)(16 in − 2.286 in) = 608.56 in-kips

Stresses in the weld are critical at point A.

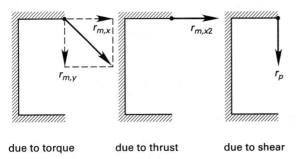

due to torque   due to thrust   due to shear

Use the parallel axis theorem to calculate the centroidal polar moment of inertia. Disregard higher order terms.

$$I_x = (2)\left((8 \text{ in})t(6 \text{ in})^2\right) + \frac{t(12)^3}{12} = 720t$$

$$I_y = (12t)(2.286 \text{ in})^2 + \left(\tfrac{1}{12}t\right)(8 \text{ in})^3(2 \text{ in})$$
$$\quad + (8t)(2)(4 \text{ in} - 2.286 \text{ in})^2$$
$$= 195t$$

$$I_p = I_x + I_y = 720t + 195t$$
$$= 915t$$

$$r_{m,x} = \frac{Mc}{I} = \frac{(608.56 \text{ in-kips})(6 \text{ in})}{915t} = \frac{3.991}{t}$$

$$r_{m,x2} = \frac{V}{A} = \frac{10 \text{ kips}}{(8 \text{ in} + 8 \text{ in} + 12 \text{ in})t} = \frac{0.357}{t}$$

$$r_{m,y} = \frac{Mc}{I} = (608.56 \text{ in-kips})\left(\frac{8 \text{ in} - 2.286 \text{ in}}{915t}\right)$$
$$= \frac{3.800}{t}$$

$$r_p = \frac{V}{A} = \frac{40 \text{ kips}}{(8 \text{ in} + 8 \text{ in} + 12 \text{ in})t} = \frac{1.429}{t}$$

$$\sum r_x = \frac{3.991}{t} + \frac{0.357}{t} = \frac{4.348}{t}$$

$$\sum r_y = \frac{3.800}{t} + \frac{1.429}{t} = \frac{5.229}{t}$$

The resultant is

$$r_a = \sqrt{\left(\frac{4.348}{t}\right)^2 + \left(\frac{5.229}{t}\right)^2}$$
$$= \frac{6.801}{t}$$

The design strength for E70 electrodes, F_w/Ω, is 21 kips/in^2.

$$\frac{6.801}{t} = 21 \frac{\text{kips}}{\text{in}^2}$$
$$t = 0.324 \text{ in}$$

The nominal weld size is

$$w_n = \sqrt{2}t = \sqrt{2}\,(0.324 \text{ in})$$
$$= 0.458 \text{ in}$$

Try $^1/_2$ in fillets.

Check the minimum and maximum limits, using *AISC Specification* Table J2.4 and Sec. J2.2b.

$$\text{minimum} = \frac{1}{4} \text{ in} \quad [\text{OK}]$$

$$\text{maximum} = \frac{5}{8} \text{ in} - \frac{1}{16} \text{ in}$$
$$= \frac{9}{16} \text{ in} \quad [\text{OK}]$$

$$\boxed{\text{Use a } ^1/_2 \text{ in fillet.}}$$

SOLUTION 9

The lower vertical load is

$$10 \text{ kips} + \left(2 \frac{\text{kips}}{\text{ft}}\right)(3 \text{ ft}) = 16 \text{ kips}$$

The lower moment is

$$(3 \text{ ft})(10 \text{ kips}) + \left(\frac{1}{2}\right)\left(2 \, \frac{\text{kips}}{\text{ft}}\right)(3 \text{ ft})^2 = 39 \text{ ft-kips}$$

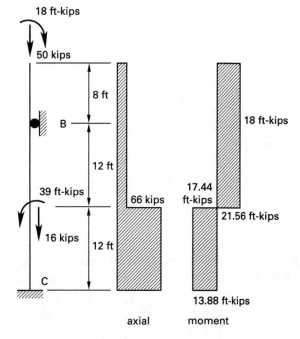

axial   moment

Obtain the bending moments.

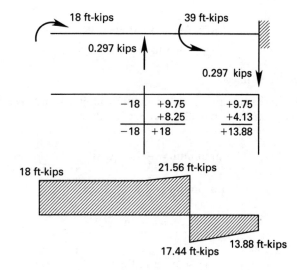

Obtain the effective length for segment AB. Neglect second-order effects in segment BC.

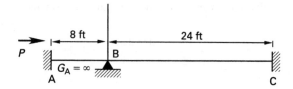

$$G_B = \frac{\left(\dfrac{I}{L}\right)_{\text{column}}}{\left(\dfrac{I}{L}\right)_{\text{girder}}} = \frac{\left(\dfrac{I}{8 \text{ ft}}\right)}{\left(\dfrac{I}{24 \text{ ft}}\right)}$$

$$= 3$$

From *AISC Commentary* Fig. C-A-7.2, $K \approx 3$.

$$KL = (3 \text{ ft})(8 \text{ ft}) = 24 \text{ ft}$$

Try a W8 × 48.

$$A = 14.1 \text{ in}^2$$
$$r_x = 3.61 \text{ in}$$
$$r_y = 2.18 \text{ in} \quad [\text{controls}]$$

For segment BC,

$$P_r = 50 \text{ kips}$$
$$M_r = 18 \text{ ft-kips}$$
$$\frac{KL}{r_y} = \frac{(24 \text{ ft})\left(12 \, \dfrac{\text{in}}{\text{ft}}\right)}{2.08 \text{ in}}$$
$$= 138$$

From *AISC Manual* Table 4-22,

$$\frac{F_{\text{cr}}}{\Omega} = 7.89 \text{ kips/in}^2$$
$$P_c = \frac{P_n}{\Omega} = \left(\frac{F_{\text{cr}}}{\Omega}\right) A$$
$$= \left(7.89 \, \frac{\text{kips}}{\text{in}^2}\right)(14.1 \text{ in}^2)$$
$$= 111.2 \text{ kips}$$

From *AISC Manual* Table 3-10, for the unbraced length of 8 ft,

$$M_c = 122 \text{ ft-kips}$$

Check combined axial compression plus bending.

$$\frac{P_r}{P_c} = \frac{50 \text{ kips}}{111.2 \text{ kips}}$$
$$= 0.45 > 0.2$$

Use *AISC Specification* Eq. H1-1a.

$$\frac{P_r}{P_c} + \left(\frac{8}{9}\right)\left(\frac{M_r}{M_c}\right) \leq 1.0$$
$$= 0.45 + \left(\frac{8}{9}\right)\left(\frac{18 \text{ ft-kips}}{122 \text{ ft-kips}}\right) \leq 1.0$$
$$0.58 \leq 1.0 \quad [\text{OK}]$$

The W8 × 48 is OK for segment AB. Next, check segment BC. For segment BC,

$$P_r = 66 \text{ kips}$$
$$M_r = M_2 = 21.56 \text{ ft-kips}$$
$$KL = 24 \text{ ft}$$
$$P_c = 111.2 \text{ kips}$$

The laterally unsupported length of 24 ft is three times larger than for AB. From *AISC Manual* Table 3-10, with a conservative value of $C_b = 1.0$,

$$M_c = 93 \text{ ft-kips}$$

Check combined axial compression and bending.

$$\frac{P_r}{P_c} = \frac{66 \text{ kips}}{111.2 \text{ kips}}$$
$$= 0.59 > 0.2$$

Use *AISC Specification* Eq. H1-1a.

$$\frac{P_r}{P_c} + \left(\frac{8}{9}\right)\left(\frac{M_r}{M_c}\right) \leq 1.0$$
$$= 0.59 + \left(\frac{8}{9}\right)\left(\frac{21.56 \text{ ft-kips}}{93 \text{ ft-kips}}\right) \leq 1.0$$
$$0.80 \leq 1.0 \quad [\text{OK}]$$

The W8 × 48 is OK. The relatively low value indicates that a lighter section might also work. Since the controlling segment, BC, is primarily limited by axial compression, try a W8 × 40.

$$A = 11.7 \text{ in}^2$$
$$r_x = 3.53 \text{ in}$$
$$r_y = 2.04 \text{ in} \quad [\text{controls}]$$

For segment BC,

$$P_r = 50 \text{ kips}$$
$$M_r = 18 \text{ ft-kips}$$
$$\frac{KL}{r_y} = \frac{(24 \text{ ft})\left(12 \frac{\text{in}}{\text{ft}}\right)}{2.04 \text{ in}}$$
$$= 141$$

From *AISC Manual* Table 4-22,

$$\frac{F_{cr}}{\Omega} = 7.56 \text{ kips/in}^2$$
$$P_c = \frac{P_n}{\Omega} = \left(\frac{F_{cr}}{\Omega}\right)A$$
$$= \left(7.56 \frac{\text{kips}}{\text{in}^2}\right)(11.7 \text{ in}^2)$$
$$= 88.5 \text{ kips}$$

Check combined axial compression plus bending.

$$\frac{P_r}{P_c} = \frac{66 \text{ kips}}{88.5 \text{ kips}}$$
$$= 0.75 > 0.2$$

Use *AISC Specification* Eq. H1-1a.

$$\frac{P_r}{P_c} + \left(\frac{8}{9}\right)\left(\frac{M_r}{M_c}\right) \leq 1.0$$
$$= 0.75 + \left(\frac{8}{9}\right)\left(\frac{21.56 \text{ ft-kips}}{71.5 \text{ ft-kips}}\right) \leq 1.0$$
$$1.02 > 1.0$$

Although there is a slight overstress, the *AISC Manual* tables use a conservative value of $C_b = 1.0$. For this case (i.e., a reversal of moment at midspan), from *AISC Manual* Table 3-1, $C_b = 1.32$.

So,

$$M_c = C_b(71.5 \text{ ft-kips})$$
$$= (1.32)(71.5 \text{ ft-kips})$$
$$= 94.4 \text{ ft-kips}$$

Use *AISC Specification* Eq. H1-1a.

$$\frac{P_r}{P_c} + \left(\frac{8}{9}\right)\left(\frac{M_r}{M_c}\right) \leq 1.0$$
$$= 0.75 + \left(\frac{8}{9}\right)\left(\frac{21.56 \text{ ft-kips}}{94.4 \text{ ft-kips}}\right) \leq 1.0$$
$$0.95 < 1.0 \quad [\text{OK}]$$

Use a W8 × 40.

11 Timber

PROBLEM 1

Lumber is used to form and shore up a cast-in-place, monolithic concrete slab during curing. The concrete has a specific weight of 150 lbf/ft^3, and the form lumber must also support a construction live load of 50 lbf/ft^2. The forms will be removed after 7 days.

The 6 in thick slab is poured on a layer of 1 in × 6 in (nominal) lumber (shown end-on in *Illustration for Prob. 1*). The lumber is, in turn, supported by joists. The joists sit on the ledger area of soffit beams with a minimal actual width of 15$^{1}/_{2}$ in each. 4 in × 8 in (nominal) posts support the entire assembly. The 9 in × 12 in concrete beams are precast, and no formwork is required for them.

To prevent cross-grain bending of the soffit beam, 4 in × 4 in headers are placed on top of the posts to extend the post support to the full width of the soffit beam.

The concrete compressive strength is 3000 lbf/in^2. The reinforcing steel has a yield strength of 40,000 lbf/in^2. The following characteristics can be used for all lumber.

$$E = 1.6 \times 10^6 \text{ lbf/in}^2$$

$$F_b = 1200 \text{ lbf/in}^2$$

$$F_v = 125 \text{ lbf/in}^2$$

$$F_c = 1200 \text{ lbf/in}^2$$

$$F_{c\perp} = 400 \text{ lbf/in}^2$$

$$\Delta_{\text{allowable}} = l/360$$

Illustration for Prob. 1

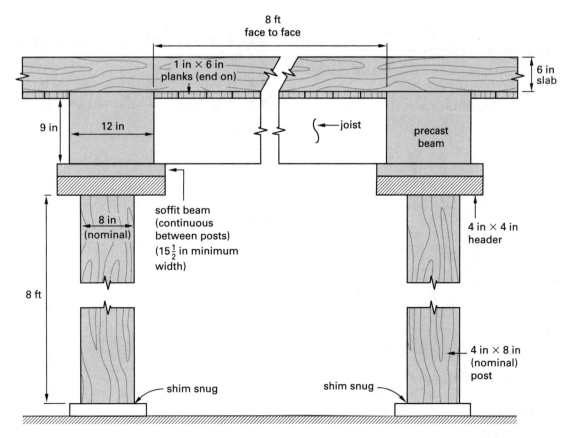

not to scale

1.1. Find the maximum spacing between joists as limited by the 1 in × 6 in lumber.

1.2. Given a 2 in (nominal) joist material and joist spacing of 2 ft 6 in, find the required joist size. (Joists can be doubled if additional joist capacity is required.)

1.3. What is the maximum post spacing?

1.4. If the posts are spaced every 5 ft and joist spacing is 2$\frac{1}{2}$ ft, what size soffit beam is required?

PROBLEM 2

A retaining wall is constructed by pouring type 1 normal-weight concrete between temporary wood forms. The concrete is at 70°F, has a setting time of 1.5 hr, and is placed at the rate of 2 ft (vertical) per hour. No admixtures are used. The concrete is vibrated internally to ensure all voids are filled. The wall sheathing is constructed of 1 in × 6 in (nominal) boards, wales, and studs. $\frac{1}{2}$ in diameter threaded-steel (yield strength = 36 kips/in^2) tie rods (tie backs) are used to connect the wales between the two walls. Wales consist of double 2 in × 6 in boards. The wales are spaced 2 ft 6 in apart vertically. The following characteristics can be used for all lumber.

$$E = 1.6 \times 10^6 \text{ lbf/in}^2$$
$$F_b = 1700 \text{ lbf/in}^2$$
$$F_v = 100 \text{ lbf/in}^2$$
$$F_c = 1200 \text{ lbf/in}^2$$

2.1. Determine the spacing of the studs.

2.2. Determine the size of the studs.

2.3. If the studs are spaced every 2 ft, determine if one tie rod between the studs is sufficient.

2.4. Determine the size of the wales.

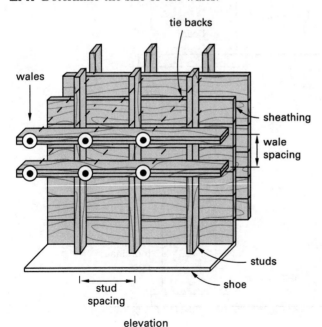

elevation

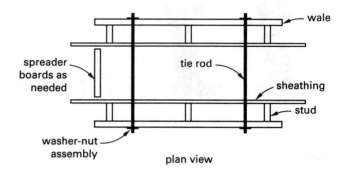

plan view

PROBLEM 3

A 6$\frac{3}{4}$ in × 24 in southern pine glulam beam 25 ft in length is uniformly loaded with 1000 lbf/ft. Its radius of curvature is 100 ft. The beam is used indoors.

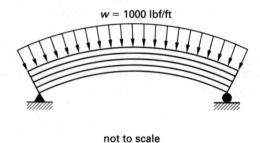

not to scale

3.1. What is the difference between a 22F-V3 and a 22F-E3 beam?

3.2. What is the volume factor for the beam described?

3.3. What is the curvature factor for the beam?

3.4. What is the maximum shear stress in the beam?

3.5. What is the maximum bending stress in the beam?

3.6. Given that the modulus of elasticity, E, is 1.6×10^6 lbf/in^2, what is the short-term deflection?

3.7. What moisture content would you expect the finished glulam beam to have?

PROBLEM 4

Design a 25 ft high solid square timber member to serve as a signpost that carries the sign and wind loads indicated. The column is set in a concrete pier at its base but is free to rotate and translate at the top. Assume the southern pine timber is dry and has been tested to determine the following properties.

$$E = 1.8 \times 10^6 \text{ lbf/in}^2$$
$$F_b = 1600 \text{ lbf/in}^2$$
$$F_v = 95 \text{ lbf/in}^2$$
$$F_c = 1200 \text{ lbf/in}^2$$
$$F_{c\perp} = 600 \text{ lbf/in}^2$$

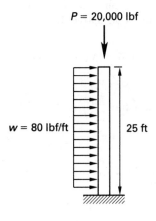

P = 20,000 lbf

w = 80 lbf/ft   25 ft

SOLUTION 1

1.1. The uniform load acting on the boards is

$$\text{concrete slab weight} \left(150 \ \frac{\text{lbf}}{\text{ft}^3}\right)\left(\frac{6 \text{ in}}{12 \ \frac{\text{in}}{\text{ft}}}\right)$$

	= 75 lbf/ft²
construction live load	50 lbf/ft²
estimated weight of boards	4 lbf/ft²
total	129 lbf/ft²

The load per foot on a single board is

$$q = \left(\frac{5.5 \text{ in}}{12 \ \frac{\text{in}}{\text{ft}}}\right)\left(129 \ \frac{\text{lbf}}{\text{ft}^2}\right) = 59.1 \text{ lbf/ft}$$

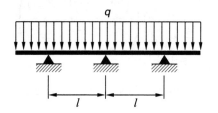

q

Assuming planking is continuous, the approximate maximum moment per board is

$$M = 0.10ql^2 = (0.10)\left(59.1 \ \frac{\text{lbf}}{\text{ft}}\right)l^2$$
$$= 5.91l^2 \text{ ft-lbf}$$

(This equation is specified in ACI 318 Sec. 8.3.3 as $w_u l_n^2/10$.)

The section modulus of 1 in × 6 in lumber is

$$S = \tfrac{1}{6}bh^2 = \left(\tfrac{1}{6}\right)(5.5 \text{ in})(0.75 \text{ in})^2$$
$$= 0.52 \text{ in}^3$$
$$F_b' = F_b C_M C_D$$

$C_M = 0.85$ for wet use   [*NDS Supplement* Table 4A]

$C_D = 1.25$ for construction loads

$$F_b' = \left(1200 \ \frac{\text{lbf}}{\text{in}^2}\right)(0.85)(1.25)$$
$$= 1275 \text{ lbf/in}^2$$

The adjusted moment capacity in one board is

$$M' = F_b'S = \frac{\left(1275 \ \frac{lbf}{in^2}\right)(0.52 \ in^3)}{12 \ \frac{in}{ft}}$$

$$= 55.3 \ \text{ft-lbf}$$

$$M' = M_{max}$$

$$55.3 \ \text{ft-lbf} = 5.91l^2 \ \text{ft-lbf}$$

$$l = \sqrt{\frac{55.3 \ \text{ft-lbf}}{5.91 \ \frac{lbf}{ft}}}$$

$$= 3.06 \ \text{ft based on stress}$$

Check the shear using ACI 318 Sec. 8.3.3. For the first interior support,

$$V = 1.15 \frac{ql}{2}$$

$$= (1.15) \left(\frac{\left(59.1 \ \frac{lbf}{ft}\right)(3.06 \ ft)}{2} \right)$$

$$= 104 \ \text{lbf}$$

Check the maximum shear stress (NDS Eq. 3.4-2).

$$f_v = \frac{3V}{2A}$$

$$= \frac{(3)(104 \ lbf)}{(2)(5.5 \ in)(0.75 \ in)}$$

$$= 37.8 \ \text{lbf/in}^2$$

Check.

$$C_D = 1.25 \quad [\text{NDS App. B}]$$

$$C_M = 0.97 \quad [NDS \ Supplement \ \text{Table 4B}]$$

$$F_v' = F_v C_D C_M$$

$$= \left(125 \ \frac{lbf}{in^2}\right)(1.25)(0.97)$$

$$= 151.56 \ \text{lbf/in}^2 \quad [f_v < F_v', \text{so OK}]$$

Check the bearing. The reaction over the first interior joist is the largest.

$$R = V + \frac{ql}{2}$$

$$= 104 \ \text{lbf} + \frac{\left(59.1 \ \frac{lbf}{ft}\right)(3.06 \ ft)}{2}$$

$$= 194.4 \ \text{lbf}$$

Assuming the nominal joist width is 2 in, the bearing area is

$$A_b = (5.5 \ in)(1.5 \ in)$$

$$= 8.25 \ in^2$$

The bearing stress is

$$f = \frac{R}{A_b}$$

$$= \frac{194.4 \ lbf}{8.25 \ in^2}$$

$$= 23.6 \ \text{lbf/in}^2$$

Check. C_D is not applicable to compression perpendicular to the grain [NDS Table 4.3.1].

$$C_M = 0.67 \quad [NDS \ Supplement \ 4B]$$

$$F_{c\perp}' = F_{c\perp} C_M$$

$$= \left(400 \ \frac{lbf}{in^2}\right)(0.67)$$

$$= 268 \ \text{lbf/in}^2 \quad [f < F_{c\perp}', \text{so OK}]$$

Check the deflection. For the assumed continuous beam with uniform loading,

$$\Delta = \frac{ql^4}{145EI}$$

Equate the deflection to the allowable value of $l/360$.

$$\Delta = \frac{ql^4}{145EI} = \frac{l}{360}$$

$$I = \frac{bh^3}{12} = \frac{(5.5 \ in)(0.75 \ in)^3}{12} = 0.19 \ in^4$$

$$l = \sqrt[3]{\frac{145EI}{360q}}$$

$$= \sqrt[3]{\frac{(145)\left(1 \times 10^6 \ \frac{lbf}{in^2}\right)(0.19 \ in^4)\left(12 \ \frac{in}{ft}\right)}{(360)\left(59.1 \ \frac{lbf}{ft}\right)}}$$

$$= 24.95 \ \text{in}$$

Deflection controls.

> The maximum spacing between the joists as limited by the 1 in × 6 in lumber is 25 in.

1.2. Calculate the load on the joists based on 2 ft 6 in spacing. Normally, the joist dead load is disregarded. The load from the boards is

$$(1.15)\left(\frac{wl}{2}\right) + \frac{wl}{2} = (2.15)\left(\frac{wl}{2}\right)$$

$$(2.15)\left(\frac{\left(129 \ \frac{\text{lbf}}{\text{ft}^2}\right)(2.5 \ \text{ft})}{2}\right) = 346.7 \ \text{lbf/ft}$$

Find the maximum moment. (For a conservative estimate, use the total span length.)

$$M = \tfrac{1}{8}wl^2 = \left(\tfrac{1}{8}\right)\left(346.7 \ \frac{\text{lbf}}{\text{ft}}\right)(8 \ \text{ft})^2$$

$$= 2773.6 \ \text{ft-lbf}$$

The required section modulus based on a single joist is

$$S_{\text{req}} = \frac{M}{F_b'} = \frac{(2773.6 \ \text{ft-lbf})\left(12 \ \frac{\text{in}}{\text{ft}}\right)}{\left(1200 \ \frac{\text{lbf}}{\text{in}^2}\right)(1.25)}$$

$$= 22.2 \ \text{in}^3$$

For a 2 in × 8 in joist,

$$S_{\text{provided}} = 13.14 \ \text{in}^3 \quad \text{[NDS Table 1B]}$$

For a 2 in × 10 in joist,

$$S_{\text{provided}} = 21.39 \ \text{in}^3$$

Try double 2 in × 8 in joists.

$$S_{\text{provided}} = (2)(13.14 \ \text{in}^3) = 26.28 \ \text{in}^3$$

Check the shear. (Since loading is symmetric, consider half of the joist.)

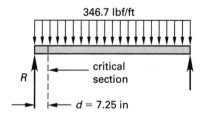

$$V = \left(346.7 \ \frac{\text{lbf}}{\text{ft}}\right)\left(4 \ \text{ft} - \frac{7.25 \ \text{in}}{12 \ \frac{\text{in}}{\text{ft}}}\right) = 1177 \ \text{lbf}$$

$$f_v = \frac{3V}{2A}$$

$$= \frac{(3)(1177 \ \text{lbf})}{(2)((2)(1.5 \ \text{in})(7.25 \ \text{in}))} \quad \text{[with two joists]}$$

$$= 81.2 \ \text{lbf/in}^2$$

Check.

$$F_v' = F_v C_D$$

$$= \left(125 \ \frac{\text{lbf}}{\text{in}^2}\right)(1.25)$$

$$= 156 \ \text{lbf/in}^2 \quad [f_v < F_v', \text{so OK}]$$

Check the bearing.

$$R = \left(346.7 \ \frac{\text{lbf}}{\text{ft}}\right)(4 \ \text{ft})$$

$$= 1386.8 \ \text{lbf}$$

The bearing area is

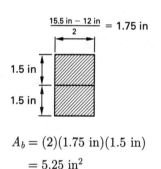

$$A_b = (2)(1.75 \ \text{in})(1.5 \ \text{in})$$

$$= 5.25 \ \text{in}^2$$

The bearing stress is

$$f = \frac{1386.8 \ \text{lbf}}{5.25 \ \text{in}^2}$$

$$= 264.2 \ \text{lbf/in}^2 \quad [< F_{c\perp} = 400 \ \text{lbf/in}^2, \text{so OK}]$$

For the 2 ft 6 in prescribed spacing, use 2 in × 8 in double joists.

1.3. First determine post capacity, then calculate spacing. Check the limit imposed by the capacity of the post as a column.

$$F_c = 1200 \ \text{lbf/in}^2$$

$$E = 1.6 \times 10^6 \ \text{lbf/in}^2$$

$$C_D = 1.25 \quad [\text{7 days, construction load}]$$

$$F_c^* = F_c C_D = \left(1200 \ \frac{\text{lbf}}{\text{in}^2}\right)(1.25)$$

$$= 1500 \ \text{lbf/in}^2$$

$$F_c' = F_c^* C_p = \left(1500 \ \frac{\text{lbf}}{\text{in}^2}\right) C_p$$

$$E' = E = 1.6 \times 10^6 \ \text{lbf/in}^2$$

For a 4 in × 8 in nominal post,

$$b = 7.25 \ \text{in}$$

$$d = 3.5 \ \text{in}$$

$$l = 8 \ \text{ft}$$

Assume pinned ends.

$$K_e = 1.0$$

$$l_e = K_e l = (1.0)(8 \ \text{ft})\left(12 \ \frac{\text{in}}{\text{ft}}\right)$$

$$= 96 \ \text{in}$$

Check the slenderness ratio.

$$\frac{l_e}{d} = \frac{96 \ \text{in}}{3.5 \ \text{in}} = 27.4 \quad [< 50, \ \text{so OK}]$$

$$c = 0.8 \quad [\text{for sawn lumber}]$$

$$F_{cE} = \frac{0.822 E'}{\left(\frac{l_e}{d}\right)^2} = \frac{(0.822)\left(1.6 \times 10^6 \ \frac{\text{lbf}}{\text{in}^2}\right)}{(27.4)^2}$$

$$= 1751.8 \ \text{lbf/in}^2$$

$$\frac{F_{cE}}{F_c^*} = \frac{1751.8 \ \frac{\text{lbf}}{\text{in}^2}}{1500 \ \frac{\text{lbf}}{\text{in}^2}} = 1.17$$

From NDS Eq. 3.7-1,

$$C_p = \frac{1 + \dfrac{F_{cE}}{F_c^*}}{2c} - \sqrt{\left(\frac{1 + \dfrac{F_{cE}}{F_c^*}}{2c}\right)^2 - \frac{\dfrac{F_{cE}}{F_c^*}}{c}}$$

$$= \frac{1 + 1.17}{(2)(0.8)} - \sqrt{\left(\frac{1 + 1.17}{(2)(0.8)}\right)^2 - \frac{1.17}{0.8}}$$

$$= 0.75$$

$$F_c' = F_c^* C_p = \left(1500 \ \frac{\text{lbf}}{\text{in}^2}\right)(0.75) = 1125 \ \text{lbf/in}^2$$

Calculate the post capacity based on column behavior.

$$P_{\text{column crushing}} = F_c' bd = \frac{\left(1125 \ \frac{\text{lbf}}{\text{in}^2}\right)(3.5 \ \text{in})(7.25 \ \text{in})}{1000 \ \frac{\text{lbf}}{\text{kip}}}$$

$$= 28.5 \ \text{kips}$$

Calculate the post capacity as limited by the bearing of the post on the 4 in × 4 in header.

$$C_b = 1.0 \quad [\text{conservatively}]$$

$$F_{c\perp}' = F_{c\perp} C_b$$

$$= \left(400 \ \frac{\text{lbf}}{\text{in}^2}\right)(1.0)$$

$$= 400 \ \text{lbf/in}^2$$

The bearing area is

$$A_b = (3.5 \ \text{in})(7.25 \ \text{in}) = 25.4 \ \text{in}^2$$

The bearing capacity is

$$P_{\text{header bearing}} = F_{c\perp} A = \frac{\left(400 \ \frac{\text{lbf}}{\text{in}^2}\right)(25.4 \ \text{in}^2)}{1000 \ \frac{\text{lbf}}{\text{kip}}}$$

$$= 10.16 \ \text{kips}$$

Post capacity is limited by bearing to 10.16 kips.

Calculate post spacing, l_p. The reaction on the post from the precast beam self-weight is

$$R = \left(\frac{(9 \ \text{in} + 6 \ \text{in})(12 \ \text{in})}{\left(12 \ \frac{\text{in}}{\text{ft}}\right)^2}\right)\left(150 \ \frac{\text{lbf}}{\text{ft}^3}\right) l_p$$

$$= (187.5 \ \text{lbf/ft}) l_p$$

Ignoring the self-weight of the joists, the total load on one post is

$$P_{\text{tot}} = w_{\text{slab}} + w_{\text{beam}}$$

$$= \left(129 \ \frac{\text{lbf}}{\text{ft}^2}\right)(8 \ \text{ft}) l_p + \left(187.5 \ \frac{\text{lbf}}{\text{ft}}\right) l_p$$

$$\approx (1220 \ \text{lbf/ft}) l_p$$

$$P_{\text{header bearing}} = P_{\text{tot}}$$

$$10{,}160 \ \text{lbf} = \left(1220 \ \frac{\text{lbf}}{\text{ft}}\right) l_p$$

$$l_p = \boxed{8.3 \ \text{ft}}$$

1.4. Calculate loads on soffit beams.

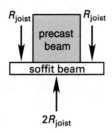

Disregarding joist self-weight,

$$R_{joist} = \left(129 \; \frac{lbf}{ft^2}\right)(2.5 \; ft)(4 \; ft) = 1290 \; lbf$$

Assuming the joists are placed over the posts and the soffit beams are fixed at supports,

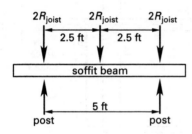

$$M = \tfrac{1}{8}Pl \quad [AISC \; Manual \; Table \; 3\text{-}23]$$
$$= \left(\tfrac{1}{8}\right)2R_{joist} \, l_{span}$$
$$= \left(\tfrac{1}{8}\right)(2)(1290 \; lbf)(5 \; ft)$$
$$= 1612.5 \; ft\text{-}lbf$$

$$S_{required} = \frac{M}{F_b'} = \frac{(1612.5 \; ft\text{-}lbf)\left(12 \; \frac{in}{ft}\right)}{(1.25)\left(1200 \; \frac{lbf}{in^2}\right)}$$
$$= 12.9 \; in^3$$

Assuming the minimum width of 15.5 in,

$$S = 12.9 \; in^3 = \frac{bt^2}{6} = \frac{(15.5 \; in)t^2}{6}$$
$$t = 2.23 \; in$$

Try a 15.5 in × 2.5 in (actual dimensions) soffit beam. Check deflection.

$$I = \frac{bt^3}{12} = \frac{(15.5 \; in)(2.5 \; in)^3}{12} = 20.2 \; in^4$$

$$\Delta = \frac{Pl^3}{192EI}$$

$$= \frac{(2)(1290 \; lbf)(5 \; ft)^3\left(12 \; \frac{in}{ft}\right)^3}{(192)\left(1.6 \times 10^6 \; \frac{lbf}{in^2}\right)(20.2 \; in^4)}$$

$$= 0.09 \; in$$

$$\Delta_{allowable} = \frac{l}{360} = \frac{(5 \; ft)\left(12 \; \frac{in}{ft}\right)}{360} = 0.17 \; in$$

$$\Delta_{actual} < \Delta_{allowable} \quad [OK]$$

$$\boxed{\text{Use a 15.5 in} \times \text{2.5 in soffit beam.}}$$

SOLUTION 2

2.1. Calculate the maximum pressure exerted by the concrete. Use the ACI recommended equation. (This equation is not dimensionally consistent.) (Reference: ACI 347.)

For regular (type I) concrete without admixtures with 4 in slump or less, ordinary workmanship, and internal vibration, and used in a wall,

$$R = 2 \; ft/hr$$
$$T = 70°F$$

unit weight coefficient, $C_w = 1.0$

chemistry coefficient, $C_c = 1.0$

$$p = (1.0)(1.0)\left(150 \; \frac{lbf}{ft^2} + 9000\left(\frac{R}{T}\right)\right) \quad [R \leq 7 \; ft/hr]$$

$$= 150 \; \frac{lbf}{ft^2} + (9000)\left(\frac{2 \; \frac{ft}{hr}}{70F}\right)$$

$$= 407.1 \; lbf/ft^2$$

There is no need to account for variations of the pressure with height.

Assume flexure in sheathing controls the stud spacing. Find the load on one board.

$$w_L = \left(407.1 \; \frac{lbf}{ft^2}\right)\left(\frac{5.5 \; in}{12 \; \frac{in}{ft}}\right) = 186.6 \; lbf/ft$$

From *AISC Manual* Table 3-23, the maximum moment for the continuous span is

$$M_{max} = \tfrac{1}{10}w_L s^2$$
$$= \left(\tfrac{1}{10}\right)\left(186.6 \; \frac{lbf}{ft}\right)s^2$$
$$= 18.7s^2$$

The section modulus of the 1 in × 6 in (nominal) sheathing is

$$S = 0.516 \text{ in}^3 \quad \text{[NDS Table 1B]}$$

The sheathing is used wet; therefore, $C_M = 0.85$. The load duration is short term during construction; therefore, $C_D = 1.25$.

Calculate the adjusted moment capacity.

$$M' = F_b C_M C_D S$$

$$= \frac{\left(1700 \dfrac{\text{lbf}}{\text{in}^2}\right)(0.85)(1.25)(0.516 \text{ in}^3)}{12 \dfrac{\text{in}}{\text{ft}}}$$

$$= 77.6 \text{ ft-lbf}$$

Set the maximum moment equal to the moment capacity.

$$M_{\text{max}} = 18.7s^2 = 77.6 \text{ ft-lbf} = M'$$

$$s^2 = 4.15$$

$$s = 2.03 \text{ ft} \quad \text{[use 2 ft]}$$

Check the shear.

$$V = \frac{1.15 w_L s}{2} = \frac{(1.15)\left(186.6 \dfrac{\text{lbf}}{\text{ft}}\right)(2 \text{ ft})}{2}$$

$$= 214.6 \text{ lbf}$$

$$f_v = \frac{3V}{2A} = \frac{(3)(214.6 \text{ lbf})}{(2)(0.75 \text{ in})(5.5 \text{ in})}$$

$$= 78.0 \text{ lbf/in}^2$$

For shear, the wet-use factor, C_M, is 0.97 (NDS Table 4A). The adjusted shear design value is

$$F_v' = F_v C_M C_D = \left(100 \dfrac{\text{lbf}}{\text{in}^2}\right)(0.97)(1.25)$$

$$= 121.3 \text{ lbf/in}^2 \quad [f_v < F_v', \text{ so OK}]$$

$$\boxed{\text{Use studs every 2 ft.}}$$

2.2. The studs act as continuous beams supported on the wales. Studs are dry and are used edgewise.

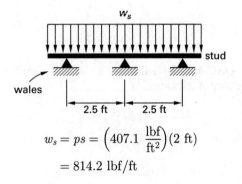

$$w_s = ps = \left(407.1 \dfrac{\text{lbf}}{\text{ft}^2}\right)(2 \text{ ft})$$

$$= 814.2 \text{ lbf/ft}$$

The maximum moment is

$$M = \tfrac{1}{10} w_s s_{\text{wale}}^2$$

$$= \left(\tfrac{1}{10}\right)\left(814.2 \dfrac{\text{lbf}}{\text{ft}}\right)(2.5 \text{ ft})^2$$

$$= 508.9 \text{ ft-lbf}$$

$$S_{\text{required}} = \frac{M}{F_b'} = \frac{(508.9 \text{ ft-lbf})\left(12 \dfrac{\text{in}}{\text{ft}}\right)}{(1.25)\left(1700 \dfrac{\text{lbf}}{\text{in}^2}\right)}$$

$$= 2.87 \text{ in}^3$$

A 2 in × 4 in stud has a section modulus, S, of 3.06 in³ (NDS Table 1B). Try 2 in × 4 in studs. Check the shear.

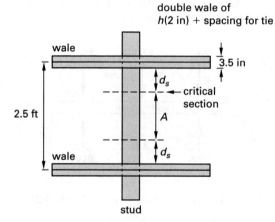

The critical section occurs at a distance of $d_s + 3.5 \text{ in}/2$ from the center of the wale.

$$V_{\text{max}} = w_s\left(\frac{s_{\text{wale}}}{2}\right) = \left(814.2 \dfrac{\text{lbf}}{\text{ft}}\right)\left(\frac{2.5 \text{ ft}}{2}\right)$$

$$= 1017.8 \text{ lbf}$$

Shear at the critical section is

$$V_{d_s} = V_{\text{max}} - w_s\left(d_s + \frac{3.5 \text{ in}}{2}\right)\left(\frac{1}{12 \dfrac{\text{in}}{\text{ft}}}\right)$$

$$= 1017.8 \text{ lbf} - \left(814.2 \dfrac{\text{lbf}}{\text{ft}}\right)\left(3.5 \text{ in} + \frac{3.5 \text{ in}}{2}\right)$$

$$\times \left(\frac{1}{12 \dfrac{\text{in}}{\text{ft}}}\right)$$

$$= 661.6 \text{ lbf}$$

$$f_v = \frac{3V_{d_s}}{2A} = \frac{(3)(661.6 \text{ lbf})}{(2)(1.5 \text{ in})(3.5 \text{ in})} = 189.0 \text{ lbf/in}^2$$

Increase F_v by 1.25 for load duration.

$$F_v' = 1.25F_v = (1.25)\left(100\ \frac{\text{lbf}}{\text{in}^2}\right)$$
$$= 125\ \text{lbf/in}^2 \quad [\text{no good}]$$

Try 2 in × 6 in studs. Conservatively assume the same shear as for the 2 in × 4 in studs.

$$f_v = \frac{3V_{d_s}}{2A} = \frac{(3)(661.6\ \text{lbf})}{(2)(1.5\ \text{in})(5.5\ \text{in})}$$
$$= 120.3\ \text{lbf/in}^2 \quad [< F_v' = 125\ \text{lbf/in}^2,\ \text{so OK}]$$

$$\boxed{\text{Use 2 in × 6 in studs.}}$$

2.3. The available strength in the tie is

$$\text{available strength} = \frac{F_y A_\text{tie}}{1.67}$$
$$= \frac{\left(36\ \frac{\text{kips}}{\text{in}^2}\right)\pi\left(\frac{0.5\ \text{in}}{2}\right)^2\left(1000\ \frac{\text{lbf}}{\text{kip}}\right)}{1.67}$$
$$= 4233\ \text{lbf}$$

The load per tie is

$$P_\text{tie} = ps_\text{stud}s_\text{wale}$$
$$= \left(407.1\ \frac{\text{lbf}}{\text{ft}^2}\right)(2\ \text{ft})(2.5\ \text{ft})$$
$$= 2035.5\ \text{lbf} \quad [< 4233\ \text{lbf tie capacity, so OK}]$$

$$\boxed{\text{One tie between studs is sufficient.}}$$

2.4. Wales are likely to be controlled by shear.

$$V = \frac{P_\text{tie}}{2} = \frac{2035.5\ \text{lbf}}{2} = 1017.8\ \text{lbf}$$

Assuming a 2 in × 6 in double wale,

$$V_\text{wale} = \frac{V}{2} = \frac{1017.8\ \text{lbf}}{2} = 508.9\ \text{lbf}$$
$$f_v = \frac{3V}{2A} = \frac{(3)(508.9\ \text{lbf})}{(2)(1.5\ \text{in})(5.5\ \text{in})}$$
$$= 92.5\ \text{lbf/in}^2 \quad [< F_v' = 125\ \text{lbf/in}^2,\ \text{so OK}]$$

$$\boxed{\text{Use double 2 in × 6 in wales.}}$$

SOLUTION 3

3.1. A 22F-V3 beam is a visually graded member, while a 22F-E3 beam is a mechanically graded member. Typically, mechanically graded members have higher allowable stresses for some properties.

3.2. Find the volume factor. (This equation is not dimensionally consistent.)

$$C_V = \left(\frac{21}{L}\right)^{1/x}\left(\frac{12}{d}\right)^{1/x}\left(\frac{5.125}{b}\right)^{1/x} \le 1.0 \quad [\text{NDS Eq. 5.3-1}]$$
$$L = 25\ \text{ft}$$
$$b = 6.75\ \text{in}$$
$$d = 23.375\ \text{in}$$
$$x = 20\ \text{for southern pine}$$

$$C_V = (1.0)\left(\frac{21}{25\ \text{ft}}\right)^{1/20}\left(\frac{12}{23.375\ \text{in}}\right)^{1/20}\left(\frac{5.125}{6.75\ \text{in}}\right)^{1/20} \le 1.0$$
$$= \boxed{0.946}$$

3.3. The curvature factor is

$$C_C = 1 - 2000\left(\frac{t}{R}\right)^2 \quad [\text{NDS Eq. 5.3-2}]$$
$$t = 1.5\ \text{in}$$
$$R = (100\ \text{ft})\left(12\ \frac{\text{in}}{\text{ft}}\right) = 1200\ \text{in}$$
$$\frac{t}{R} = \frac{1.5\ \text{in}}{1200\ \text{in}} = 0.00125 < 0.01 \quad [\text{OK}]$$
$$C_C = 1 - (2000)(0.00125)^2$$
$$= \boxed{0.997}$$

3.4. The maximum shear stress, f_v, is $3V/2bd$.

$$w = \text{applied load} + \text{self-weight}$$
$$= 1000\ \frac{\text{lbf}}{\text{ft}}$$
$$+ \left(35\ \frac{\text{lbf}}{\text{ft}^3}\right)\left(\frac{(6.75\ \text{in})(23.375\ \text{in})}{\left(12\ \frac{\text{in}}{\text{ft}}\right)^2}\right)$$
$$= 1038.35\ \text{lbf/ft} \quad (1040\ \text{lbf/ft})$$
$$V = w\left(\frac{L}{2}\right)$$
$$f_v = \frac{(3)\left(1040\ \frac{\text{lbf}}{\text{ft}}\right)\left(\frac{25\ \text{ft}}{2}\right)}{(2)(6.75\ \text{in})(23.375\ \text{in})}$$
$$= \boxed{124\ \text{lbf/in}^2}$$

3.5. The maximum bending stress in the beam, simply supported as shown, is

$$f_b = \frac{M}{S}$$

$$S = 614.7 \text{ in}^3 \quad \text{[NDS Table 1D]}$$

$$M = \frac{wl^2}{8} = \frac{\left(1040 \; \frac{\text{lbf}}{\text{ft}}\right)(25 \text{ ft})^2}{8} = 81{,}250 \text{ ft-lbf}$$

$$f_b = \frac{M}{S} = \frac{(81{,}250 \text{ ft-lbf})\left(12 \; \frac{\text{in}}{\text{ft}}\right)}{614.7 \text{ in}^3}$$

$$= \boxed{1586 \text{ lbf/in}^2}$$

3.6. Deflection can be calculated using the following values.

$$E = 1.6 \times 10^6 \text{ lbf/in}^2$$

$$I = 7184 \text{ in}^4 \quad \text{[NDS Table 1D]}$$

From *AISC Manual* Table 3-23,

$$\Delta = \left(\frac{5}{384}\right)\left(\frac{wl^4}{EI}\right)$$

$$= \left(\frac{5}{384}\right)\left(\frac{\left(1040 \; \frac{\text{lbf}}{\text{ft}}\right)(25 \text{ ft})^4\left(12 \; \frac{\text{in}}{\text{ft}}\right)^3}{\left(1.6 \times 10^6 \; \frac{\text{lbf}}{\text{in}^2}\right)(7184 \text{ in}^4)}\right)$$

$$= \boxed{0.80 \text{ in}}$$

3.7. Moisture content depends on ambient conditions and, for dry use, must be less than 16%. Otherwise, the wet service factor, C_M, must be used.

SOLUTION 4

The member is a beam-column. Calculate the effective length for slenderness. For a fixed-free column, NDS App. G specifies $K_e = 2.1$.

$$l_e = K_e l = (2.1)(25 \text{ ft})\left(12 \; \frac{\text{in}}{\text{ft}}\right) = 630 \text{ in}$$

The column slenderness ratio should be less than 50.

$$\frac{l_e}{d} = \frac{630 \text{ in}}{d} < 50$$

$$d \geq 12.6 \text{ in}$$

Try a 14 in × 14 in beam-column. (A larger size could also be selected.)

$$b = 13.5 \text{ in}$$

$$d = 13.5 \text{ in}$$

$$S = 410.1 \text{ in}^3 \quad \text{[NDS Table 1B]}$$

Calculate the adjustment factors.

$$C_D = 1.6 \text{ for wind load}$$

$$C_D = 1.0 \text{ for long-term load}$$

$$C_M = 1.0 \text{ for dry conditions}$$

$$C_F = \left(\frac{12}{d}\right)^{1/9} = \left(\frac{12}{13.5 \text{ in}}\right)^{1/9} = 0.987 \text{ for bending, } F_b$$

$$C_F = 1.0 \text{ for compression, } F_C \quad \text{[NDS Table 4D]}$$

In compression, the column stability factor, C_p, can be calculated from the following properties, given in NDS Sec. 3.7.

$$E' = \left(1.8 \times 10^6 \; \frac{\text{lbf}}{\text{in}^2}\right)(1.0) = 1.8 \times 10^6 \text{ lbf/in}^2$$

$$c = 0.8 \text{ for sawn lumber}$$

$$F_{cE} = \frac{0.822 E'}{\left(\frac{l_e}{d}\right)^2} = \frac{(0.822)\left(1.8 \times 10^6 \; \frac{\text{lbf}}{\text{in}^2}\right)}{\left(\frac{630 \text{ in}}{13.5 \text{ in}}\right)^2}$$

$$= 679.4 \text{ lbf/in}^2$$

$$F_c = 1200 \text{ lbf/in}^2$$

$$F_c^* = 1.0 F_c = 1200 \text{ lbf/in}^2 \text{ for axial sign load}$$

$$\frac{F_{cE}}{F_c^*} = \frac{679.4 \; \frac{\text{lbf}}{\text{in}^2}}{1200 \; \frac{\text{lbf}}{\text{in}^2}}$$

$$= 0.566$$

From NDS Eq. 3.7-1,

$$C_p = \frac{1 + \frac{F_{cE}}{F_c^*}}{2c} - \sqrt{\left(\frac{1 + \frac{F_{cE}}{F_c^*}}{2c}\right)^2 - \frac{\frac{F_{cE}}{F_c^*}}{c}}$$

$$= \frac{1 + 0.566}{(2)(0.8)} - \sqrt{\left(\frac{1 + 0.566}{(2)(0.8)}\right)^2 - \frac{0.566}{0.8}}$$

$$= 0.479$$

In bending, the beam stability factor, C_L, is 1.0 when $b = d$ (NDS Sec. 3.3.3.1).

Calculate the stresses.

$$M_{\max} = \frac{wl^2}{2} = \frac{\left(80 \ \frac{\text{lbf}}{\text{ft}}\right)(25 \ \text{ft})^2}{2}$$
$$= 25{,}000 \ \text{ft-lbf}$$

$$f_b = \frac{M}{S} = \frac{(25{,}000 \ \text{ft-lbf})\left(12 \ \frac{\text{in}}{\text{ft}}\right)}{410.1 \ \text{in}^3}$$
$$= 731.5 \ \text{lbf/in}^2$$

$$f_c = \frac{P}{bd} = \frac{20{,}000 \ \text{lbf}}{(13.5 \ \text{in})(13.5 \ \text{in})} = 109.7 \ \text{lbf/in}^2$$

$$f_v = \frac{3V}{2bd} = \frac{(3)\left(80 \ \frac{\text{lbf}}{\text{ft}}\right)(25 \ \text{ft})}{(2)(13.5 \ \text{in})(13.5 \ \text{in})} = 16.46 \ \text{lbf/in}^2$$

$$F_c' = F_c^* C_p = \left(1200 \ \frac{\text{lbf}}{\text{in}^2}\right)(0.479)$$
$$= 575 \ \text{lbf/in}^2$$

$$F_b^* = F_b C_D C_F = \left(1600 \ \frac{\text{lbf}}{\text{in}^2}\right)(1.6)(0.987)$$
$$= 2527 \ \text{lbf/in}^2$$

$$F_b' = F_b^* C_L = F_b^*(1.0)$$
$$= 2527 \ \text{lbf/in}^2 \quad \text{[for square members]}$$

Check the combined bending and axial compression (NDS Sec. 3.9.2).

$$\left(\frac{f_c}{F_c'}\right)^2 + \frac{f_b}{F_b'\left(1 - \dfrac{f_c}{F_{cE}}\right)} \leq 1.0 \quad \text{[NDS Eq. 3.9-3]}$$

$$\left(\frac{109.7 \ \frac{\text{lbf}}{\text{in}^2}}{575 \ \frac{\text{lbf}}{\text{in}^2}}\right)^2 + \frac{731.5 \ \frac{\text{lbf}}{\text{in}^2}}{\left(2527 \ \frac{\text{lbf}}{\text{in}^2}\right)\left(1 - \dfrac{109.7 \ \frac{\text{lbf}}{\text{in}^2}}{679.4 \ \frac{\text{lbf}}{\text{in}^2}}\right)}$$

$$= 0.38 \quad [< 1.0, \text{ so OK}]$$

Check shear.

$$f_v = 16.46 \ \text{lbf/in}^2$$

$$F_v' = F_v C_D$$
$$= \left(95 \ \frac{\text{lbf}}{\text{in}^2}\right)(1.6)$$
$$= 152 \ \text{lbf/in}^2 \quad [f_v < F_v', \text{ so OK}]$$

Use a 14 in × 14 in column.

12 Masonry[1]

PROBLEM 1

A short masonry column has a height of 10 ft and is constructed of concrete block with four no. 7 steel bars as shown. Adequate lateral reinforcing ties are provided. The column is subjected to a live load moment of 100 in-kips. Sidesway is permitted. What is the maximum axial load the column can support in addition to the applied moment? Solve using ACI 530.

| masonry compressive strength | 1500 lbf/in^2 |
| steel yield stress | 60,000 lbf/in^2 |

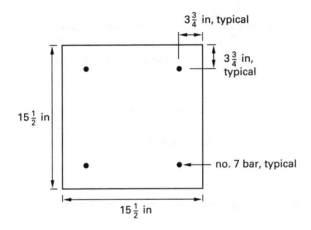

3$\frac{3}{4}$ in, typical

3$\frac{3}{4}$ in, typical

15$\frac{1}{2}$ in

no. 7 bar, typical

15$\frac{1}{2}$ in

PROBLEM 2

A square column has a nominal 12 in outside dimension. It is constructed of concrete block, mortar, and four no. 8 bars. Solve using ACI 530.

2.1. What is the maximum axial concentric load this column can support if the effective height is 20 ft?

2.2. What is the maximum eccentricity that would be permitted in this column if the axial load is 40 kips?

masonry compressive strength	1500 lbf/in^2
steel yield stress	60,000 lbf/in^2
effective height	20 ft

PROBLEM 3

A masonry wall is eccentrically loaded by a steel beam as shown. Design the support and connection between the steel beam and wall. Solve using ACI 530.

$$f'_m = 2000 \text{ lbf/in}^2$$

Use 1 in diameter anchor bolts and $f_y = 60,000$ lbf/in^2.

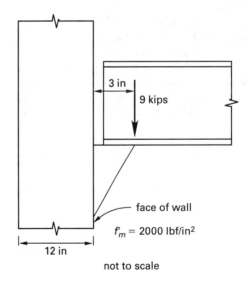

3 in

9 kips

face of wall

$f'_m = 2000$ lbf/in^2

12 in

not to scale

---

[1]Strength design solutions are generally not permitted by NCEES for masonry problems, but they are presented in this chapter for your reference only.

PROBLEM 4

A two-story (20 ft high) masonry wall is constructed of 8 in concrete masonry blocks with solid grouting. Reinforcing steel consists of no. 5 bars spaced every 36 in. The wall is supported laterally by steel beams at the bottom, top, and midpoint as illustrated. Is the wall adequate with a 35 lbf/ft^2 wind load? Solve using ACI 530. State your assumptions.

masonry compressive strength	1500 lbf/in^2
steel yield stress	40,000 lbf/in^2

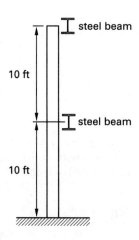

PROBLEM 5

A 24 in square brick column 32 ft high is subjected to a uniformly distributed lateral load in addition to an axial force. The service moment from the uniform load is 45 ft-kips, and the axial force (unfactored) is 65 kips. The column is supported laterally only at its top and bottom, and the ends are free to rotate in the plane of bending. Reinforcing steel consists of four vertical no. 8 bars.

brick compressive strength	4200 lbf/in^2

What is the moment of inertia of the cracked section?

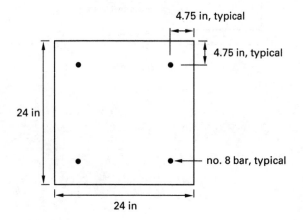

SOLUTION 1

ACI 530 (ASD) Solution

Assume lateral reinforcing ties are provided as required per ACI 530 Sec. 1.14.1.4.

$$A_b = 0.6 \text{ in}^2 \quad \text{[no. 7 bar]}$$

Calculate h/r.

$$h = (120 \text{ in})k = (120 \text{ in})(2) = 240 \text{ in}$$

$$r = \frac{15.5 \text{ in}}{\sqrt{12}} = 4.475 \text{ in}$$

$$\frac{h}{r} = \frac{240 \text{ in}}{4.475 \text{ in}} = 53.6 < 99$$

$$A_n = (15.5 \text{ in})^2 = 240.25 \text{ in}^2$$

$$s = \frac{b^3}{6} = \frac{(15.5 \text{ in})^3}{6} = 620.6 \text{ in}^3$$

$$A_s = A_s' = 2A_b = (2)(0.6 \text{ in}^2)$$

$$= 1.20 \text{ in}^2$$

$$F_s = 32,000 \text{ lbf/in}^2 \quad \text{[ACI 530 Sec. 2.3.3.1]}$$

For combined axial and flexure loading, use ACI 530 Sec. 2.3.4.2.2.

$$f_m \le 0.45 f_m'$$

Assume an uncracked section.

$$f_m = \frac{P}{A} + \frac{M}{S} = \frac{P}{240 \text{ in}^2} + \frac{(100 \text{ in-kips})\left(1000 \frac{\text{lbf}}{\text{kip}}\right)}{620.6 \text{ in}^3}$$

$$= 0.45 f_m'$$

$$\frac{P}{240 \text{ in}^2} + 161 \frac{\text{lbf}}{\text{in}^2} = (0.45)\left(1500 \frac{\text{lbf}}{\text{in}^2}\right) = 675 \text{ lbf/in}^2$$

$$P = \left(514 \frac{\text{lbf}}{\text{in}^2}\right)(240 \text{ in}^2)$$

$$= 123,360 \text{ lbf}$$

Determine whether the section is uncracked: $f_a > f_b$.

$$f_a = \frac{123,360 \text{ lbf}}{240 \text{ in}^2} = 514 \text{ lbf/in}^2$$

$$f_b = \frac{M}{S} = 161 \text{ lbf/in}^2$$

Therefore, the section is uncracked.

Check the compressive force in the reinforced masonry due to axial load only. From ACI 530 Eq. 2-21, for $h/r < 99$,

$$P_a = (0.25f'_m A_n + 0.65 A_{st} F_s)\left(1 - \left(\frac{h}{140r}\right)^2\right)$$

$$= \left((0.25)\left(1500 \ \frac{\text{lbf}}{\text{in}^2}\right)(240.25 \ \text{in}^2)\right.$$

$$\left. + (0.65)(2.40 \ \text{in}^2)\left(32,000 \ \frac{\text{lbf}}{\text{in}^2}\right)\right)$$

$$\times \left(1 - \left(\frac{53.6}{140}\right)^2\right)$$

$$= 119,490 \ \text{lbf}$$

Check ACI 530 Eq. 2-16, per ACI 530 Sec. 2.3.4.2.2.

$$F_a = 0.25f'_m\left(1 - \left(\frac{h}{140r}\right)^2\right)$$

$$= (0.25)\left(1500 \ \frac{\text{lbf}}{\text{in}^2}\right)\left(1 - \left(\frac{53.6}{140}\right)^2\right)$$

$$= 320 \ \text{lbf/in}^2$$

$$P_a = F_a A_n = \left(320 \ \frac{\text{lbf}}{\text{in}^2}\right)(240.25 \ \text{in}^2)$$

$$= 76,880 \ \text{lbf}$$

$$\boxed{\text{Therefore, } P_{\max} = 76,880 \ \text{lbf.}}$$

ACI 530 (Strength Design) Solution

Solve using a simplified approach and the unity equation.

$$\frac{P_u}{\phi P_n} + \frac{M_u}{\phi M_n} \leq 1.0$$

Use ACI 530 Eq. 3-18.

$$\phi P_n = \phi(0.80)\left((0.80)f'_m(A_n - A_{st}) + f_y A_{st}\right)$$

$$\times \left(1 - \left(\frac{h}{140r}\right)^2\right)$$

$$\frac{h}{r} = 53.6$$

$$A_{st} = 4A_b = (4)(0.6 \ \text{in}^2)$$

$$= 2.40 \ \text{in}^2$$

From ACI 530 Sec. 3.1.4.4, ϕ is 0.9 for combinations of flexure and axial load.

$$\phi P_n = (0.90)(0.80)\left((0.80)\left(1500 \ \frac{\text{lbf}}{\text{in}^2}\right)\right.$$

$$\times (240.25 \ \text{in}^2 - 2.40 \ \text{in}^2)$$

$$\left. + \left(60,000 \ \frac{\text{lbf}}{\text{in}^2}\right)(2.40 \ \text{in}^2)\right)$$

$$\times \left(1 - \left(\frac{53.6}{140}\right)^2\right)$$

$$= 263,860 \ \text{lbf}$$

$$\phi M_n = \phi T\left(d - \frac{a}{2}\right)$$

$$a = \frac{A_s f_y}{0.80 f'_m b}$$

$$= \frac{(1.20 \ \text{in}^2)\left(60,000 \ \frac{\text{lbf}}{\text{in}^2}\right)}{(0.80)\left(1500 \ \frac{\text{lbf}}{\text{in}^2}\right)(15.5 \ \text{in})}$$

$$= 3.87 \ \text{in}$$

$$T = A_s f_y = (1.20 \ \text{in}^2)\left(60,000 \ \frac{\text{lbf}}{\text{in}^2}\right)$$

$$= 72,000 \ \text{lbf}$$

$$\phi M_n = (0.9)(72,000 \ \text{lbf})\left(11.75 \ \text{in} - \frac{3.87 \ \text{in}}{2}\right)$$

$$= 636,000 \ \text{in-lbf}$$

$$\frac{P_u}{\phi P_n} + \frac{M_u}{\phi M_n} = \frac{P_u}{263,860 \ \text{lbf}} + \frac{100,000 \ \text{in-lbf}}{636,000 \ \text{in-lbf}} = 1.0$$

$$P_u = \frac{222,370 \ \text{lbf}}{1000 \ \frac{\text{lbf}}{\text{kip}}}$$

$$= \boxed{222.4 \ \text{kips}}$$

SOLUTION 2

ACI 530 (ASD) Solution

Assume the specified dimension of a 12 in nominal column is 11.5 in.

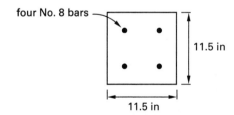

four No. 8 bars

11.5 in

11.5 in

2.1. Calculate h/r.

$$h = 20 \ \text{ft}$$

$$\frac{h}{r} = \frac{(20 \ \text{ft})\left(12 \ \frac{\text{in}}{\text{ft}}\right)}{\dfrac{11.5 \ \text{in}}{\sqrt{12}}} = 72.3 < 99$$

Structural

From ACI 530 Eq. 2-21,

$$P_a = (0.25 f'_m A_n + 0.65 A_{st} F_s)\left(1 - \left(\frac{h}{140r}\right)^2\right)$$

$A_n = (11.5 \text{ in})^2 = 132.25 \text{ in}^2$

$A_{st} = (4)(0.79 \text{ in}^2) = 3.16 \text{ in}^2$

$f'_m = 1500 \text{ lbf/in}^2$

$F_s = 32{,}000 \text{ lbf/in}^2$ [ACI 530 Sec. 2.3.3.1]

$$P_a = \frac{\left(\begin{array}{c}(0.25)\left(1500\,\frac{\text{lbf}}{\text{in}^2}\right)(132.25 \text{ in}^2) \\[4pt] + (0.65)(3.16 \text{ in}^2)\left(32{,}000\,\frac{\text{lbf}}{\text{in}^2}\right)\end{array}\right) \times \left(1 - \left(\frac{72.3}{140}\right)^2\right)}{1000\,\frac{\text{lbf}}{\text{kip}}}$$

$= 84.6 \text{ kips}$

2.2. Given $P = 40$ kips, find the maximum eccentricity using ACI 530 Sec. 2.3.4.2.2.

$$\frac{P}{P_a} + \frac{M}{M_a} \le 1.0$$

$$\frac{40 \text{ kips}}{84.6 \text{ kips}} + \frac{e(40 \text{ kips})}{M_a} = 1.0$$

$$M_a = F_b S = 0.45 f'_m \frac{bh^2}{6}$$

$$= \frac{(0.45)\left(1500\,\frac{\text{lbf}}{\text{in}^2}\right)\left(\frac{(11.5 \text{ in})^3}{6}\right)}{1000\,\frac{\text{lbf}}{\text{kip}}}$$

$$= 171.1 \text{ in-kips}$$

$$\frac{40 \text{ kips}}{84.6 \text{ kips}} + \frac{e(40 \text{ kips})}{171.1 \text{ in-kips}} = 1.0$$

$$e = \boxed{2.26 \text{ in}}$$

Check minimum eccentricity. Per ACI 530 Sec. 2.3.4.3, the minimum eccentricity of the load is 0.1 multiplied by the side length.

$$(0.1)(12 \text{ in}) = 1.2 \text{ in}$$

$$e > 1.2 \text{ in} \quad [\text{OK}]$$

ACI 530 (Strength Design) Solution

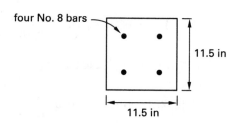

four No. 8 bars

11.5 in

11.5 in

2.1. ACI 530 Eq. 3-18 applies.

$$\frac{h}{r} = \frac{(20 \text{ ft})\left(12\,\frac{\text{in}}{\text{ft}}\right)}{\frac{11.5 \text{ in}}{\sqrt{12}}}$$

$$= 72.3 < 99$$

$$A_{st} = 4A_b = (4)(0.79 \text{ in}^2)$$

$$= 3.16 \text{ in}^2$$

$$A_n = b^2 = (11.5 \text{ in})^2$$

$$= 132.25 \text{ in}^2$$

$$\phi P_n = \phi(0.8)\left(0.8 f'_m (A_n - A_{st}) + f_y A_{st}\right)\left(1 - \left(\frac{h}{140r}\right)^2\right)$$

$$= \frac{\begin{array}{c}(0.65)(0.8) \\[4pt] \times \left(\begin{array}{c}(0.8)\left(1500\,\frac{\text{lbf}}{\text{in}^2}\right)(132.25 \text{ in}^2 - 3.16 \text{ in}^2) \\[4pt] + \left(60{,}000\,\frac{\text{lbf}}{\text{in}^2}\right)(3.16 \text{ in}^2)\end{array}\right) \\[4pt] \times \left(1 - \left(\frac{72.3}{140}\right)^2\right)\end{array}}{1000\,\frac{\text{lbf}}{\text{kip}}}$$

$$= \boxed{131 \text{ kips}}$$

2.2. The maximum eccentricity is limited by the bending moment capacity.

$$\frac{P_u}{\phi P_n} + \frac{M}{M_a} \le 1.0$$

$$\frac{40 \text{ kips}}{131 \text{ kips}} + \frac{e(40 \text{ kips})}{M_a} = 1.0$$

$$M_a = \phi M_n$$

$$M_n = A_s f_y \left(d - \frac{a}{2} \right)$$

$$a = \frac{P_u + A_s f_y}{0.85 f'_m b} = \frac{40{,}000 \text{ lbf} + (1.58 \text{ in}^2)\left(60{,}000 \, \frac{\text{lbf}}{\text{in}^2}\right)}{(0.85)\left(1500 \, \frac{\text{lbf}}{\text{in}^2}\right)(11.5 \text{ in})}$$

$$= 9.2 \text{ in}$$

$$M_n = (1.58 \text{ in}^2)\left(60{,}000 \, \frac{\text{lbf}}{\text{in}^2}\right)\left(8 \text{ in} - \frac{9.2 \text{ in}}{2}\right)$$

$$= 322{,}320 \text{ in-lbf}$$

$$M_a = \phi M_n = \frac{(0.80)(322{,}320 \text{ in-lbf})}{1000 \, \frac{\text{lbf}}{\text{kip}}}$$

$$= 257.8 \text{ in-kips}$$

$$\frac{40 \text{ kips}}{131 \text{ kips}} + \frac{e(40 \text{ kips})}{258.1 \text{ in-kips}} = 1.0$$

$$e = \boxed{4.48 \text{ in}}$$

SOLUTION 3

ACI 530 (ASD) Solution

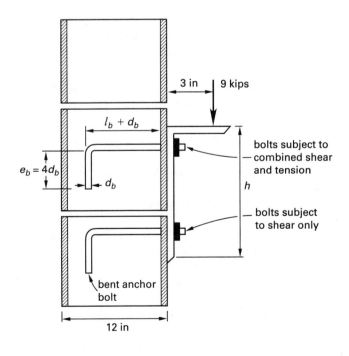

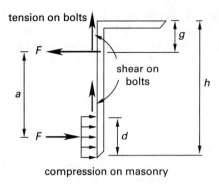

Estimate the number of bolts based on shear. From ACI 530 Sec. 2.1.4.3.2, the shear capacity, B_v, is the smallest of B_{vb}, B_{vc}, B_{vpry}, and B_{vs}.

For a 1 in diameter anchor bolt,

$$d_b = 1 \text{ in}$$

$$A_b = \frac{\pi d^2}{4} = \frac{\pi (1 \text{ in})^2}{4} = 0.79 \text{ in}^2$$

Either the embedment length or effective embedment length, l_b, must be assumed in this problem. Since the masonry units have widths of 12 in, the embedment length must be less than 12 in. Given the interior dimensions, an embedment of 9 in seems reasonable. ACI 530 Sec. 1.17.5 specifies that the effective embedment length, l_b, is the actual length of embedment minus one anchor bolt diameter.

$$l_b = l - d_b = 9 \text{ in} - 1 \text{ in} = 8 \text{ in} \quad [\text{controls}]$$

Check minimum l_b per ACI 530 Sec. 1.17.6.

$$\text{greater of} \begin{cases} 4d_b = (4)(1 \text{ in}) = 4 \text{ in} \\ 2 \text{ in} \end{cases}$$

A common anchor bolt minimum spacing is $4d_b$, and this dimension is essentially the maximum leg extension. Calculate the projected leg extension of the anchor bolt as

$$e_b = 4d_b = (4)(1 \text{ in}) = 4 \text{ in}$$

The anchor bolt edge distance, l_{be}, in this problem is difficult to determine. For an isolated bolt in an infinite wall without adjacent edges, the distance to an edge is infinite. However, only the embedment material within a 45° cone is effective in resisting the load, so l_{be} is taken as the distance from the bolt outside diameter to the perimeter of the pull-out cone. That distance depends on the effective embedment length, l_b.

$$l_{be} = \frac{2l_b}{\sqrt{2}} \approx \frac{l_b}{0.707}$$

Structural

The projected area for an isolated anchor bolt in shear is calculated simply from ACI 530 Eq. 1-5, essentially one-half of the area of a circle with radius l_{be}. (This is 50% of the projected area of an anchor bolt in tension, which is given by ACI 530 Eq. 1-4.)

$$A_{pv} = \frac{\pi l_{be}^2}{2}$$

For an anchor bolt placed close to an edge, l_{be} is reduced to the distance from the bolt outside diameter to the edge, and the projected area is again calculated simply from ACI 530 Eq. 1-5.

For an anchor bolt placed near other bolts, the project area is reduced by areas attributable to other bolts resisting the same loading. In practice, the entire area of a load group resisting shear (or pull-out) would be determined and proportioned to all of the resisting bolts, resulting in an "equivalent" value of l_{be}. In this problem, there is no information on edge distance, and the number of bolts and their installation grid pattern are not known. All that can be said is that there are (or, will be) interactions with other anchor bolts, and the projected area will need to be determined iteratively. The second iteration will depend on the first iteration's determination of the number of bolts and their layout.

For this problem, the theoretical projected area of an isolated bolt will be used in the first iteration. Other assumptions may be more rational, although any actual fraction of the theoretical projected area would still be an assumption.

$$A_{pv} = \frac{\pi l_{be}^2}{2} = \frac{\pi \left(\dfrac{l_b}{0.707}\right)^2}{2} \quad \text{[ACI 530 Eq. 1-5]}$$

$$= \frac{\pi \left(\dfrac{8 \text{ in}}{0.707}\right)^2}{2}$$

$$= 201 \text{ in}^2$$

From ACI 530 Eq. 2-6, the shear load based on masonry breakout is

$$B_{vb} = 1.25 A_{pv} \sqrt{f'_m}$$

$$= (1.25)(201 \text{ in}^2)\sqrt{2000 \ \frac{\text{lbf}}{\text{in}^2}}$$

$$= 11{,}240 \text{ lbf}$$

From ACI 530 Eq. 2-7, the allowable shear load governed by masonry crushing is

$$B_{vc} = 350 \sqrt[4]{f'_m A_b}$$

$$= (350)\sqrt[4]{\left(2000 \ \frac{\text{lbf}}{\text{in}^2}\right)(0.79 \text{ in}^2)}$$

$$= 2207 \text{ lbf} \quad \text{[governs]}$$

From ACI 530 Eq. 2-8, the allowable shear load governed by anchor bolt pryout is

$$B_{vpry} = 2.5 A_{pt} \sqrt{f'_m}$$

$$= (2.5)(201 \text{ in}^2)\sqrt{2000 \ \frac{\text{lbf}}{\text{in}^2}}$$

$$= 22{,}470 \text{ lbf}$$

From ACI 530 Eq. 2-9, the allowable shear load governed by steel yielding is

$$B_{vs} = 0.36 A_b f_y$$

$$= (0.36)(0.79 \text{ in}^2)\left(60{,}000 \ \frac{\text{lbf}}{\text{in}^2}\right)$$

$$= 17{,}060 \text{ lbf}$$

Therefore, $B_v = 2207$ lbf.

The number of bolts is greater than or equal to

$$n = \frac{9000 \text{ lbf}}{2207 \text{ lbf}} = 4.1$$

Try six bolts.

The allowable tension load of the bent bar anchor, B_a, is set by ACI 530 Sec. 2.1.4.3.1.2 as the minimum value calculated from ACI 530 Eq. 2-3, Eq. 2-4, and Eq. 2-5. The projected areas of the six bolts overlap and ACI 530 Sec. 1.17.2 requires an adjustment so that area resisting tension is not counted twice. Arbitrarily adjust ACI 530 Eq. 1-4 by 50%.

$$A_{pt} = 0.5\pi l_b^2 = (0.5)\pi(8 \text{ in})^2$$

$$= 100.5 \text{ in}^2$$

From ACI 530 Eq. 2-3, the allowable axial tensile load governed by masonry breakout is

$$B_{ab} = 1.25 A_{pt} \sqrt{f'_m}$$

$$= (1.25)(100.5 \text{ in}^2)\sqrt{2000 \ \frac{\text{lbf}}{\text{in}^2}}$$

$$= 5618 \text{ lbf} \quad \text{[governs]}$$

From ACI 530 Eq. 2-4, the allowable axial tensile load governed by anchor bolt pullout is

$$B_{ap} = 0.6 f'_m e_b d_b + 120\pi(l_b + e_b + d_b)d_b$$

$$= (0.6)\left(2000 \ \frac{\text{lbf}}{\text{in}^2}\right)(4.0 \text{ in})(1.0 \text{ in})$$

$$+ (120)\pi(8 \text{ in} + 4.0 \text{ in} + 1.0 \text{ in})(1.0 \text{ in})$$

$$= 9700 \text{ lbf}$$

From ACI 530 Eq. 2-5, the allowable axial tensile load governed by steel yielding is

$$B_{as} = 0.6A_b f_y$$
$$= (0.6)(0.79 \text{ in}^2)\left(60{,}000 \frac{\text{lbf}}{\text{in}^2}\right)$$
$$= 28{,}440 \text{ lbf}$$

Therefore, $B_a = B_{ab} = 5618$ lbf.

The allowable load in tension, B_a, is 5618 lbf.

The allowable load in shear, B_v, is 2207 lbf.

The top row of bolts is in shear and tension.

The bottom row of bolts is in shear.

From ACI 530 Eq. 2-10,

$$\frac{b_a}{B_a} + \frac{b_v}{B_v} \le 1$$

Calculate the tension in the bolts. Try $a = 4$ in and $g = 2.5$ in.

Summing moments,

$$Ve + F(g + a) = Fg$$
$$Ve + Fa = 0$$
$$(9 \text{ kips})(3 \text{ in}) + F(4 \text{ in}) = 0 \text{ in-kips}$$
$$F = 6.75 \text{ kips}$$
$$F_{\text{bolt}} = \left(\frac{6.75 \text{ kips}}{3 \text{ bolts}}\right)\left(1000 \frac{\text{lbf}}{\text{kip}}\right)$$
$$= 2250 \text{ lbf/bolt}$$

For the top row of bolts,

$$\frac{2250 \text{ lbf}}{5618 \text{ lbf}} + \frac{b_v}{2207 \text{ lbf}} = 1$$
$$b_v = 1323 \text{ lbf/bolt}$$
$$b_a = 2250 \text{ lbf/bolt}$$

For the bottom row of bolts,

$$b_v = 2207 \text{ lbf/bolt}$$
$$b_a = 0 \text{ lbf/bolt}$$

The total shear capacity is

$$(3)(1323 \text{ lbf}) + (3)(2207 \text{ lbf})$$
$$= 10{,}590 \text{ lbf} \quad [> 9000 \text{ lbf, so OK}]$$

Try $a = 5$ in.

Check the compression on the face of the masonry.

$$F = 2250 \text{ lbf/bolt}$$
$$\text{total compression} = (2250 \text{ lbf})(3) \quad [\text{three bolts}]$$
$$= 6750 \text{ lbf}$$

From ACI 530 Sec. 2.1.6, the bearing stress is

$$f_p \le 0.33 f_m' = (0.33)\left(2000 \frac{\text{lbf}}{\text{in}^2}\right) = 660 \text{ lbf/in}^2$$
$$\frac{6750 \text{ lbf}}{A} \le 660 \text{ lbf/in}^2$$
$$A \ge 10.23 \text{ in}^2$$

If the plate is 8 in wide based on 2 in bolt spacing, the height of the bearing area, d, is 1.28 in.

The total height of the plate is

$$h \ge a + g + \frac{d}{2}$$
$$h \ge 4 \text{ in} + 2.5 \text{ in} + \frac{1.28 \text{ in}}{2} = 7.14 \text{ in}$$

> Use an 8 in × 8 in plate with six 1 in diameter bolts in two rows 4 in apart with 2 in bolt spacing.

ACI 530 (Strength Design) Solution

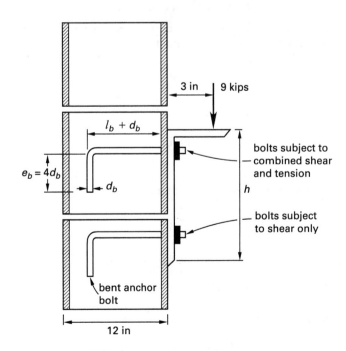

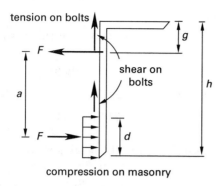

ACI 530 Sec. 1.17.6 requires a minimum spacing of $4d_b$ between anchor bolts.

Try 1 in bolts, spaced 4 in apart.

$$A_b = 0.79 \text{ in}^2$$

$$l_b = 8 \text{ in}$$

$$l_{be} \geq 12d_b$$

$$A_{pv} = \frac{\pi l_{be}^2}{2} = \frac{\pi \left(\dfrac{l_b}{0.707}\right)^2}{2} \quad \text{[ACI 530 Eq. 1-5]}$$

$$= \frac{\pi \left(\dfrac{8 \text{ in}}{0.707}\right)^2}{2}$$

$$= 201 \text{ in}^2$$

Estimate the number of bolts based on shear. From ACI 530 Sec. 3.1.6.3.2, the shear capacity, B_{vn}, is the smallest of B_{vnb}, B_{vnc}, B_{vpry}, and B_{vns}.

From ACI 530 Eq. 3-6, the nominal shear strength governed by masonry breakout is

$$B_{vnb} = 4A_{pv}\sqrt{f'_m}$$

$$= (4)(201 \text{ in}^2)\sqrt{2000 \ \dfrac{\text{lbf}}{\text{in}^2}}$$

$$= 35{,}960 \text{ lbf}$$

From ACI 530 Eq. 3-7, the nominal shear strength governed by masonry crushing is

$$B_{vnc} = 1050\sqrt[4]{f'_m A_b}$$

$$= 1050\sqrt[4]{\left(2000 \ \dfrac{\text{lbf}}{\text{in}^2}\right)(0.79 \text{ in}^2)}$$

$$= 6620 \text{ lbf} \quad \text{[governs]}$$

From ACI 530 Eq. 1-4,

$$A_{pt} = \pi l_b^2 = \pi(8 \text{ in})^2 = 201 \text{ in}^2$$

From ACI 530 Eq. 3-8, the nominal shear strength governed by anchor bolt pryout is

$$B_{vpry} = 8A_{pt}\sqrt{f'_m}$$

$$= (8)(201 \text{ in}^2)\sqrt{2000 \ \dfrac{\text{lbf}}{\text{in}^2}}$$

$$= 71{,}911 \text{ lbf}$$

From ACI 530 Eq. 3-9, the nominal shear strength governed by steel yielding is

$$B_{vns} = 0.6A_b f_y$$

$$= (0.60)(0.79 \text{ in}^2)\left(60{,}000 \ \dfrac{\text{lbf}}{\text{in}^2}\right)$$

$$= 28{,}440 \text{ lbf}$$

Therefore, $B_{vn} = B_{vnc} = 6620$ lbf. The strength reduction factor, ϕ, is set by ACI 530 Sec. 3.1.4.1 as 0.5 for strength controlled by masonry crushing. Therefore,

$$\phi B_{vn} = (0.5)(6620 \text{ lbf}) = 3310 \text{ lbf}$$

Assume 9 kips is the factored load. The number of bolts is

$$\frac{9000 \ \text{lbf}}{3310 \ \dfrac{\text{lbf}}{\text{bolt}}} = 2.7 \text{ bolts}$$

Try four 1 in diameter bolts.

The nominal tensile capacity of the bent bar anchor, B_{an}, is set by ACI 530 Sec. 3.1.6.3.1.2 as the minimum value calculated by Eq. 3-3, Eq. 3-4, and Eq. 3-5. The projected areas of the six bolts overlap, and ACI 530 Sec. 1.17.2 requires an adjustment so that area resisting tension is not counted twice. Therefore,

$$A_{pt} = 0.5\pi l_b^2 = (0.5)\pi(8 \text{ in})^2$$

$$= 100.5 \text{ in}^2$$

From ACI 530 Eq. 3-3, the nominal axial tensile strength governed by masonry breakout is

$$B_{ab} = 4A_{pt}\sqrt{f'_m}$$

$$= (4)(100.5 \text{ in}^2)\sqrt{2000 \ \dfrac{\text{lbf}}{\text{in}^2}}$$

$$= 17{,}980 \text{ lbf} \quad \text{[governs]}$$

$$e_b = 4d_b = (4)(1.0 \text{ in})$$

$$= 4.0 \text{ in}$$

From ACI 530 Eq. 3-4, the nominal axial tensile strength governed by anchor bolt pullout is

$$B_{anp} = 1.5f'_m e_b d_b + 300\pi(l_b + e_b + d_b)d_b$$

$$= (1.5)\left(2000 \; \frac{\text{lbf}}{\text{in}^2}\right)(4.0 \text{ in})(1.0 \text{ in})$$

$$+ (300)\pi(8 \text{ in} + 4.0 \text{ in} + 1.0 \text{ in})(1.0 \text{ in})$$

$$= 24{,}252 \text{ lbf}$$

From ACI 530 Eq. 3-5, the nominal axial tensile strength governed by steel yielding is

$$B_{ans} = A_b f_y$$

$$= (0.79 \text{ in}^2)\left(60{,}000 \; \frac{\text{lbf}}{\text{in}^2}\right)$$

$$= 47{,}400 \text{ lbf}$$

Therefore, $B_{an} = B_{ab} = 17{,}980$ lbf.

The strength reduction factor, ϕ, is set by ACI 530 Sec. 3.1.4.1 as 0.5 for strength controlled by masonry breakout. Therefore,

$$\phi B_{an} = (0.5)(17{,}980 \text{ lbf}) = 8990 \text{ lbf}$$

The design strength is $\phi B_{vn} = 3310$ lbf and $\phi B_{an} = 8990$ lbf.

The top row of bolts is in shear and tension.

The bottom row of bolts is in shear.

From ACI 530 Eq. 3-10,

$$\frac{b_{af}}{\phi B_{an}} + \frac{b_{vf}}{\phi B_{vn}} \leq 1.0$$

Calculate the tension in bolts. Try $a = 4$ in and $g = 2.5$ in.

Summing moments gives

$$Ve + F(g + a) = Fg$$

$$Ve + Fa = 0$$

$$(9 \text{ kips})(3 \text{ in}) = F(4 \text{ in}) = 0 \text{ in-kips}$$

$$F = 6.75 \text{ kips}$$

$$F_{\text{bolt}} = b_{af} = \left(\frac{6.75 \text{ kips}}{2 \text{ bolts}}\right)\left(1000 \; \frac{\text{lbf}}{\text{kip}}\right)$$

$$= 3375 \text{ lbf/bolt}$$

For the top row of bolts,

$$\frac{3375 \text{ lbf}}{8990 \text{ lbf}} + \frac{b_{vf}}{3310 \text{ lbf}} = 1.0$$

$$b_{vf} = 2067 \text{ lbf}$$

$$b_{af} = 3375 \text{ lbf/bolt}$$

For the bottom row of bolts,

$$b_{su} = 3310 \text{ lbf/bolt}$$

$$b_{tu} = 0 \text{ lbf/bolt}$$

The total shear capacity is

$$(2)(2067 \text{ lbf}) + (2)(3310 \text{ lbf})$$

$$= 10{,}754 \text{ lbf} \quad [> 9000 \text{ lbf, so OK}]$$

Check compression in the masonry. The strength reduction factor, ϕ, is set by ACI 530 3.1.4.4 as 0.9 for reinforced masonry subjected to compression, and the compression strength is given by ACI 530 Sec. 3.3.2(g) as $0.8f'_m$. The compression force must be equal and opposite to the bolt tension. Therefore,

$$C = 2b_{af} = (2)(3375 \text{ lbf})$$

$$= 6750 \text{ lbf}$$

For an 8 in wide plate based on bolt spacing of 4 in, the required depth of the compression zone is

$$a = \frac{C}{\phi 0.8f'_m b}$$

$$= \frac{6750 \text{ lbf}}{(0.9)(0.8)\left(2000 \; \frac{\text{lbf}}{\text{in}^2}\right)(8 \text{ in})}$$

$$= 0.59 \text{ in}$$

The depth to the neutral axis is

$$c = \frac{a}{0.8} = \frac{0.59 \text{ in}}{0.8}$$

$$= 0.74 \text{ in}$$

The total height of the plate is

$$h \geq a + g + \frac{d}{2}$$

$$h \geq 4 \text{ in} + 2.5 \text{ in} + \frac{0.74 \text{ in}}{2} = 6.87 \text{ in}$$

> Use an 8 in × 8 in plate with four 1 in diameter bolts in two rows spaced 4 in apart in each direction.

SOLUTION 4

ACI 530 (ASD) Solution

Assume that steel has been placed in the center of the wall.

$$d = \frac{7.625 \text{ in}}{2} = 3.81 \text{ in}$$

Consider a 1 ft wide strip.

$$A_s = \left(\frac{0.31 \text{ in}^2}{3 \text{ ft}}\right)(1 \text{ ft}) = 0.10 \text{ in}^2$$

Assume pinned supports.

From the *AISC Manual* Table 3-23, case 12,

$$+M = \left(\frac{9}{128}\right)wl^2$$

$$= \left(\frac{9}{128}\right)\left(35 \ \frac{\text{lbf}}{\text{ft}^2}\right)(1 \text{ ft})(10 \text{ ft})^2$$

$$= 246.1 \text{ ft-lbf}$$

$$-M = \tfrac{1}{8}wl^2$$

$$= \left(\tfrac{1}{8}\right)\left(35 \ \frac{\text{lbf}}{\text{ft}^2}\right)(1 \text{ ft})(10 \text{ ft})^2$$

$$= 437.5 \text{ ft-lbf}$$

Using ACI 530 Sec. 2.3.3.1, check the steel.

$$\rho = \frac{A_s}{bd} = \frac{0.10 \text{ in}^2}{(12 \text{ in})(3.81 \text{ in})} = 0.0022$$

$$n = \frac{E_s}{E_m} = \frac{29 \times 10^6 \ \frac{\text{lbf}}{\text{in}^2}}{900 f_m'}$$

$$= \frac{29 \times 10^6 \ \frac{\text{lbf}}{\text{in}^2}}{(900)\left(1500 \ \frac{\text{lbf}}{\text{in}^2}\right)}$$

$$= 21.5$$

$$\rho n = (0.0022)(21.5) = 0.047$$

$$k = -\rho n + \sqrt{2\rho n + (\rho n)^2}$$

$$= -0.047 + \sqrt{(2)(0.047) + (0.047)^2}$$

$$= 0.263$$

$$j = 1 - \frac{k}{3} = 1 - \frac{0.263}{3}$$

$$= 0.912$$

$$f_s = \frac{M}{A_s j d}$$

$$= \frac{(437.5 \text{ ft-lbf})\left(12 \ \frac{\text{in}}{\text{ft}}\right)}{(0.10 \text{ in}^2)(0.912)(3.18 \text{ in})}$$

$$= 18{,}102 \text{ lbf/in}^2 \quad [< F_s = 32{,}000 \text{ lbf/in}^2, \text{ so OK}]$$

Using ACI 530 Sec. 2.3.4.2.2, check the masonry.

$$f_m = \frac{2M}{bjkd^2} = \frac{(2)(437.5 \text{ ft-lbf})\left(12 \ \frac{\text{in}}{\text{ft}}\right)}{(12 \text{ in})(0.912)(0.263)(3.81 \text{ in})^2}$$

$$= 251 \text{ lbf/in}^2$$

$$F_b = 0.45 f_m' = (0.45)\left(1500 \ \frac{\text{lbf}}{\text{in}^2}\right)$$

$$= 675 \text{ lbf/in}^2 \quad [> 251 \text{ lbf/in}^2, \text{ so OK}]$$

$$\boxed{\text{The wall is adequate.}}$$

ACI 530 (Strength Design) Solution

Assume that steel has been placed in the center of the wall.

$$d = \frac{7.625 \text{ in}}{2} = 3.81 \text{ in}$$

$$A_s = \frac{0.31 \text{ in}^2}{4 \text{ ft}} = 0.0775 \text{ in}^2/\text{ft}$$

Assume pinned supports. Consider a 1 ft wide strip.

From the *AISC Manual* Table 3-23, case 12,

$$+M = \left(\frac{9}{128}\right)wl^2 = \left(\frac{9}{128}\right)\left(35 \ \frac{\text{lbf}}{\text{ft}^2}\right)(1 \text{ ft})(10 \text{ ft})^2$$

$$= 245 \text{ ft-lbf}$$

$$-M = \tfrac{1}{8}wl^2 = \left(\tfrac{1}{8}\right)\left(35 \ \frac{\text{lbf}}{\text{ft}^2}\right)(1 \text{ ft})(10 \text{ ft})^2$$

$$= 437.5 \text{ ft-lbf}$$

The factored maximum moment is

$$M_u = 1.6M = (1.6)(437.5 \text{ ft-lbf}) = 700 \text{ ft-lbf}$$

Calculate the design strength. From ACI 530 Commentary Sec. 3.3.5.4,

$$a = \frac{P_u + A_s f_y}{0.80 f_m' b}$$

$$= \frac{0 \text{ lbf} + (0.10 \text{ in}^2)\left(40{,}000 \ \frac{\text{lbf}}{\text{in}^2}\right)}{(0.80)\left(1500 \ \frac{\text{lbf}}{\text{in}^2}\right)(12 \text{ in})}$$

$$= 0.277 \text{ in}$$

From ACI 530 Commentary Sec. 3.3.5.4,

$$M_n = \left(A_s f_y + P_u\right)\left(d - \frac{a}{2}\right)$$

$$= \frac{\left(\left(0.10 \text{ in}^2\right)\left(40{,}000 \ \frac{\text{lbf}}{\text{in}^2}\right) + 0 \text{ lbf}\right)}{12 \ \frac{\text{in}}{\text{ft}}} \times \left(3.81 \text{ in} - \frac{0.277 \text{ in}}{2}\right)$$

$$= 1224 \text{ ft-lbf}$$

$$\phi M_n = (0.9)(1224 \text{ ft-lbf}) = 1102 \text{ ft-lbf}$$

$$[>700 \text{ ft-lbf, so OK}]$$

> The wall is adequate.

SOLUTION 5

Deflection calculations are based on unfactored (service) loads.

First calculate the masonry compressive strength, f'_m.

From ACI 530.1 Sec. 1.4B.2.a, assuming type N mortar, f'_m is 1500 lbf/in^2.

$$n = \frac{E_s}{E_m} = \frac{29{,}000{,}000 \ \frac{\text{lbf}}{\text{in}^2}}{700 f'_m} = \frac{29{,}000{,}000 \ \frac{\text{lbf}}{\text{in}^2}}{(700)\left(1500 \ \frac{\text{lbf}}{\text{in}^2}\right)}$$

$$= 27.6$$

$$A'_s = A_s = (2)(0.79 \text{ in}^2)$$

$$= 1.58 \text{ in}^2$$

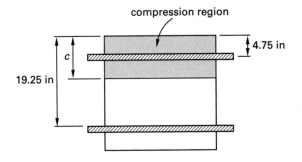

Sum moments about the neutral axis.

$$\sum M = 0 = (24 \text{ in})c\frac{c}{2} + (n-1)A'_s(c - 4.75 \text{ in}) - A_s n(19.25 \text{ in} - c)$$

$$(1.58 \text{ in}^2)(27.6)(19.25 \text{ in} - c) = \frac{(24 \text{ in})c^2}{2} + (27.6 - 1)(1.58 \text{ in}^2) \times (c - 4.75 \text{ in})$$

$$c = 6.40 \text{ in}$$

$$\begin{aligned} I_{cr} &= \tfrac{1}{3}bc^3 + (n-1)A'_s(c - 4.75 \text{ in})^2 \\ &\quad + nA_s(19.25 \text{ in} - c)^2 \\ &= \left(\tfrac{1}{3}\right)(24 \text{ in})(6.40 \text{ in})^3 \\ &\quad + (27.6 - 1)(1.58 \text{ in}^2)(6.40 \text{ in} - 4.75 \text{ in})^2 \\ &\quad + (27.6)(1.58 \text{ in}^2)(19.25 \text{ in} - 6.40 \text{ in})^2 \\ &= \boxed{9412 \text{ in}^4} \end{aligned}$$

13 Bridge Design

PROBLEM 1

A semicircular-nosed concrete pier with a width of 9 ft is located in a major stream. The pier is supported by piles and a pile cap. The velocity of the water is 5 ft/sec. The design flood water level is 16 ft above the top of the pile cap.

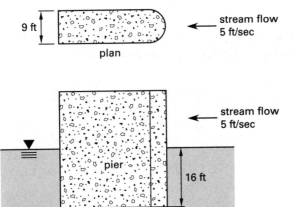

plan

elevation

What is the longitudinal drag force on the concrete pier?

PROBLEM 2

A road deck is to be supported by a prestressed concrete girder. The girder is 75 ft long, 4.0 ft deep, and 2.0 ft wide. The concrete strength during prestressing is 4 ksi. The concrete contains shrinkage-free aggregates, and no drying occurs before it has cured for 5 days. The local relative humidity is 90%. The girder is 35 days old, and the deck will be placed when the girder is 45 days old. Calculations are being performed to determine the deflection of the girder (including the effects of creep) during the time between the girder fabrication and the deck placement. What is the strain due to concrete girder shrinkage?

SOLUTION 1

Use *AASHTO LRFD Bridge Design Specifications* (AASHTO *LRFD*) Sec. 3.7.3. From AASHTO *LRFD* Table 3.7.3.1-1, the drag coefficient, C_D, is 0.7 for a semicircular-nosed pier.

Calculate the longitudinal drag pressure, p, using AASHTO *LRFD* Eq. 3.7.3.1-1. (This equation is not dimensionally consistent.)

$$p = \frac{C_D v^2}{1000}$$

$$= \frac{(0.7)\left(5 \, \frac{\text{ft}}{\text{sec}}\right)^2}{1000}$$

$$= 0.0175 \, \text{kip/ft}^2$$

Calculate the longitudinal drag force, F_D, from the pier width, d_{pier}, and the depth of water above the pile cap, h.

$$F_D = p A_{\text{pier}} = p d_{\text{pier}} h$$

$$= \left(0.0175 \, \frac{\text{kip}}{\text{ft}^2}\right)(9 \, \text{ft})(16 \, \text{ft})$$

$$= \boxed{2.52 \, \text{kips} \quad (2.5 \, \text{kips})}$$

SOLUTION 2

The effects of shrinkage are described in AASHTO *LRFD* Sec. 5.4.2.3.

Calculate the volume, V, of the girder.

$$V = LDW$$

$$= (75 \, \text{ft})(4.0 \, \text{ft})(2.0 \, \text{ft})$$

$$= 600 \, \text{ft}^3$$

Calculate the total surface area, S, of the girder. Since the beam is long, the ends do not significantly contribute to surface area.

$$S = 2(W + D)L$$

$$= (2)(2.0 \, \text{ft} + 4.0 \, \text{ft})(75 \, \text{ft})$$

$$= 900 \, \text{ft}^2$$

Calculate the volume-to-surface area ratio.

$$\frac{V}{S} = \left(\frac{600 \, \text{ft}^3}{900 \, \text{ft}^2}\right)\left(12 \, \frac{\text{in}}{\text{ft}}\right)$$

$$= 8 \, \text{in}$$

Use AASHTO *LRFD* Eq. 5.4.2.3.2-2 to calculate the factor for the effect of the volume-to-surface area ratio, k_s. k_s must be at least 1.0.

$$k_s = 1.45 - 0.13\left(\frac{V}{S}\right)$$

$$= 1.45 - (0.13)(8 \, \text{in})$$

$$= 0.41 \quad [< 1.0. \text{ Use } 1.0.]$$

Use AASHTO *LRFD* Eq. 5.4.2.3.3-2 to calculate the humidity factor for shrinkage, k_{hs}.

$$k_{\text{hs}} = 2.00 - 0.014H$$

$$= 2.00 - (0.014)(90\%)$$

$$= 0.74$$

Use AASHTO *LRFD* Eq. 5.4.2.3.2-4 to calculate the factor for the effect of concrete strength.

$$k_f = \frac{5}{1 + f'_{ci}} = \frac{5}{1 + 4 \, \frac{\text{kips}}{\text{in}^2}}$$

$$= 1.0$$

Use AASHTO *LRFD* Eq. 5.4.2.3.2-5 to calculate the factor for time development.

$$k_{\text{td}} = \frac{t}{61 - 4f'_{ci} + t}$$

$$= \frac{45 \, \text{days} - 35 \, \text{days}}{61 - (4)\left(4 \, \frac{\text{kips}}{\text{in}^2}\right) + (45 \, \text{days} - 35 \, \text{days})}$$

$$= 0.182$$

From AASHTO *LRFD* Eq. 5.4.2.3.3-1, the shrinkage strain, ϵ_{sh}, is

$$\epsilon_{\text{sh}} = k_s k_{\text{hs}} k_f k_{\text{td}}(0.48 \times 10^{-3})$$

$$= (1.0)(0.74)(1.0)(0.182)(0.48 \times 10^{-3})$$

$$= \boxed{6.5 \times 10^{-5}}$$

Topic V: Transportation

Chapter

Transportation

14 Transportation

PROBLEM 1

A suburban central business district has a maximum daytime population of 120,000. Traffic passing through a level intersection in the business district is controlled by a pretimed signal with a total cycle length of 60 sec. There are no bus stops in or near the intersection. The intersection is not part of any transit bus routes. Left turns from the west-to-east direction are made exclusively from the left-hand lane.

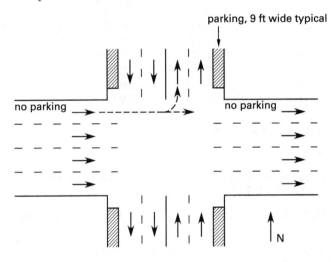

The following data have been collected on the nature of the traffic during a particular time of day.

	west-to-east	north-to-south and south-to-north
lane width	11 ft	12 ft
peak hour factor	0.80	0.70
percent right turns	0%	0%
percent left turns	20%	0%
percent trucks and local bus traffic	10%	7%
length of green and amber phases	32 sec	26 sec
length of all red phases	1 sec	1 sec
approach grade	0%	0%
parking maneuvers	0	20/hr

1.1. Find the maximum service volume for level of service (LOS) C on the west-to-east approach to the intersection.

1.2. Find the maximum service volume for LOS E on the west-to-east approach to the intersection.

1.3. Find the maximum service volume for LOS B on the north-to-south approach to the intersection.

1.4. Find the maximum service volume for LOS E on the south-to-north approach to the intersection.

PROBLEM 2

List at least eight major factors that influence the capacity of a signalized intersection.

PROBLEM 3

The average daily traffic (ADT) patterns for four different sets of traffic are shown. For each pattern, specify a type of interchange (e.g., cloverleaf, loop-connector) and briefly indicate your reason for choosing that type.

3.1.

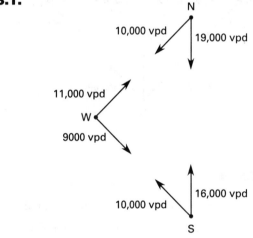

3.2.

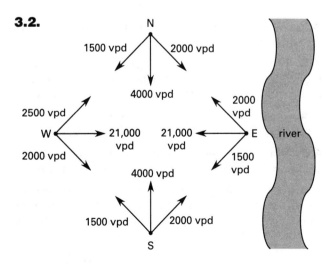

3.3.

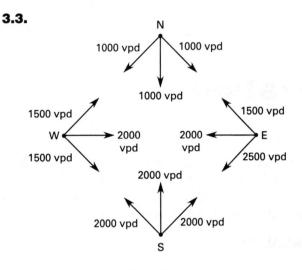

3.4.

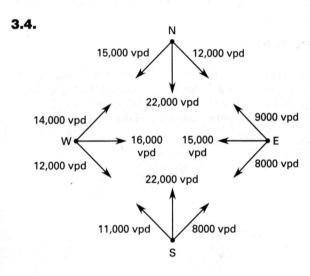

PROBLEM 4

The following questions are not necessarily related.

4.1. What is the passenger car equivalent for a truck on a 2% multilane highway grade that is 0.6 mi long?

4.2. What is the primary variable that determines LOS for automobiles?

4.3. List the major guardrail (roadside barrier) warrants. What values or conditions must these warrants demonstrate to justify the installation of guardrails?

4.4. What is the primary variable used to rate the need for making future improvements to intersections?

4.5. What dimensions should be given to a single-vehicle parking slot in a municipal parking structure?

4.6. On average (i.e., considering all other uses for space), what is the amount of space (in square feet) that should be allowed per parked vehicle in a municipal parking structure?

PROBLEM 5

A rural two-lane highway was designed for 50 mph travel. The following traffic counts were taken over a one-year period.

annualized volume (10^6 veh/yr)	cumulative percentage
0	0%
0.11	10%
0.23	20%
0.39	30%
0.77	40%
0.99	50%
1.30	60%
1.96	70%
3.37	80%
7.81	90%
9.38	93.15%
11.50	95.43%
14.46	97.72%
17.15	98.86%
19.59	99.43%
21.55	99.66%
22.62	99.77%
23.98	99.89%
28.04	100.00%

5.1. What is the 30th hour volume?

5.2. What is the maximum hourly volume?

5.3. What is the approximate average annual daily traffic (AADT)?

5.4. Given a vehicle speed of 42 mph, what is the 30th hour density?

5.5. What is the headway in each direction of flow under the conditions given in part 5.4, assuming a 50-50 directional split?

5.6. Estimate the density when the road is at capacity.

5.7. What is the space mean unit travel time?

5.8. If traffic is expected to grow at the rate of 2% per year, what will be the percentage increase in traffic after four years?

PROBLEM 6

The accident report of a highway patrol officer reads as follows.

Responding to a radio message, I reached the scene of the accident at 18:34 hours and observed a truck that had, apparently, just skidded into a parked car. There were no injuries. The road was dry. The truck left 180 ft of skid marks up to the point of impact. The skid was down (south) on Scenic Drive, which has a uniform downward grade of −5.0% in the direction of travel. Based on the limited amount of damage to both vehicles, I estimate that the impact speed was 20 mph.

6.1. Using the following coefficient of friction values, determine the initial velocity of the truck just prior to locking the brakes.

speed prior to skidding	coefficient of friction
20 mph	0.46
30 mph	0.38
40 mph	0.32
50 mph	0.30
60 mph	0.28

6.2. How would your answer to part 6.1 change if the accident report stated that the road was wet?

PROBLEM 7

A new undivided four-lane highway is being designed to cross an existing two-lane road. Lanes are 12 ft wide. There is no median. The intersection is in a level wooded area and trees will have to be cleared to provide proper sight distance. The four-lane highway will be posted for 50 mph travel; the two-lane road will be posted for 40 mph travel. The design speed should be 10 mph greater than the posted speeds. The intersection will be controlled by a stop sign on each approach of the minor road to the two-lane highway. There is little or no truck traffic.

Draw a plan view of the intersection and show the areas you would clear to provide sight distance. Label and dimension the sight triangles. Use AASHTO specifications.

PROBLEM 8

An airport runway is being redesigned to current FAA regulations.

8.1. What are the reasonable values of longitudinal and transverse gradients?

8.2. Draw two intersecting runways. What are the VFR and IFR maximum loads?

8.3. What other factors must be considered when designing airport runway pavement sections?

PROBLEM 9

The following questions are not necessarily related.

9.1. What arrival type represents highly favorable progression at a signalized intersection? How is arrival type determined?

9.2. How many trips will be generated by a shopping center with 100,000 ft^2 of leasable space (1) per weekday, and (2) during the peak hour?

9.3. What is the definition of the peak hour factor?

9.4. What is the definition of (1) the gap, (2) the critical gap, and (3) the trip distribution?

9.5. What are typical warrants and numerical criteria for signalizing an intersection?

PROBLEM 10

A compacted asphalt pavement with a bulk specific gravity (SG) contains the following components. (Percentages are by weight of total mixture.)

asphalt cement	SG = 1.01	7% by weight
coarse aggregate	SG = 2.61	52% by weight
fine aggregate	SG = 2.71	34% by weight
mineral filler	SG = 2.70	7% by weight

The maximum SG of a paving mixture sample (as measured according to ASTM D2041) was 2.455.

The bulk SG of the compacted paving mixture sample (as measured according to ASTM D2726) was 2.360.

10.1. What is the bulk SG of the aggregate?

10.2. What is the effective SG of the aggregate?

10.3. What would the maximum (zero voids) SG of the paving mixture be if the asphalt mixture is increased to 8%?

10.4. What is the asphalt absorption (by weight)?

10.5. What is the effective asphalt content (by weight) of the paving mixture?

10.6. What is the percent voids in the mineral aggregate (VMA)?

10.7. What is the percent air voids in the compacted mixture?

10.8. Define bleeding in asphalt concrete. What is (are) its primary cause(s)?

10.9. What are the elements comprising Superpave mix design?

10.10. What are typical air-void requirements for asphalt concrete?

SOLUTION 1

(All references in this solution are from the Transportation Research Board's *Highway Capacity Manual* (*HCM*) Chap. 18.)

1.1. Geometry and timing are given. Find the saturation flow rate and capacity.

The lanes can be numbered from 1 to 4, with the rightmost lane being assigned lane number 1. On the west-to-east approach, 20% of the flow turns left. With four approach lanes and a large proportion of left turns, assume that the leftmost lane (lane 4) acts as an exclusive left-turn lane, other than as an occasional insignificant straight-through movement. Each of the through lanes, 1, 2, and 3, has the same characteristics for adjustment. Therefore, the saturation flow rate is the same for each of these lanes. Find the adjusted saturation flow rate for the through lanes. (Use a default value of 1.0 for all factors for which no information was given.)

$$s = s_o f_w f_{HV} f_g f_p f_{bb} f_a f_{LU} f_{LT} f_{RT} f_{Lpb} f_{Rpb} \quad \text{[Eq. 18-5]}$$

$$f_w = 1.0 \quad [HCM \text{ Exh. 18-13}]$$

$$f_{LU} = 1.0 \quad [\text{exclusive lane; p. 18-38}]$$

$$f_{LT} = 1.0 \quad [\text{no turns in lanes 1–3; p. 18-38}]$$

$$f_{Lpb} = 1.0 \quad [\text{no pedestrians; p. 18-38}]$$

$$f_{Rpb} = 1.0 \quad [\text{no bicycles; p. 18-38}]$$

$$s_o = 1900 \text{ veh/hr} \quad [\text{Exh. 18-28}]$$

$$s_{1,2,3} = s_o f_w f_{HV} f_g f_p f_{bb} f_a f_{LU} f_{LT} f_{RT} f_{Lpb} f_{Rpb}$$

$$= \left(1900 \ \frac{\text{veh}}{\text{hr}}\right)(1.0)(0.909)(1.0)(1.0)(1.0)$$

$$\times (0.90)(1.0)(1.0)(1.0)(1.0)(1.0)$$

$$= 1554 \text{ veh/hr}$$

Lane 4 acts as an exclusive left-turn lane. There is no pedestrian interference, and the left turn is unopposed.

$$f_{LT} = \frac{1}{E_L} = \frac{1}{1.05} = 0.95 \quad \text{[Eq. 18-11]}$$

Find the g/C ratio. The cycle length, C, is 60 sec. The effective green time is

$$g_{eff} = D_p - l_1 - l_2$$

$$= g_s + g_e + e \quad \text{[Eq. 18-14]}$$

The length of the queue is unknown, so the queue service time and green extension time are unknown. Since the all-red clearance intervals and cycle length are given,

the given green times must be considered effective green times. For the east-to-west movement,

$$g_{eff,EW} = 32 \text{ sec}$$

$$\left(\frac{g_{eff}}{C}\right)_{EW} = \frac{32 \text{ sec}}{60 \text{ sec}}$$

$$= 0.533$$

For the north-to-south and south-to-north movements,

$$g_{eff,NS-SN} = 26 \text{ sec}$$

$$\left(\frac{g_{eff}}{C}\right)_{NS-SN} = \frac{26 \text{ sec}}{60 \text{ sec}}$$

$$= 0.433$$

Find the capacity. The left-turn adjustment factor has already been applied to each lane group. Therefore, the capacity can be determined from the total g/C ratio applied to the total service volume. For the west-to-east movement,

$$c_{WE} = s\left(\frac{g}{C}\right)_{WE} \quad \text{[Eq. 18-15]}$$

$$= \left(6139 \ \frac{\text{veh}}{\text{hr}}\right)(0.533)$$

$$= 3274 \text{ vph}$$

For the north-to-south and south-to-north movements,

$$c_{NS-SN} = s\left(\frac{g}{C}\right)_{NS-SN} \quad \text{[Eq. 18-15]}$$

$$= \left(3182 \ \frac{\text{veh}}{\text{hr}}\right)(0.433)$$

$$= 1379 \text{ vph}$$

The control delay, d, is found from the following equations.[1] Use $v_i/c_i = 1.0$ for values greater than 1.0.

$$d = d_1 + d_2 + d_3 \quad \text{[Eq. 18-19]}$$

$$d_1 = \frac{0.5 C\left(1 - \dfrac{g}{C}\right)^2}{1 - \left(\min(1, X)\dfrac{g}{C}\right)} \quad \text{[Eq. 18-20]}$$

$$d_2 = 900 T\left((X - 1) + \sqrt{(X - 1)^2 + \frac{8kIX}{cT}}\right)$$

$$T = 1 \text{ hr}$$

$$k = 1.0 \quad [\text{pretimed conditions}]$$

$$I = 1.0 \quad [\text{isolated intersections}]$$

$$d_3 = 0 \text{ sec} \quad [\text{no initial queue}]$$

---

[1]The 2010 *HCM* has removed the equation for incremental delay, d_2. The equation presented for d_2 is *HCM* 2000 Eq. 16-12, and it is still valid for computations.

For the west-to-east movement,

$$d_{1,\text{WE}} = \frac{0.5C\left(1 - \dfrac{g}{C}\right)^2_{\text{WE}}}{1 - \left(\min(1, X)\left(\dfrac{g}{C}\right)_{\text{WE}}\right)}$$

$$= \frac{(0.5)(60 \text{ sec})(1 - 0.533)^2}{1 - 0.533}$$

$$= 14 \text{ sec}$$

For the north-to-south and south-to-north movements,

$$d_{1,\text{NS}} = \frac{0.5C\left(1 - \dfrac{g}{C}\right)^2_{\text{NS-SN}}}{1 - \left(\min(1, X)\left(\dfrac{g}{C}\right)_{\text{NS-SN}}\right)}$$

$$= \frac{(0.5)(60 \text{ sec})(1 - 0.433)^2}{1 - 0.433}$$

$$= 17 \text{ sec}$$

Level of service is determined by the control delay per vehicle on each approach as shown in *HCM* Exh. 18-4. Results are tabulated.

case	g/C	c (vph)	LOS	d_{target} (sec)	$X = v_i/c_i$	d_{calc} (sec)	adjusted flow (vph)
Sol. 1.1	0.533	3274	C	35	0.9850	34.5	3225
Sol. 1.2	0.533	3274	E	80	1.0278	79.5	3365
Sol. 1.3	0.433	1379	B	20	0.6854	19.9	950
Sol. 1.4	0.433	1379	E	80	1.018	79.8	1398

1.2. See the table in Sol. 1.1.

1.3. See the table in Sol. 1.1.

1.4. See the table in Sol. 1.1.

SOLUTION 2

According to Eq. 18-5 of the *Highway Capacity Manual*, there are 12 primary factors that influence capacity. They are

1. base saturation flow rate

2. lane width

3. percentage of heavy vehicles in the traffic stream

4. grade of the approach

5. presence of a parking lane and number of parking maneuvers per hour

6. blockage caused by local buses making stops

7. area type (capacity is less in central business districts than elsewhere)

8. lane utilization

9. percentage of right-turning vehicles in the traffic stream

10. percentage of left-turning vehicles in the traffic stream

11. percentages of pedestrians and bicyclists turning left

12. percentages of pedestrians and bicyclists turning right

Key secondary factors that interact with items 8 through 12 are

- whether turns are made from an exclusive or shared lane

- whether turning movements are protected, permitted (unprotected), or both

- volume of pedestrians using the conflicting crosswalk for both left and right turns

- opposing flow rate when the signal phasing includes permitted left turns

- bicycle traffic, especially bicycle lanes that are to the right of right-turning vehicles

- number of lanes

SOLUTION 3

3.1. From AASHTO's *A Policy on Geometric Design of Highways and Streets*, the practical capacity of loop ramps (270° turns) is 800–1200 vph regardless of width (loop ramps are not amenable to double lanes). Assuming a *k*-factor (ratio of peak hour flow to ADT) of between 0.12 (urban) and 0.15 (rural), peak hour flow for the lower volume left turn is at least 1200 vph. A directional three-leg interchange is recommended.

3.2. The low flows on all except the freeway mainline dictate a simple diamond with stop- or signal-controlled intersections on the north-to-south road. If the river prevents ramps on the east, a half cloverleaf with diagonal and loop ramps in the two western quadrants is recommended, with stop or signal control at the minor street terminals.

3.3. The relatively low flows suggest that no interchange will be needed. The high proportion of turning movements suggests a rotary intersection. If an interchange is desired, a loop-connector type is recommended, with the east-to-west street at a separate grade from a loop that carries all other movements. This is an inexpensive type of interchange that works well with low flows. Because it treats the north-to-south flows as turning flows (they must enter the loop and then leave it), it is most appropriate when the through volume is small compared to the turning flows.

3.4. The interchange is between two very high-volume roads, presumably freeways. With its high turning volumes, it requires a fully directional or semidirectional interchange that provides the best level of service (i.e., high speeds for all turning flows). The lowest volume left turn movement (east-to-south) will have a peak hour flow of at least 1000–1200 vph, so it could possibly be accommodated by a loop ramp; however, including a single loop ramp in a directional interchange is impractical, and the level of service on the loop would be poor.

SOLUTION 4

4.1. From *Highway Capacity Manual* (*HCM*) Exh. 14-13, $E_T = 1.5$ for all percentages of truck traffic.

4.2. (Reference: *HCM* Exh. 2-2.)

- On basic freeway segments and multilane highways, density is the primary variable.

- On freeway weaving sections, speed for weaving and nonweaving vehicles is the primary variable.

- On freeway ramp sections, density is the primary variable.

- On two-lane rural highways, speed, percent time spent following is the primary variable.

- At signalized intersections, average control delay is the primary variable.

- At unsignalized intersections, control delay is the primary variable.

- On urban/suburban arterials, average through-vehicle travel speed is the primary variable.

4.3. (Reference: AASHTO, *Roadside Design Guide*, Chap. 5, Chap. 6, Chap. 10.) (Guardrail warrants are not given in AASHTO's *A Policy on Geometric Design of Highways and Streets* or the *HCM*.)

Traditionally, the two situations that have been used to justify (warrant) roadside guardrails are *embankments* and *roadside hazards*. Depending on the interpretation of "roadside guardrail," three additional situations that require *separation* may be combined with the roadside hazards situation.

- Embankment slope and height: Various combinations warrant a roadside barrier, as illustrated (from AASHTO's *Roadside Design Guide* Fig. 5.1 (b)). Modifications for traffic and embankment length may also be included in the warrant.

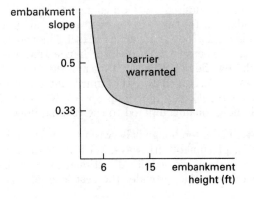

- Presence of roadside obstacles: A barrier is warranted when removal or modification of the obstacle is not possible or cost effective, and when the barrier is less of a hazard than the obstacle. Examples of obstacles calling for a barrier are

  - wood poles or posts with an area greater than or equal to 50 in^2

  - metal shapes with a moment of inertia greater than or equal to 3.0 in^4 for steel and 4.5 in^4 for aluminum

  - concrete bases extending 6 in or more above the ground

  - sign bridge supports

  - breakaway light poles, with a breakaway linear impulse greater than or equal to 110 lbf-sec

  - trees with diameter greater than or equal to 6 in

- Protection of bystanders, cyclists, and pedestrians: No simple numerical value is given, but under normal circumstances, a raised curb is a sufficient barrier to protect pedestrians for speeds up to 25 mph. Special circumstances (e.g., accident experience and the presence of schoolchildren) may apply.

- Median barriers: These may be needed, based on roadside warrants. In addition, there is a median warrant to prevent head-on collisions, based on various combinations of ADT and median width (from AASHTO's *Roadside Design Guide* Fig. 6-1).

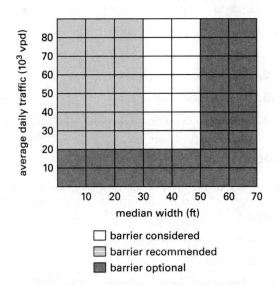

- Barriers in work areas: No numerical values are given. Work-area barriers are warranted when needed to

  - protect traffic from entering unsafe areas

  - protect workers

  - separate two-way traffic

  - protect exposed construction works (e.g., falsework)

4.4. The primary variable is the total number of accidents. Other measures used are the accident rate (accidents per 100,000 veh) and a weighted sum of personal injury accidents and property damage only (PDO) accidents, in which personal injury accidents count as four or five PDO accidents.

4.5. For most purposes, a 90° parking stall can be 8.5 ft wide × 17.5 ft long. Recommended width varies from 9 ft for high turnover uses to 8.25 ft for low turnover uses. "Small car only" stalls can, of course, be smaller.

4.6. A typical parking module is 8.5 ft wide × 61 ft long, including two stalls and an aisle. Adding 10% for losses at corners and such, the area per vehicle is

$$A_{\text{veh}} = (1.1)(8.5 \text{ ft})\left(\frac{61 \text{ ft}}{2}\right)$$

$$= \boxed{285 \text{ ft}^2}$$

SOLUTION 5

5.1. There are (24 hr/day)(365 days/yr) = 8760 hr in a year, so the 30th busiest hour is the $(1 - 30/8760) \times (100) = 99.66$ percentile value, which (from the problem statement) is 21,550,000 veh/yr. Converting to an hourly volume, the 30th hour volume is

$$V_{30\text{th}} = \frac{21{,}550{,}000 \ \dfrac{\text{veh}}{\text{yr}}}{8760 \ \dfrac{\text{hr}}{\text{yr}}} = \boxed{2460 \text{ vph}}$$

5.2. The maximum hourly volume is

$$V_{\text{max}} = \frac{28{,}040{,}000 \ \dfrac{\text{veh}}{\text{yr}}}{8760 \ \dfrac{\text{hr}}{\text{yr}}}$$

$$= \boxed{3201 \text{ vph}}$$

5.3. An approximate average is found using the 50th percentile volume. Converting to average annual daily traffic volume (AADT),

$$\text{AADT} \approx \frac{990{,}000 \ \dfrac{\text{veh}}{\text{yr}}}{365 \ \dfrac{\text{days}}{\text{yr}}}$$

$$= \boxed{2712 \text{ vpd}}$$

5.4. Density is equal to volume divided by speed. Using the volume from Sol. 5.1,

$$D = \frac{V}{S} = \frac{2460 \ \dfrac{\text{veh}}{\text{hr}}}{42 \ \dfrac{\text{mi}}{\text{hr}}}$$

$$= \boxed{58.6 \text{ veh/mi}}$$

5.5. Assuming a 50–50 directional split, the volume in each direction is

$$V = \frac{2460 \ \dfrac{\text{veh}}{\text{hr}}}{2}$$

$$= 1230 \text{ vph}$$

$$\text{headway} = \frac{1}{V} = \frac{(1)\left(3600 \ \dfrac{\text{sec}}{\text{hr}}\right)}{1230 \ \dfrac{\text{veh}}{\text{hr}}}$$

$$= \boxed{2.93 \text{ sec/veh} \quad (2.93 \text{ sec})}$$

5.6. Because the road is sometimes congested, its flow reaches capacity. An estimate of the capacity is the maximum hourly flow as determined in Sol. 5.2. At capacity, the average speed is approximately 30 mph.

$$D = \frac{V}{S}$$

$$\approx \frac{3201 \ \dfrac{\text{veh}}{\text{hr}}}{30 \ \dfrac{\text{mi}}{\text{hr}}}$$

$$= \boxed{107 \text{ veh/mi}}$$

5.7. Space mean speed differs little from time mean speed, so the given speed can be equated to a space mean speed. The space mean unit travel time is

$$t = \frac{1}{S_{\text{mean}}}$$

$$= \left(\frac{1}{42 \ \dfrac{\text{mi}}{\text{hr}}}\right)\left(60 \ \dfrac{\text{min}}{\text{hr}}\right)$$

$$= \boxed{1.43 \text{ min/mi}}$$

5.8. $(1.02)^4 = 1.0824$. The increase will be $\boxed{8.24\%}$.

SOLUTION 6

6.1. Since the coefficient of friction factor, μ, must be considered with the initial speed, v_0, solve using trial and error. The *Green Book* equation for braking distance (where $d_B = v_0$) can be used to find the initial speed. Use interpolation, and start by assuming an initial speed of 45 mph.

$$
d_B = \frac{v^2}{30\left(\dfrac{a}{32.2} \pm G\right)} = \frac{v_i^2 - v_o^2}{30\left(\dfrac{a}{32.2} \pm G\right)} \quad \left[\begin{array}{c} \textit{Green Book} \\ \text{Eq. 3-3} \end{array}\right]
$$

$$
= \frac{\left(45\ \dfrac{\text{mi}}{\text{hr}}\right)^2 - \left(20\ \dfrac{\text{mi}}{\text{hr}}\right)^2}{(30)\left(\dfrac{(0.31)\left(32.2\ \dfrac{\text{ft}}{\text{sec}^2}\right)}{32.2\ \dfrac{\text{ft}}{\text{sec}^2}} - 0.05\right)}
$$

$$
= 208\ \text{ft} \quad \text{[inconsistent with skid length]}
$$

Assume an initial speed of 43 mph.

$$
d_B = \frac{v^2}{30\left(\dfrac{a}{32.2} \pm G\right)} = \frac{v_i^2 - v_o^2}{30\left(\dfrac{a}{32.2} \pm G\right)}
$$

$$
= \frac{\left(43\ \dfrac{\text{mi}}{\text{hr}}\right)^2 - \left(20\ \dfrac{\text{mi}}{\text{hr}}\right)^2}{(30)\left(\dfrac{(0.314)\left(32.2\ \dfrac{\text{ft}}{\text{sec}^2}\right)}{32.2\ \dfrac{\text{ft}}{\text{sec}^2}} - 0.05\right)}
$$

$$
= 182\ \text{ft} \quad \text{[consistent with skid length]}
$$

The speed of the truck at the beginning of the skid was 43 mph.

6.2. The answer given in Sol. 6.1 would not change appreciably because the given coefficients of friction are valid for most wet pavements.

SOLUTION 7

(All references in this solution are from AASHTO's *A Policy on Geometric Design of Highways and Streets*.)

This solution requires the determination of sight triangles as covered in Chap. 9, in the section "Intersection Sight Distance." Since the intersection is controlled by signs, departure sight triangles control the design. Stop signs are used as opposed to yield signs, so this problem corresponds to case B, intersections with stop control on the minor road. Within case B are three conditions that must be evaluated: case B1, left turn from the minor road; case B2, right turn from the minor road; and case B3, crossing maneuver from the minor road.

Using the AASHTO methodology, tables and graphs have essentially eliminated the need to factor in reaction and acceleration or deceleration times for movements relative to single-lane highways, although the tabulated values have been derived from certain assumed values of these times. For movements relative to multilane highways, the time gap must be used with the AASHTO equations.

(a) case B1, left turn from the minor road:

AASHTO specifies that the decision point, one of the vertices on the departure sight triangle, is located 14.5 ft back from the edge of the traveled way, and that a value of 18 ft should be used where practical (essentially a matter of economics). The length of the shorter orthogonal leg is this distance plus the distance from the edge of the traveled way to the location of the oncoming traffic, taken as one-half of the lane width.

$$
\text{short leg} = 14.5\ \text{ft} + \frac{12\ \text{ft}}{2} = 20.5\ \text{ft}
$$

The length of the longer orthogonal leg is a function of the time required by the stopped vehicle to cross all lanes of opposing traffic, and this time is greater for multilane highways than single-lane highways. Since truck traffic is considered to be insignificant, from Table 9-5, the time gap, t_g, for a passenger car is 7.5 seconds. From the footnote in Table 9-5, the correction for an additional lane is 0.5 sec. Since the terrain is level, no adjustment for grades is required. Therefore, the time gap is

$$
t_g = 7.5\ \text{sec} + 0.5\ \text{sec} = 8\ \text{sec}
$$

The design speed for the major road is 10 mph greater than the posted limit.

$$
v_{\text{major}} = 50\ \text{mph} + 10\ \text{mph} = 60\ \text{mph}
$$

From Eq. 9-1, the intersection sight distance is

$$
\text{ISD} = 1.47 v_{\text{major}} t_g = (1.47)\left(60\ \frac{\text{mi}}{\text{hr}}\right)(8\ \text{sec})
$$

$$
= 705.6\ \text{ft} \quad \text{[use 710 ft]}
$$

If this distance is provided for opposing traffic approaching in the near (i.e., outside or "slow") lane, the sight distance of opposing traffic approaching in the far (i.e., inside or "fast") lane will be adequate.

(b) case B2, right turn from the minor road:

AASHTO specifies that the time gap can be reduced by 1 sec for right turns from the minor road. Either Table 9-7 can be used with Eq. 9-1, or the intersection sight distance can be read directly from Table 9-8 as 575 ft (design value).

Transportation

This is less than the distance calculated for case B1.

(c) case B3, crossing maneuver from the minor road:

AASHTO specifies that intersection sight distances for crossing maneuvers generally do not control the design. The conditions where the time for the crossing maneuver should be checked are not present in this exercise.

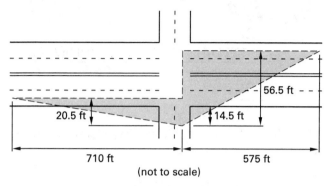

710 ft   575 ft
(not to scale)

SOLUTION 8

8.1. Reasonable gradients are 0–2% longitudinal and no less than 1.5% transverse (crown) for draining.

8.2. There is more than one way to lay out two intersecting runways. One way is

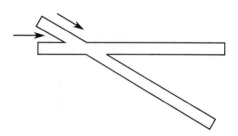

PHOCAP (practical hourly capacity) under VFR (visual flight rules) is 72–98 operations/hr, depending on the mix of aircraft. For IFR (instrument flight rules), PHOCAP is about 60 operations/hr. (Reference: FAA Advisory Circular 150/5060-5, "Airport Capacity and Delay," cited in Haefner, L.E., *Introduction to Transportation Systems*, Holt, Rinehart and Winston.)

8.3. Other factors to consider are the weight of the largest expected aircraft, the strength of the base and subbase, drainage, resistance to petroleum-based fuels, location of underground utilities, and frost interaction.

SOLUTION 9

9.1. From the *Highway Capacity Manual* p. 18-13, a very good (i.e., highly favorable) progression is arrival type 5. There are no definitive parameters used to quantify arrival type. It is best determined using field observations, but approximations can also be obtained by reviewing time-space diagrams for the street in question. The platoon ratio, R_p, may also be used.

9.2. The *ITE Transportation Planning Handbook* Table 4-5 provides the value of 42.92 daily trips/1000 ft^2 for a shopping center and 3.74 afternoon peak trips/1000 ft^2. Therefore, the number of trips generated each weekday is

$$j_{\text{weekday}} = \left(\frac{42.92 \ \dfrac{\text{trips}}{\text{day}}}{1000 \ \text{ft}^2} \right) (100{,}000 \ \text{ft}^2)$$

$$= \boxed{4292 \ \text{trips/day}}$$

The number of trips during the peak hour is

$$j_{\text{peak}} = \left(\frac{3.74 \ \dfrac{\text{trips}}{\text{hr}}}{1000 \ \text{ft}^2} \right) (100{,}000 \ \text{ft}^2)$$

$$= \boxed{374 \ \text{trips/hr}}$$

9.3. The peak hour factor, PHF, is

$$\boxed{\text{PHF} = \frac{V_{\text{peak hour}}}{4 \, V_{\text{peak 15 min}}}}$$

9.4. From the *Highway Capacity Manual* Chap. 9, a *gap* (or headway) is the time (in seconds) it takes the front bumper of the second of two successive vehicles to reach the starting point of the front bumper of the first vehicle. The *critical gap* (critical headway) is the minimum time (in seconds) between successive major-street vehicles in which a minor-street vehicle can make a maneuver.

Trip distribution is the result of matching productions to attractions. Given estimates of the number of trips beginning at each origin, i, and the number of trips ending at each destination, j, trip distribution is the art of estimating the number of trips that go from i to j.

9.5. The U.S. Department of Transportation, Federal Highway Administration's *Manual on Uniform Traffic Control Devices* (MUTCD) identifies nine warrants for signalizing an intersection. They include

warrant 1, eight-hour vehicular volume: This warrant comprises two conditions. Either condition A or condition B, or a combination of both, may be met.

- condition A, minimum vehicular volume: For example, at an intersection of two two-lane roads, the major street volume (total for both approaches) should be at least 500 vph and the minor street volume at least 150 vph. These volumes should hold for eight hours.

• condition B, interruption of continuous traffic: This applies when the delay on the minor street would be excessive without a signal due to high volume on the major street. For example, at an intersection of two two-lane roads, the major street volume (total for both approaches) should be at least 750 vph and the minor street volume at least 75 vph. These volumes should hold for eight hours.

warrant 2, four-hour vehicular volume: This warrant applies if for any four hours of an average day, the plotted points (representing vehicles per hour on the major street and the corresponding vehicles per hour on the higher-volume minor street) fall above the applicable curve in MUTCD Fig. 4C-1. (MUTCD Fig. 4C-2 may be used in place of Fig. 4C-1.)

warrant 3, peak hour: This warrant applies to intersections where conditions are such that for one hour of the day, minor street traffic suffers undue delay. Total delay on minor streets must exceed 4 veh-hr on one-lane approaches and 5 veh-hr for two-lane approaches, and street vehicle volume must be greater than 100 veh for single lanes and 150 veh for two lanes. The total volume at the intersection must be greater than 800 vph.

warrant 4, pedestrian volume: This warrant applies to intersections where traffic volume on a major street is so heavy that pedestrians experience excessive delay in crossing a major street.

warrant 5, school crossing: This warrant applies to intersections where schoolchildren (elementary through high school students) cross a major street. There must be at least 20 schoolchildren during the highest crossing hour.

warrant 6, coordinated signal system: This warrant applies to intersections that require a traffic control signal where it would be needed to maintain proper platooning of vehicles.

warrant 7, crash experience: This warrant applies to intersections where there are at least five accidents in 12 months, if less-restrictive remedies have failed to reduce accident frequency, and if there are traffic volumes that meet 80% of the first three given warrants.

warrant 8, roadway network: This warrant applies to intersections where installing a traffic control signal would encourage concentration and organization of traffic flow on a roadway network.

warrant 9, intersection near a grade crossing: This warrant applies to intersections where no single warrant is satisfied, but the proximity to the intersection of a grade crossing on an intersection approach controlled by a "stop" or "yield" sign is cause for installing a traffic control signal.

SOLUTION 10

(All references in this solution are from the Asphalt Institute's *The Asphalt Handbook*.)

10.1. The bulk specific gravity, G_{sb}, is a weighted average of the aggregate and mineral specific gravities.

$$
\begin{aligned}
G_{sb} &= \frac{\sum x_i G_i}{\sum x_i} \\
&= \frac{(0.52)(2.61) + (0.34)(2.71) + (0.07)(2.70)}{0.52 + 0.34 + 0.07} \\
&= \boxed{2.653}
\end{aligned}
$$

10.2. The effective specific gravity, G_{se}, of the aggregate is

$$
\begin{aligned}
G_{se} &= \frac{1 - x_b}{\dfrac{1}{G_{mm}} - \dfrac{x_b}{G_b}} \\
&= \frac{1 - 0.07}{\dfrac{1}{2.455} - \dfrac{0.07}{1.01}} \\
&= \boxed{2.751}
\end{aligned}
$$

10.3. The maximum specific gravity, G_{mm}, under the given conditions is

$$
\begin{aligned}
G_{mm} &= \frac{1}{\dfrac{1 - x_b}{G_{se}} + \dfrac{x_b}{G_b}} \\
&= \frac{1}{\dfrac{1 - 0.08}{2.751} + \dfrac{0.08}{1.01}} \\
&= \boxed{2.418}
\end{aligned}
$$

10.4. The asphalt absorption fraction, $x_{ba} = P_{ba}/100\%$, is

$$
\begin{aligned}
x_{ba} &= G_b\left(\frac{G_{se} - G_{sb}}{G_{sb} G_{se}}\right) \\
&= (1.01)\left(\frac{2.751 - 2.653}{(2.751)(2.653)}\right) \\
&= \boxed{0.0136 \quad (1.36\%)}
\end{aligned}
$$

10.5. The effective asphalt content fraction, $x_{be} = P_{be}/100\%$, is

$$
\begin{aligned}
x_{be} &= x_b - x_{ba}(1 - x_b) \\
&= 0.07 - (0.0136)(1 - 0.07) \\
&= \boxed{0.0574 \quad (5.74\%)}
\end{aligned}
$$

10.6. The voids in the mineral aggregate fraction, $x_{\text{VMA}} = \text{VMA}/100\%$, is

$$x_{\text{VMA}} = 1 - \frac{G_{\text{mb}}(1 - x_b)}{G_{\text{sb}}}$$
$$= 1 - \frac{(2.360)(1 - 0.07)}{2.653}$$
$$= \boxed{0.1727 \quad (17.27\%)}$$

10.7. The air voids fraction, $x_{\text{a}} = P_{\text{a}}/100\%$, is

$$x_a = \frac{G_{\text{mm}} - G_{\text{mb}}}{G_{\text{mm}}}$$
$$= \frac{2.455 - 2.360}{2.455}$$
$$= \boxed{0.0387 \quad (3.87\%)}$$

10.8. Bleeding is the migration of free asphalt to the pavement surface, resulting in a loss of skid resistance. Bleeding typically occurs under loading on hot days in mixtures that have too few air voids.

10.9. There are three elements comprising *Superpave* (Superior Performing Asphalt Pavement) mix design. *Level 1 mix design* includes specifying asphalts by a set of performance-based binder specifications (e.g., stiffness, dynamic shear, bending). Level 1 is used by most highway agencies and in most new construction, replacing older methods of asphalt call-outs.

Level 2 mix design includes performance-based tests that measure primary mixture performance factors (e.g., fatigue, cracking, permanent deformation, aging, water sensitivity).

Level 3 mix design is a computer-aided volumetric mix design and analysis that incorporates test results, geographic location, and climatological data. Level 3 also incorporates mix-testing technology such as the Superpave gyratory compactor (SGC), which is a laboratory compaction device that can create laboratory densities equivalent to field densities.

10.10. Air voids are generally 3–5% of the mixture by volume.

15 Surveying

PROBLEM 1

Two highway tangents intersect at sta 60+50 as shown. The roadway follows a horizontal circular curve constructed between the two tangents, and it passes through point A. Other data are given in the illustration.

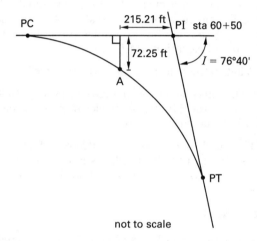

not to scale

1.1. What is the radius of the horizontal circular curve that passes through the PC, PT, and point A?

1.2. Measuring from the intersection point, PI, what are the stations of the PC and PT?

1.3. Assume the curve radius is 660 ft and the design speed is 40 mph. Use the current AASHTO standards to determine if the radius is acceptable.

PROBLEM 2

The PI of two tangents of a highway is located in a lake. Stations A and B are selected to replace the inaccessible PI.

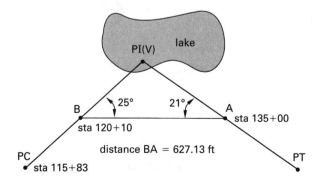

2.1. What is the radius of the circular curve between the PC and PT?

2.2. What is the degree of the curve?

2.3. What is the length of the curve?

2.4. What is the station of the PT?

2.5. If the instrument is at the PC, what deflection angle should be used to locate staking sta 116+00 on the curve?

2.6. If the instrument is at the PC, what deflection angle should be used to locate staking sta 117+00 on the curve?

2.7. If the instrument is at sta 116+00, what deflection angle should be used to locate staking sta 118+00 on the curve?

PROBLEM 3

The bearings of two tangents connected by a horizontal circular curve are N 50° E and S 35° E, respectively. The tangents intersect at sta 37+00. The curve radius is 800 ft.

3.1. What is the length of the curve?

3.2. What is the station of the PC?

3.3. What is the station of the PT?

3.4. What is the interior angle at the PI?

3.5. What is the tangent distance from the PI to the PC?

3.6. What is the long chord distance?

3.7. What is the external distance?

3.8. What is the degree of the curve (arc basis)?

3.9. What is the degree of the curve (chord basis)?

3.10. What is the chord length of a 100 ft arc (arc basis)?

PROBLEM 4

Numerous accidents have been occurring at several horizontal highway curves in a county. It is suspected that the geometries of the curves are contributing to the accidents.

4.1. What are the geometric and trigonometric relationships that need to be considered when replacing a compound curve with a simple circular curve?

4.2. What are the geometric and trigonometric relationships that need to be considered when replacing a simple circular curve with a simple spiral curve?

4.3. What are the geometric and trigonometric relationships that need to be considered when replacing a broken-back curve with a simple circular curve?

PROBLEM 5

A road contains a vertical curve starting at sta 60+00 and ending at sta 68+00. The elevation at the beginning of the curve (BVC) is 562 ft. The grade prior to the BVC is −1.5%. Just after the end of the curve (EVC), the grade is +2.5%.

5.1. What are the centerline elevations of the road every 50 ft from the BVC to the EVC?

5.2. The road is constructed on marshy ground subject to flooding. At sta 63+00, the road becomes elevated and is supported by five-pile bents every 50 ft. The piles are separated from each other by 10 ft (perpendicular to the road centerline) and are cut 3 ft below the finished road grade. A 4° horizontal circular curve to the right starts at sta 64+50 and finishes at sta 66+00. Superelevation on the curve is 12% and is fully developed around the curve. What are the cutoff elevations for each of the five piles in the four bents in the horizontal curve?

PROBLEM 6

The grade into a vertical sag curve is −2%. The curve length is 1400 ft. The grade out of the curve is +4%. The elevation and station of the grade intersection are 226.88 ft and 7+20, respectively. The curve goes through a flood plain where the 50 yr flood elevation is 240 ft.

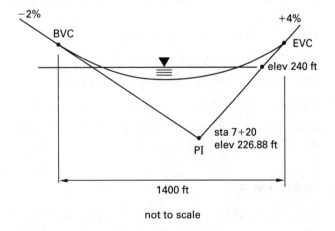

not to scale

6.1. Find the stations where the curve drops below and emerges from the 50 yr flood plain.

6.2. What length of curve will be submerged in a 50 yr flood?

6.3. How will increasing the length of curve change the length of the curve that will be submerged?

6.4. How will decreasing the length of curve change the length of the curve that will be submerged?

6.5. What can be done to the pavement so that flooding will not degrade it?

PROBLEM 7

A vertical curve connects a −1.3% tangent and a +1.8% tangent. The tangents intersect at sta 74+00 and elevation 310 ft. A railroad line at grade crosses the curve at sta 75+20 and elevation 314 ft.

7.1. What is the length of the vertical curve?

7.2. If the curve length is changed to exactly 1300 ft, what will be the difference in elevation between the railroad line and the roadway?

7.3. The road is curbed and guttered. If the curve length is exactly 1300 ft, at what station and elevation would a single grate inlet be placed on the curve?

PROBLEM 8

During a survey, lengths and bearings are collected for legs of a traverse.

leg	bearing	distance
AB	N 35.15° W	905.21 ft
BC	N 81.28° E	1135.76 ft
CD	S 7.19° E	1207.92 ft
DE	S 15.25° W	800.25 ft
EF	N 48.17° W	1100.85 ft
FA	N 40.73° E	429.53 ft

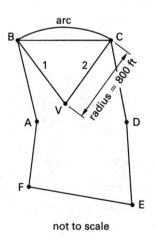

not to scale

8.1. What is the precision of the survey?

8.2. Adjust the traverse by the compass rule.

8.3. Using the adjusted traverse, what is the area of the figure ABCDEF?

8.4. What is the area of the circular segment bounded by the chord BC?

8.5. What is the length of the arc BC?

PROBLEM 9

The partial results of a survey of a traverse are shown in the following illustration.

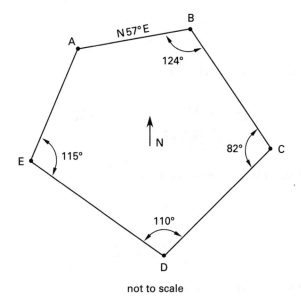

not to scale

9.1. What is the interior angle at A?

9.2. What is the bearing of line BC?

9.3. What is the bearing of line CD?

9.4. What is the bearing of line DE?

9.5. What is the bearing of line EA?

SOLUTION 1

1.1.

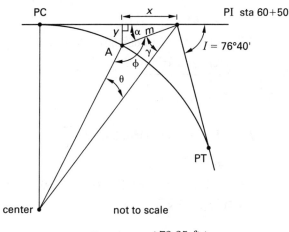

center   not to scale

$$\alpha = \arctan\frac{y}{x} = \arctan\left(\frac{72.25 \text{ ft}}{215.21 \text{ ft}}\right)$$
$$= 18.558°$$

$$m = \sqrt{x^2 + y^2} = \sqrt{(215.21 \text{ ft})^2 + (72.25 \text{ ft})^2}$$
$$= 227.01 \text{ ft}$$

$$I = 76°40' = 76.667°$$

$$\gamma = 90° - \frac{I}{2} - \alpha$$
$$= 90° - \frac{76.667°}{2} - 18.558°$$
$$= 33.109°$$

$$\phi = 180° - \arcsin\left(\frac{\sin\gamma}{\cos\dfrac{I}{2}}\right)$$
$$= 180° - \arcsin\left(\frac{\sin 33.109°}{\cos\dfrac{76.667°}{2}}\right)$$
$$= 135.864°$$

$$\theta = 180° - \phi - \gamma$$
$$= 180° - 135.864° - 33.109°$$
$$= 11.027°$$

From the law of sines,

$$\frac{\sin\theta}{m} = \frac{\sin\phi}{R\sec\dfrac{I}{2}} = \frac{\sin\phi\cos\dfrac{I}{2}}{R}$$

$$R = \frac{(227.01 \text{ ft})\sin 135.864°\cos\dfrac{76.667°}{2}}{\sin 11.027°}$$
$$= \boxed{648.30 \text{ ft}}$$

1.2. Find the stations for the PC and PT.

$$T = R \tan \frac{I}{2} = (648.30 \text{ ft}) \tan \frac{76.667°}{2}$$
$$= 512.61 \text{ ft}$$
$$PC = PI - T = (\text{sta } 60{+}50) - 512.61 \text{ ft}$$
$$= \boxed{\text{sta } 55{+}37.39}$$

The length of the curve is

$$L = RI = (648.30 \text{ ft})(76.667°)\left(\frac{2\pi}{360°}\right)$$
$$= 867.48 \text{ ft}$$
$$\text{sta PT} = \text{sta PC} + L = (\text{sta } 55{+}37.39) + 867.48 \text{ ft}$$
$$= \boxed{\text{sta } 64{+}4.87}$$

1.3. Calculate the minimum radius length. e is the superelevation, and f is the side-friction factor.

$$R_{\min} = \frac{v^2}{15(e+f)}$$

The minimum value of e is 0.04, and for a speed of 40 mph, $f = 0.16$. (Reference: AASHTO, *A Policy on Geometric Design of Highways and Streets*, Exh. 3-15.)

$$R_{\min} = \frac{(40 \text{ mph})^2}{(15)(0.04 + 0.16)}$$
$$= 533 \text{ ft} < 660 \text{ ft} \quad [\text{OK}]$$

$$\boxed{\text{The radius is acceptable.}}$$

SOLUTION 2

2.1. Angle BVA is

$$\angle BVA = 180° - \angle VBA - \angle VAB$$
$$= 180° - 25° - 21°$$
$$= 134°$$
$$I = 180° - \angle BVA$$
$$= 180° - 134°$$
$$= 46°$$

From the law of sines,

$$\frac{BV}{\sin 21°} = \frac{BA}{\sin 134°}$$
$$BV = (627.13 \text{ ft})\left(\frac{\sin 21°}{\sin 134°}\right) = 312.43 \text{ ft}$$

The distance between the PC and point B is

$$(\text{sta } 120{+}10) - (\text{sta } 115{+}83) = 427 \text{ ft}$$
$$T = 427 \text{ ft} + BV = 427 \text{ ft} + 312.43 \text{ ft}$$
$$= 739.43 \text{ ft}$$

So,

$$T = R \tan \frac{I}{2}$$
$$R = \frac{739.43 \text{ ft}}{\tan \dfrac{46°}{2}} = \boxed{1742.0 \text{ ft}}$$

2.2. The degree of the curve is

$$D = \frac{5729.6}{R} = \frac{5729.6}{1742 \text{ ft}} = \boxed{3.289°}$$

2.3. The length of the curve is

$$L = (100 \text{ ft})\frac{I}{D} = (100 \text{ ft})\left(\frac{46°}{3.289°}\right)$$
$$= \boxed{1398.60 \text{ ft}}$$

2.4. The station of the PT is

$$\text{sta PT} = \text{sta PC} + L$$
$$= (\text{sta } 115{+}83) + 1398.60 \text{ ft}$$
$$= \boxed{\text{sta } 129{+}81.60}$$

2.5. The deflection angle for the whole curve is

$$\frac{I}{2} = \frac{46°}{2} = 23°$$

The deflection angle per foot of curve is

$$\frac{\dfrac{I}{2}}{L} = \frac{\dfrac{46°}{2}}{1398.60 \text{ ft}} = 0.016445°/\text{ft}$$

For sta 116+00,

$$\text{deflection angle} = \big((\text{sta } 116{+}00) - (\text{sta } 115{+}83)\big)$$
$$\times \left(100 \ \frac{\text{ft}}{\text{sta}}\right)(0.016445°/\text{ft})$$
$$= \boxed{0.2796°}$$

2.6. The deflection angle for locating sta 117+00 on the curve is

$$\text{deflection angle} = \big((\text{sta } 117{+}00) - (\text{sta } 115{+}83)\big)$$
$$\times \left(100 \ \frac{\text{ft}}{\text{sta}}\right)(0.016445°/\text{ft})$$
$$= \boxed{1.9241°}$$

Transportation

2.7. The instrument at K (sta 116+00) is as shown.

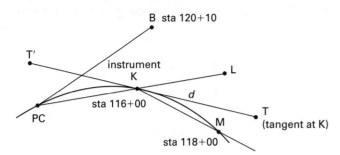

With the instrument at K (sta 116+00), locate direction KL (along line PC-L). The total deflection angle, ∠LKM, will be ∠LKT + ∠TKM. M is at sta 118+00 and TT′ is tangent to the curve at point K.

$$\angle LKT = \angle T'KPC$$
$$= \text{deflection of sta } 116{+}00$$
$$= 0.2796°$$
$$\angle TKM = \big((\text{sta } 118{+}00) - (\text{sta } 116{+}00)\big)$$
$$\times \left(\begin{array}{c}\text{deflection angle} \\ \text{per ft of curve}\end{array}\right)$$
$$= (200 \text{ ft})(0.016445°/\text{ft})$$
$$= 3.289°$$

The deflection angle for M (sta 118+00), as measured from bearing PC-L, is

$$\angle LKT + \angle TKM = 0.2796° + 3.289° = \boxed{3.5686°}$$

SOLUTION 3

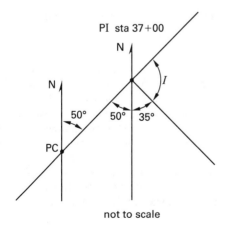

not to scale

3.1. Calculate the intersection angle to find the length of the curve.

$$I = 180° - 50° - 35° = 95°$$
$$L = RI = (800 \text{ ft})(95°)\left(\frac{2\pi}{360°}\right)$$
$$= \boxed{1326.45 \text{ ft}}$$

3.2. Calculate the tangent length to find the station of the PC.

$$T = R \tan\frac{I}{2}$$
$$= (800 \text{ ft})\tan\frac{95°}{2}$$
$$= 873.05 \text{ ft} \quad [\text{sta } 8{+}73.05]$$
$$PC = PI - T$$
$$= (\text{sta } 37{+}00) - (\text{sta } 8{+}73.05)$$
$$= \boxed{\text{sta } 28{+}26.95}$$

3.3. The station of the PT is

$$\text{sta PT} = \text{sta PC} + L$$
$$= (\text{sta } 28{+}26.95) + (\text{sta } 13{+}26.45)$$
$$= \boxed{\text{sta } 41{+}53.40}$$

3.4. The interior angle at the PI is

$$I = \boxed{95°}$$

3.5. T was calculated in Sol. 3.2.

$$T = \boxed{873.05 \text{ ft}}$$

3.6. The long chord distance is

$$C = 2R \sin\frac{I}{2}$$
$$= (2)(800 \text{ ft}) \sin\frac{95°}{2}$$
$$= \boxed{1179.64 \text{ ft}}$$

3.7. The external distance is

$$E = R \tan \frac{I}{2} \tan \frac{I}{4}$$

$$= (800 \text{ ft}) \tan \frac{95°}{2} \tan \frac{95°}{4}$$

$$= \boxed{384.15 \text{ ft}}$$

3.8. The arc basis is

$$D = \frac{5729.6}{R} = \frac{5729.6}{800 \text{ ft}}$$

$$= \boxed{7.162°}$$

3.9. The chord basis is

$$\sin \frac{D}{2} = \frac{50}{R} = \frac{50}{800 \text{ ft}}$$

$$= 0.0625$$

$$D = 2 \arcsin 0.0625$$

$$= \boxed{7.167°}$$

3.10. The chord length is

$$2R \sin \frac{D}{2} = (2)(800 \text{ ft}) \sin \frac{7.162°}{2}$$

$$= \boxed{99.935 \text{ ft}}$$

SOLUTION 4

4.1. Compound curves often fit the topography better than simple curves due to the inequality of their tangent lengths. However, compound curves also create operating disadvantages (e.g., variations in sight distances) that can lead to unsafe driving conditions. As such, a compound curve should be replaced by a simple curve where possible.

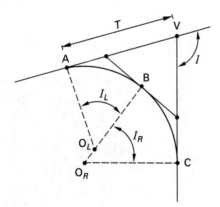

The objective as shown in the illustration is to fit a simple circular curve between two tangents: AV and CV. These are the relationships governing circular curves.

$$I = I_L + I_R$$

$$T = R \tan \frac{I}{2}$$

$$E = R \left(\sec \frac{I}{2} - 1 \right)$$

$$D = \frac{5729.6}{R}$$

$$M = R \left(1 - \cos \frac{I}{2} \right)$$

$$L = 2R \sin \frac{I}{2}$$

4.2. When used with proper transitions, a central circular curve with a smaller radius R and angle I may be used over a simple curve with no spiral transitions to increase operating safety and performance. The most common way is to use spirals to transition into and out of a central circular curve. Using AASHTO recommendations of spiral length, the governing equations are as follows.

$$L_s = \frac{1.6v^3}{R}$$

$$I_s = \frac{L_s D}{200}$$

$$D = \frac{5729.6}{R}$$

$$L_D = 100 \left(\frac{I - 2I_s}{D} \right)$$

L_D is the length of the circular curve with the degree of curve, D.

4.3. Short sections of tangents between two curves often cause driving difficulties, as multiple steering transitions are necessary to stay within the lane. This is known as a *broken-back alignment* or *broken-back curve*, and is shown in line ACDB in the following illustration. Therefore, the elimination of the tangent between curves (such as in line CD) is a common practice. If topography and clearances allow, one simple curve can be inserted between the two outer tangents that starts at point A and finishes at point X as shown.

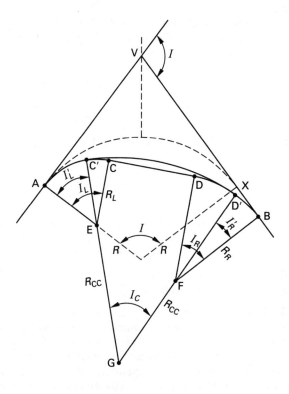

If a simple curve cannot be inserted, it may be possible to replace the short tangent section with a curve of broader radius (shown as curved line C'D') to connect the two shorter radius curves and make a three-centered curve. AASHTO recommends this center radius be no greater than twice the sharpest end radius where possible.

The radius of the new single curve, R, and the distance BX (B is the ending point of the original longer-radius curve) are given by the following equations.

$$R = \frac{R_L(\cos I_R - \cos I) + CD \sin I_R + R_R(1 - \cos I_R)}{1 - \cos I}$$

$$BX = (R_R - R_L)\sin I_R - (R - R_L)\sin I + CD \cos I_R$$

The offset, BX, between the original and proposed layouts may be so large that it prohibits the replacement.

The advantage that a three-centered curve has over a broken-back alignment is that it increases driver comfort during the superelevation transition through the center of the curve. In a three-centered curve, points C' and D' become points of a continuing curve that lie on the line through the center of the compound curves, C'EG and D'FG. The sum of the deflections equals the total deflection of the original tangents.

$$I = I_L + I_C + I_R$$

SOLUTION 5

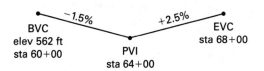

5.1. Determine the equation to find the elevations.

$$L = (\text{sta } 68{+}00) - (\text{sta } 60{+}00) = 800 \text{ ft}$$

$$y = \left(\frac{g_2 - g_1}{2LC}\right)x^2 + g_1 x + y_{\text{BVC}}$$

$$= \left(\frac{2.5\% - (-1.5\%)}{(2)(8 \text{ sta})}\right)x^2 - 1.5x + 562 \text{ ft}$$

$$= 0.25x^2 - 1.5x + 562 \text{ ft}$$

Elevations of the road at 50 ft intervals can be calculated by inserting values of $x = 0.5, 1.0, 1.5, \ldots, 8.0$ stations in the preceding equation.

$$y_{64+50} = (0.25)(4.5 \text{ sta})^2 - (1.5)(4.5 \text{ sta}) + 562 \text{ ft}$$
$$= \boxed{560.31 \text{ ft}}$$

$$y_{65+00} = (0.25)(5 \text{ sta})^2 - (1.5)(5.0 \text{ sta}) + 562 \text{ ft}$$
$$= \boxed{560.75 \text{ ft}}$$

$$y_{65+50} = (0.25)(5.5 \text{ sta})^2 - (1.5)(5.5 \text{ sta}) + 562 \text{ ft}$$
$$= \boxed{561.31 \text{ ft}}$$

$$y_{66+00} = (0.25)(6 \text{ sta})^2 - (1.5)(6 \text{ sta}) + 562 \text{ ft}$$
$$= \boxed{562.0 \text{ ft}}$$

5.2. Label the piles A through E. C is the centerline pile.

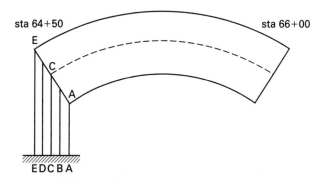

The elevation of the centerline pile (pile C) is found from Sol. 5.1, less 3 ft. There is a 12% superelevation, so for each 10 ft run, the rise is 1.2 ft.

For the first bent at sta 64+50,

pile C top $= 560.31$ ft $- 3.0$ ft $= 557.31$ ft

pile B top $= 557.31$ ft $- 1.2$ ft $= 556.11$ ft

pile A top $= 557.31$ ft $- (2)(1.2$ ft$) = 554.91$ ft

pile D top $= 557.31$ ft $+ 1.2$ ft $= 558.51$ ft

pile E top $= 557.31$ ft $+ (2)(1.2$ ft$) = 559.71$ ft

bent at station	A (ft)	B (ft)	C (ft)	D (ft)	E (ft)
64+50	554.91	556.11	557.31	558.51	559.71
65+00	555.35	556.55	557.75	558.95	560.15
65+50	555.91	557.11	558.31	559.51	560.71
66+00	556.6	557.8	559.0	560.2	561.4

SOLUTION 6

6.1. The stations of the BVC and EVC are

$$\text{sta}_{BVC} = \text{sta}_V - \frac{L}{2}$$
$$= (\text{sta } 7{+}20) - \frac{1400 \text{ ft}}{2}$$
$$= \text{sta } 0{+}20$$
$$\text{sta}_{EVC} = (\text{sta } 0{+}20) + L$$
$$= (\text{sta } 0{+}20) + 1400 \text{ ft}$$
$$= \text{sta } 14{+}20$$
$$\text{elevation}_{BVC} = \text{elevation}_V + g_1 \frac{L}{2}$$
$$= 226.88 \text{ ft} + (0.02)\left(\frac{1400 \text{ ft}}{2}\right)$$
$$= 240.88 \text{ ft}$$
$$y = \left(\frac{g_2 - g_1}{2L}\right)x^2 + g_1 x + \text{elevation}_{BVC}$$
$$= \left(\frac{4\% - (-2\%)}{(2)(14 \text{ sta})}\right)x^2 - 2x + 240.88 \text{ ft}$$
$$= 0.21429 x^2 - 2x + 240.88 \text{ ft}$$

At $y = 240$ ft, calculate x.

$$0.21429 x^2 - 2x + 240.88 = 240$$
$$x_1 = 8.871 \text{ sta} \quad (887.1 \text{ ft})$$
$$x_2 = 0.462 \text{ sta} \quad (46.2 \text{ ft})$$
$$\text{sta}_1 = (\text{sta } 0{+}20) + 46.2 \text{ ft}$$
$$= \boxed{\text{sta } 0{+}66.2}$$
$$\text{sta}_2 = (\text{sta } 0{+}20) + 887.1 \text{ ft}$$
$$= \boxed{\text{sta } 9{+}07.1}$$

Between sta 0+66.2 and sta 9+07.1, the curve drops below the flood plain. The stations dropping below this 50 yr flood plain are

> 1+00, 2+00, 3+00, 4+00, 5+00, 6+00, 7+00, 8+00, and 9+00.

6.2. The length of curve that will be submerged in a 50 yr flood is

$$907.1 \text{ ft} - 66.2 \text{ ft} = \boxed{840.9 \text{ ft}}$$

6.3. Increasing the length of curve will decrease the length of the portion that will be submerged because elevations on the curve will generally increase.

6.4. Decreasing the length of curve will increase the length of the portion that will submerge.

6.5. When the road is flooded, the embankment and the subgrade may deteriorate, which in turn degrades the pavement, regardless of pavement material. Depending on its importance, the highway may have to be rerouted. In addition, providing an adequate drainage system, paving the shoulders, and using high-quality, plant-mixed hot bituminous pavement material can help.

SOLUTION 7

7.1. The curve passes through point E at sta 75+20 and elevation 314 ft.

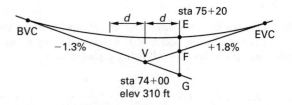

not to scale

$$EG = \text{elevation}_E - \text{elevation}_V + |dg_1|$$
$$= 314 \text{ ft} - 310 \text{ ft} + |(7520 \text{ ft} - 7400 \text{ ft})(-0.013)|$$
$$= 5.56 \text{ ft}$$
$$EF = \text{elevation}_E - \text{elevation}_V - dg_2$$
$$= 314 \text{ ft} - 310 \text{ ft} - ((7520 \text{ ft} - 7400 \text{ ft})(0.018))$$
$$= 1.84 \text{ ft}$$

The length of the curve is

$$L = 2d\left(\frac{\sqrt{\dfrac{EG}{EF}}+1}{\sqrt{\dfrac{EG}{EF}}-1}\right) = (2)(120 \text{ ft})\left(\frac{\sqrt{\dfrac{5.56 \text{ ft}}{1.84 \text{ ft}}}+1}{\sqrt{\dfrac{5.56 \text{ ft}}{1.84 \text{ ft}}}-1}\right)$$
$$= \boxed{890.13 \text{ ft}}$$

7.2. The curve length is changed to $L = 1300$ ft.

$$\text{sta}_{\text{BVC}} = \text{sta}_{\text{V}} + \frac{L}{2}$$

$$= (\text{sta } 74{+}00) - \frac{1300 \text{ ft}}{2}$$

$$= \text{sta } 67{+}50$$

$$\text{elevation}_{\text{BVC}} = \text{sta}_{\text{V}} + \frac{L}{2} g_1$$

$$= 310 \text{ ft} + \left(\frac{1300 \text{ ft}}{2}\right)(0.013)$$

$$= 318.45 \text{ ft}$$

$$y = \left(\frac{g_2 - g_1}{2L}\right) x^2 + g_1 x + \text{elevation}_{\text{BVC}}$$

$$= \left(\frac{1.8\% - (-1.3\%)}{(2)(13 \text{ sta})}\right) x^2 - 1.3x + 318.45 \text{ ft}$$

$$= 0.1192 x^2 - 1.3x + 318.45 \text{ ft}$$

At point E,

$$x_{\text{E}} = (\text{sta } 75{+}20) - (\text{sta } 67{+}50) = 7.70 \text{ sta}$$

$$y_{\text{E}} = (0.1192)(7.70 \text{ sta})^2 - (1.3)(7.70 \text{ sta}) + 318.45 \text{ ft}$$

$$= 315.51 \text{ ft}$$

The difference in elevation is

$$315.51 \text{ ft} - 314 \text{ ft} = \boxed{1.51 \text{ ft}}$$

7.3. The grate inlet should be placed at the low point on the curve.

$$y = 0.1192 x^2 - 1.3x + 318.45 \text{ ft}$$

At the low point, $dy/dx = 0$.

$$\frac{dy}{dx} = 0.2384x - 1.3 = 0$$

$$x = 5.453 \text{ sta} \quad (543.3 \text{ ft})$$

$$\text{sta}_{\text{grate inlet}} = \text{sta}_{\text{BVC}} + x$$

$$= (\text{sta } 67{+}50) + (\text{sta } 5{+}45.3)$$

$$= \boxed{\text{sta } 72{+}95.3}$$

$$\text{elevation}_{\text{grate inlet}} = (0.1192)(5.453 \text{ sta})^2$$

$$- (1.3)(5.453 \text{ sta}) + 318.45 \text{ ft}$$

$$= \boxed{314.91 \text{ ft}}$$

SOLUTION 8

8.1. Calculate the departures and latitudes for each leg.

AB: departure $= (-905.21 \text{ ft}) \sin 35.15° = -521.15 \text{ ft}$

latitude $= (905.21 \text{ ft}) \cos 35.15° = 740.14 \text{ ft}$

BC: departure $= (1135.76 \text{ ft}) \sin 81.28° = 1122.63 \text{ ft}$

latitude $= (1135.76 \text{ ft}) \cos 81.28° = 172.19 \text{ ft}$

CD: departure $= (1207.92 \text{ ft}) \sin 7.19° = 151.18 \text{ ft}$

latitude $= (-1207.92 \text{ ft}) \cos 7.19° = -1198.42 \text{ ft}$

DE: departure $= (-800.25 \text{ ft}) \sin 15.25° = -210.49 \text{ ft}$

latitude $= (-800.25 \text{ ft}) \cos 15.25° = -772.07 \text{ ft}$

EF: departure $= (-1100.85 \text{ ft}) \sin 48.17° = -820.27 \text{ ft}$

latitude $= (1100.85 \text{ ft}) \cos 48.17° = 734.18 \text{ ft}$

FA: departure $= (429.53 \text{ ft}) \sin 40.73° = 280.27 \text{ ft}$

latitude $= (429.53 \text{ ft}) \cos 40.73° = 325.49 \text{ ft}$

$$\sum \text{departure} = -521.15 \text{ ft} + 1122.63 \text{ ft} + 151.18 \text{ ft}$$

$$- 210.49 \text{ ft} - 820.27 \text{ ft} + 280.27 \text{ ft}$$

$$= 2.17 \text{ ft}$$

$$\sum \text{latitude} = 740.14 \text{ ft} + 172.19 \text{ ft} - 1198.42 \text{ ft}$$

$$- 772.07 \text{ ft} + 734.18 \text{ ft} + 325.49 \text{ ft}$$

$$= 1.51 \text{ ft}$$

The length of the traverse closure is

$$\sqrt{(2.17 \text{ ft})^2 + (1.51 \text{ ft})^2} = \boxed{2.64 \text{ ft}}$$

8.2. The total traverse length is

$$905.21 \text{ ft} + 1135.76 \text{ ft} + 1207.92 \text{ ft}$$

$$+ 800.25 \text{ ft} + 1100.85 \text{ ft} + 429.53 \text{ ft} = 5579.52 \text{ ft}$$

Use the compass rule.

$$\frac{\text{leg departure correction}}{\text{closure in departure}} = -\frac{\text{leg length}}{\text{total traverse length}}$$

$$[\text{Eq. 1}]$$

$$\frac{\text{leg latitude correction}}{\text{closure in latitude}} = -\frac{\text{leg length}}{\text{total traverse length}}$$

$$[\text{Eq. 2}]$$

Transportation

Using Eq. 1 and Eq. 2 for each leg,

$$\frac{\text{AB departure correction}}{2.17 \text{ ft}} = -\frac{905.21 \text{ ft}}{5579.52 \text{ ft}}$$

$$\text{AB departure correction} = -\left(\frac{905.21 \text{ ft}}{5579.52 \text{ ft}}\right)(2.17 \text{ ft})$$
$$= -0.35 \text{ ft}$$

$$\text{AB latitude correction} = -\left(\frac{905.21 \text{ ft}}{5579.52 \text{ ft}}\right)(1.51 \text{ ft})$$
$$= -0.24 \text{ ft}$$

$$\text{BC departure correction} = -\left(\frac{1135.76 \text{ ft}}{5579.52 \text{ ft}}\right)(2.17 \text{ ft})$$
$$= -0.44 \text{ ft}$$

$$\text{BC latitude correction} = -\left(\frac{1135.76 \text{ ft}}{5579.52 \text{ ft}}\right)(1.51 \text{ ft})$$
$$= -0.31 \text{ ft}$$

$$\text{CD departure correction} = -\left(\frac{1207.92 \text{ ft}}{5579.52 \text{ ft}}\right)(2.17 \text{ ft})$$
$$= -0.47 \text{ ft}$$

$$\text{CD latitude correction} = -\left(\frac{1207.92 \text{ ft}}{5579.52 \text{ ft}}\right)(1.51 \text{ ft})$$
$$= -0.33 \text{ ft}$$

$$\text{DE departure correction} = -\left(\frac{800.25 \text{ ft}}{5579.52 \text{ ft}}\right)(2.17 \text{ ft})$$
$$= -0.31 \text{ ft}$$

$$\text{DE latitude correction} = -\left(\frac{800.25 \text{ ft}}{5579.52 \text{ ft}}\right)(1.51 \text{ ft})$$
$$= -0.22 \text{ ft}$$

$$\text{EF departure correction} = -\left(\frac{1100.85 \text{ ft}}{5579.52 \text{ ft}}\right)(2.17 \text{ ft})$$
$$= -0.43 \text{ ft}$$

$$\text{EF latitude correction} = -\left(\frac{1100.85 \text{ ft}}{5579.52 \text{ ft}}\right)(1.51 \text{ ft})$$
$$= -0.30 \text{ ft}$$

$$\text{FA departure correction} = -\left(\frac{429.53 \text{ ft}}{5579.52 \text{ ft}}\right)(2.17 \text{ ft})$$
$$= -0.17 \text{ ft}$$

$$\text{FA latitude correction} = -\left(\frac{429.53 \text{ ft}}{5579.52 \text{ ft}}\right)(1.51 \text{ ft})$$
$$= -0.12 \text{ ft}$$

Find the adjusted lengths.

AB: departure $= -521.15 \text{ ft} - 0.35 \text{ ft} = -521.50 \text{ ft}$

latitude $= 740.14 \text{ ft} - 0.24 \text{ ft} = 739.90 \text{ ft}$

BC: departure $= 1122.63 \text{ ft} - 0.44 \text{ ft} = 1122.19 \text{ ft}$

latitude $= 172.19 \text{ ft} - 0.31 \text{ ft} = 171.88 \text{ ft}$

CD: departure $= 151.18 \text{ ft} - 0.47 \text{ ft} = 150.71 \text{ ft}$

latitude $= -1198.42 \text{ ft} - 0.33 \text{ ft} = -1198.75 \text{ ft}$

DE: departure $= -210.49 \text{ ft} - 0.31 \text{ ft} = -210.80 \text{ ft}$

latitude $= -772.07 \text{ ft} - 0.22 \text{ ft} = -772.29 \text{ ft}$

EF: departure $= -820.27 \text{ ft} - 0.43 \text{ ft} = -820.70 \text{ ft}$

latitude $= 734.18 \text{ ft} - 0.30 \text{ ft} = 733.88 \text{ ft}$

FA: departure $= 280.27 \text{ ft} - 0.17 \text{ ft} = 280.10 \text{ ft}$

latitude $= 325.49 \text{ ft} - 0.12 \text{ ft} = 325.37 \text{ ft}$

8.3. Use the double meridian distance (DMD) method (see *Table for Sol. 8.3*).

$$A = \left(\frac{1}{2}\right)(2{,}890{,}823.64 \text{ ft}^2)$$
$$= \boxed{1{,}445{,}411.82 \text{ ft}^2}$$

8.4. The length of BC is

$$\text{BC} = \sqrt{(171.88 \text{ ft})^2 + (1122.19 \text{ ft})^2} = 1135.28 \text{ ft}$$

Table for Sol. 8.3

leg	latitude (ft)	departure (ft)	DMD (ft)	latitude × DMD (ft²)
AB	739.90	−521.50	−521.50	−385,857.85
BC	171.88	1122.19	−521.50 − 521.50 + 1122.19 = 79.19	13,611.18
CD	−1198.75	150.71	79.19 + 1122.19 + 150.71 = 1352.09	−1,620,817.89
DE	−772.29	−210.80	1352.09 + 150.71 − 210.80 = 1292.00	−997,798.68
EF	733.88	−820.70	1292.00 − 210.80 − 820.70 = 260.50	191,175.74
FA	325.37	280.10	−280.10	−91,136.14
				total −2,890,823.64

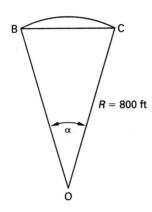

$$\sin\frac{\alpha}{2} = \frac{\dfrac{BC}{2}}{R} = \frac{\dfrac{1135.28 \text{ ft}}{2}}{800 \text{ ft}}$$
$$= 0.70955$$
$$\alpha = 2\arcsin 0.70955 = 90.397°$$

The area of circular region BC is

$$\left(\frac{90.397°}{360°}\right)\pi R^2 = \left(\frac{90.397°}{360°}\right)\pi (800 \text{ ft})^2$$
$$= 504{,}872.1 \text{ ft}^2$$

The area of triangle OBC is

$$\frac{(BC)R\cos\dfrac{\alpha}{2}}{2} = \frac{(1135.28 \text{ ft})(800 \text{ ft})\cos\dfrac{90.397°}{2}}{2}$$
$$= 319{,}991.3 \text{ ft}^2$$

The area of the circular segment bounded by the chord and arc BC is

$$504{,}870.1 \text{ ft}^2 - 319{,}991.3 \text{ ft}^2 = \boxed{184{,}878.8 \text{ ft}^2}$$

8.5. The length of curve BC is

$$BC = R\alpha = (800 \text{ ft})(90.397°)\left(\frac{2\pi}{360°}\right)$$
$$= \boxed{1262.18 \text{ ft}}$$

SOLUTION 9

9.1. The sum of the internal angles is

$$(n-2)(180°) = (5-2)(180°)$$
$$= 540°$$
$$\angle EAB = 540° - (124° + 82° + 110° + 115°)$$
$$= \boxed{109°}$$

9.2. For the bearing of line BC,

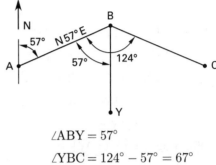

$$\angle ABY = 57°$$
$$\angle YBC = 124° - 57° = 67°$$
$$\text{bearing}_{BC} = \boxed{S\,67°\,E}$$

9.3. For the bearing of line CD,

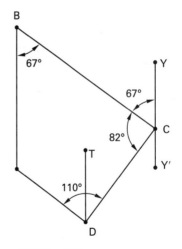

$$\angle BCY = 67°$$
$$\angle DCY' = 180° - 82° - 67° = 31°$$
$$\text{bearing}_{CD} = \boxed{S\,31°\,W}$$

9.4. For the bearing of line DE,

$$\angle TDC = \angle DCY' = 31°$$
$$\text{bearing}_{DE} = 110° - 31° = 79°$$
$$= \boxed{N\,79°\,W}$$

9.5. For the bearing of line EA,

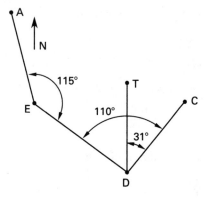

$$\angle TDE = 110° - 31° = 79°$$

$$\angle NED = 180° - 79° = 101°$$

$$\angle AEN = 115° - \angle NED$$

$$= 115° - 101°$$

$$= 14°$$

$$\text{bearing}_{EA} = \boxed{N\,14°\,W}$$

Topic VI: Systems, Management, and Professional

Chapter

Systems, Mgmt,
and Professional

16 Engineering Economic Analysis

PROBLEM 1

The grass in a public golf course turns brown each summer because the irrigation system has deteriorated to the point where the system must be replaced. Three alternatives are available to replace the existing irrigation system. The chosen alternative will be funded by tax-free bonds that pay 8% to the bearer.

Each alternative requires the installation of a pump and piping. The pump is expected to last 10 years and have no salvage value after that time. Replacement pumps of the same design will be used to extend the operation to 30 years. Each pump vendor has guaranteed to provide the replacement pumps at the same cost as the originals. The piping can be expected to last 30 years and will have no salvage value after that period of time.

alternative	A	B	C
initial pump cost	$7500	$15,000	$30,000
initial piping cost	$15,000	$15,000	$22,500
annual maintenance	$6000	$4500	$3000
pumping cost per 100 gal	$4.50	$3.75	$3.00

1.1. If the pumping cost is disregarded, which alternative should be chosen?

1.2. Is the selection sensitive to the volume of water pumped?

PROBLEM 2

A small partnership spent three years developing a product. The company spent $55,000 in the first year of development, $75,000 in the second year, and $85,000 in the third year. At the end of the third year, the product was immediately placed into production.

The partners have an average personal tax rate of 46% and want an 18% return (after taxes) on their investment. All partnership profits and losses are passed through to the partners. The development costs are to be depreciated over a three year period after the start of production using the following percentages: first year, 25%; second year, 38%; and third year, 37%. There is no (investment) tax credit. Assume all development costs are depreciable.

The company expects to produce and sell 4000 units each year for the 10 years following the beginning of production. The unit manufacturing cost, including all labor, material, and overhead (but excluding taxes and development costs) is $60. What amount should be added to the cost of each unit to recover all development costs over the 10 year period?

PROBLEM 3

Route 420 is currently the only way to get between two cities; it carries 1200 commercial vehicles per day at an average speed of 50 mph. Route 422 is being proposed as a replacement for Route 420.

	Route 420	Route 422
length	9.2 mi	6.5 mi
initial cost	$0	$1,200,000
pavement life	15 yr	15 yr
resurfacing cost	$300,000	$260,000

The vehicle operating cost is $0.22 per mile for both routes. The time savings is $0.30 per vehicle-minute. Consider costs and savings only for the commercial traffic, and neglect maintenance (vehicle and pavement) and all other factors. The average speed is unchanged. Assume an infinite life and zero salvage value. Use the incremental benefit-cost ratio method with an interest rate of 10%. Which route is superior?

PROBLEM 4

A disposal site serves a population of 100,000, and this number is not expected to change during the next 10 years. Municipal solid waste (MSW) is collected daily at the rate of 5 lbm per person. This quantity is expected to increase 5% annually. The composition of the MSW and the fraction of each component recovered for recycling are

	fraction in MSW	fraction recoverable
combustible materials (includes plastics)	50%	60%
ferrous materials	8%	90%
glass	15%	80%
aluminum	5%	70%
other (mineral matter, yard waste, other metals, etc.)	22%	0%

The disposal site sells its recoverables at the following prices.

- combustibles: 50% of the price of coal. Coal is selling at $45/ton. This price is expected to increase at the rate of 4% per year.

- ferrous: 50% of the price of steel. Steel is selling at $80/ton. This price is expected to increase at the rate of 8% per year.

- glass: $20/ton. This price is expected to increase at the rate of 2% per year.

- aluminum: $200/ton. This price is expected to increase at the rate of 12% per year.

4.1. What is the annual revenue from recoverables in the first year of operation?

4.2. What is the annual revenue from recoverables in the fifth year of operation?

4.3. What is the annual revenue from recoverables in the tenth year of operation?

PROBLEM 5

A municipality intends to purchase new pickup trucks for its public works department and drive them 17,000 mi each year. A bid has been received from a fleet dealer of $10,000 per truck, with the following guaranteed salvage values.

end of year	salvage value
1	$8000
2	$6500
3	$5000
4	$3000
5	$2500
6	$2000
7	$1500
8	$1000

Annual operating costs are expected to be $1250 for fuel and insurance. Annual maintenance costs are expected to start at $500 and increase $100 each year. An interest rate of 10% is used for comparison of alternatives.

5.1. What is the total equivalent uniform annual cost (EUAC) of ownership if a vehicle is kept for eight years?

5.2. What is the total present worth if a vehicle is kept for eight years?

5.3. What is the total equivalent uniform annual cost (EUAC) of ownership if a vehicle is kept for four years?

5.4. What is the total present worth if a vehicle is kept for two years?

5.5. What is the most economical length of time to keep a vehicle?

Use the following information for the remaining parts of this problem: A local garage has agreed to perform all maintenance during the next eight years if the municipality pays it $3500 per vehicle now. This will eliminate the annual maintenance performed by the municipality.

5.6. What is the total equivalent uniform annual cost (EUAC) of ownership if a vehicle is kept for eight years?

5.7. What is the total present worth if a vehicle is kept for eight years?

5.8. What is the most economical length of time to keep a vehicle?

PROBLEM 6

A contractor submitted a bid to an owner-developer to construct a new office building for $2,400,000, to be paid in a single payment at the completion of construction. As an alternative, the contractor agreed to give the developer three years (starting from the time of completion) to make payments of 50%, 25%, and 25% of the bid amount, with adjustments for interest (figured at 10% per year before taxes) and inflation (figured at 5% per year). The contractor's actual cost of construction is $2,000,000, which would be payable to subcontractors, employees, and suppliers at the completion of the project. The contractor's income tax rate is 45%. All other accounting conventions are to be neglected.

6.1. If the construction bid price is paid off at the time of completion, what is the contractor's after-tax rate of return (ROR)?

6.2. If the construction bid price is paid off over three years following completion, what is the contractor's before-tax rate of return (ROR)?

6.3. If the owner-developer depreciates the building over 25 years using straight-line depreciation and $0 salvage value, what is the before-tax present worth of the revenue that must be generated to recover the building cost? The owner-developer is accustomed to receiving an after-tax 15% return on his investments. Assume an income tax rate of 45% for the owner-developer, and that the owner-developer pays the contractor in a single payment.

SOLUTION 1

1.1. Alternative A:

$$P = \$7500 + \$15{,}000 + (\$6000)(P/A, 8\%, 30)$$
$$+ (\$7500)(P/F, 8\%, 10) + (\$7500)(P/F, 8\%, 20)$$
$$= \$7500 + \$15{,}000 + (\$6000)(11.2578)$$
$$+ (\$7500)(0.4632 + 0.2145)$$
$$= \$95{,}130$$

Alternative B:

$$P = \$15{,}000 + \$15{,}000 + (\$4500)(P/A, 8\%, 30)$$
$$+ (\$15{,}000)\big((P/F, 8\%, 10) + (P/F, 8\%, 20)\big)$$
$$= \$15{,}000 + \$15{,}000 + (\$4500)(11.2578)$$
$$+ (\$15{,}000)(0.4632 + 0.2145)$$
$$= \$90{,}826$$

Alternative C:

$$P = \$30{,}000 + \$22{,}500 + (\$3000)(P/A, 8\%, 30)$$
$$+ (\$30{,}000)\big((P/F, 8\%, 10) + (P/F, 8\%, 20)\big)$$
$$= \$30{,}000 + \$22{,}500 + (\$3000)(11.2578)$$
$$+ (\$30{,}000)(0.4632 + 0.2145)$$
$$= \$106{,}604$$

> Choose Alternative B.

1.2. The volume of water (in gallons) pumped per year is V.

Alternative A:

$$P = \$95{,}130 + \left(\frac{\$4.5}{100 \text{ gal}}\right) V(P/A, 8\%, 30)$$
$$= \$95{,}130 + \left(\frac{\$4.5}{100 \text{ gal}}\right) V(11.2578)$$
$$= \$95{,}130 + 0.5066V$$

Alternative B:

$$P = \$90{,}826 + \left(\frac{\$3.75}{100 \text{ gal}}\right) V(P/A, 8\%, 30)$$
$$= \$90{,}826 + \left(\frac{\$3.75}{100 \text{ gal}}\right) V(11.2578)$$
$$= \$90{,}826 + 0.4222V$$

Alternative C:

$$P = \$106{,}604 + \left(\frac{\$3.00}{100 \text{ gal}}\right) V(P/A, 8\%, 30)$$
$$= \$106{,}604 + \left(\frac{\$3.00}{100 \text{ gal}}\right) V(11.2578)$$
$$= \$106{,}604 + 0.3377V$$

The selection is sensitive to the volume of water pumped per year. B is always superior to A, but C will become superior to B if $V > 186{,}722$ gal. At $V = 186{,}722$ gal,

$$P_B = P_C = \$169{,}660$$

> If $V < 186{,}722$ gal, choose B. If $V > 186{,}722$ gal, choose C.

SOLUTION 2

The tax rate is

$$t = 46\%$$
$$\text{total investment} = \$55{,}000 + \$75{,}000 + \$85{,}000$$
$$= \$215{,}000$$

Depreciation for each year is

$$D_1 = (\$215{,}000)(0.25) = \$53{,}750$$
$$D_2 = (\$215{,}000)(0.38) = \$81{,}700$$
$$D_3 = (\$215{,}000)(0.37) = \$79{,}550$$

The production cost per year is

$$(4000)(\$60) = \$240{,}000$$

If the price per item is x, the revenue per item is $4000x$.

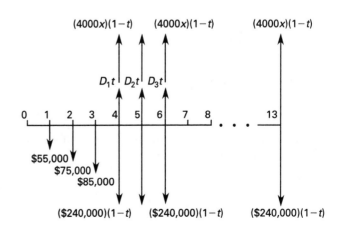

$$P = 0 = -(\$55,000)(P/F, 18\%, 1)$$
$$-(\$75,000)(P/F, 18\%, 2)$$
$$-(\$85,000)(P/F, 18\%, 3)$$
$$-(\$240,000)(1 - 0.46)$$
$$\times (P/A, 18\%, 10)(P/F, 18\%, 3)$$
$$+(\$53,750)(0.46)(P/F, 18\%, 4)$$
$$+(\$81,700)(0.46)(P/F, 18\%, 5)$$
$$+(\$79,550)(0.46)(P/F, 18\%, 6)$$
$$+(4000x)(1 - 0.46)(P/A, 18\%, 10)$$
$$\times (P/F, 18\%, 3)$$
$$P = 0 = -(\$55,000)(0.8475) - (\$75,000)(0.7182)$$
$$-(\$85,000)(0.6086)$$
$$-(\$240,000)(1 - 0.46)(4.4941)(0.6086)$$
$$+(\$53,750)(0.46)(0.5158)$$
$$+(\$81,700)(0.46)(0.4371)$$
$$+(\$79,550)(0.46)(0.3704)$$
$$+(4000)(1 - 0.46)(4.4941)(0.6086)x$$

$$5907.84x - 463,944.4 = 0$$

$$x = \boxed{\$78.53}$$

The amount to be added to the cost of each unit is

$$\$78.53 - \$60 = \boxed{\$18.53}$$

SOLUTION 3

For Route 420, the annual operating costs are

$$\left(365 \ \frac{\text{days}}{\text{yr}}\right)\left(1200 \ \frac{\text{veh}}{\text{day}}\right)\left(0.22 \ \frac{\$}{\text{veh-mi}}\right)(9.2 \ \text{mi})$$
$$= \$886,512/\text{yr}$$

For Route 422, the annual operating costs are

$$\left(365 \ \frac{\text{days}}{\text{yr}}\right)\left(1200 \ \frac{\text{veh}}{\text{day}}\right)\left(0.22 \ \frac{\$}{\text{veh-mi}}\right)(6.5 \ \text{mi})$$
$$= \$626,340/\text{yr}$$

Use capitalized cost because the project has an infinite life. The benefit-cost ratio is

$$\frac{B}{C} = \frac{P_{\Delta B}}{P_{\Delta I} + P_{\Delta M}}$$

$$\Delta = \text{Route } 422 - \text{Route } 420$$
$$P_{\Delta B} = P \text{ of increased user benefits}$$
$$P_{\Delta I} = P \text{ of increased investment costs}$$
$$P_{\Delta M} = P \text{ of increased maintenance costs}$$

Calculate users' benefits derived from using Route 422 instead of Route 420. The value of time saved is

$$\left(365 \ \frac{\text{days}}{\text{yr}}\right)\left(1200 \ \frac{\text{veh}}{\text{day}}\right)\left(\left(\frac{9.2 \ \text{mi}}{50 \ \frac{\text{mi}}{\text{hr}}}\right)\left(60 \ \frac{\text{min}}{\text{hr}}\right)\right.$$
$$\left.-\left(\frac{6.5 \ \text{mi}}{50 \ \frac{\text{mi}}{\text{hr}}}\right)\left(60 \ \frac{\text{min}}{\text{hr}}\right)\right)\left(0.30 \ \frac{\$}{\text{veh-min}}\right)$$
$$= \$425,736/\text{yr}$$

The operating costs saved per year are

$$\frac{\$886,512}{1 \ \text{yr}} - \frac{\$626,340}{1 \ \text{yr}} = \$260,172/\text{yr}$$

$$P_{\Delta B} = \frac{\$425,736 + \$260,172}{0.10}$$
$$= \$6,859,080$$
$$P_{\Delta I} = \$1,200,000 - \$0$$
$$= \$1,200,000$$

ΔM is the decrease in the cost of resurfacing. It occurs once every 15 years, and it is $\$260,000 - \$300,000 = -\$40,000$.

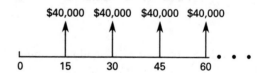

$$P_{\Delta M} = -(\$40,000)(P/F, 10\%, 15)$$
$$-(\$40,000)(P/F, 10\%, 30)$$
$$-(\$40,000)(P/F, 10\%, 45) - \cdots$$
$$= -(\$40,000)(0.2394 + 0.0573 + 0.0137 + \cdots)$$
$$= -\$12,416$$
$$\frac{B}{C} = \frac{\$6,859,080}{\$1,200,000 - \$12,416}$$
$$= 5.78 > 1.0$$

$$\boxed{\text{Choose Route 422.}}$$

SOLUTION 4

4.1. The daily MSW mass (first year) is

$$\left(5 \; \frac{\text{lbm}}{\text{person}}\right)(100{,}000 \text{ persons}) = 500{,}000 \text{ lbm/day}$$

The total MSW mass (first year) is

$$(365 \text{ days})\left(\frac{500{,}000 \; \frac{\text{lbm}}{\text{day}}}{2000 \; \frac{\text{lbm}}{\text{ton}}}\right) = 91{,}250 \text{ tons}$$

MSW composition	fraction recoverable	price/ton
combustibles	$(0.50)(0.60) = 0.30$	$(0.5)(\$45) = \$22.50$
ferrous materials	$(0.08)(0.90) = 0.072$	$(0.5)(\$80) = \$40$
glass	$(0.15)(0.80) = 0.12$	$\$20$
aluminum	$(0.05)(0.70) = 0.035$	$\$200$

Because the MSW mass increases 5% annually and prices increase in varying degrees, effective growth rates, i, must be calculated for each type of recoverable.

$$\text{combustibles:} \quad i_C = (1 + 0.05)(1 + 0.04) - 1$$
$$= 0.092$$

$$\text{ferrous materials:} \quad i_F = (1 + 0.05)(1 + 0.08) - 1$$
$$= 0.134$$

$$\text{glass:} \quad i_G = (1 + 0.05)(1 + 0.02) - 1$$
$$= 0.071$$

$$\text{aluminum:} \quad i_A = (1 + 0.05)(1 + 0.12) - 1$$
$$= 0.176$$

The first year's revenues are

combustibles:	$(91{,}250)(0.30)(\$22.50) =$	$\$615{,}937$
ferrous materials:	$(91{,}250)(0.072)(\$40) =$	$\$262{,}800$
glass:	$(91{,}250)(0.12)(\$20) =$	$\$219{,}000$
aluminum:	$(91{,}250)(0.035)(\$200) =$	$\$638{,}750$
total:		$\overline{\$1{,}736{,}487}$

The total revenues in the first year are

$$\boxed{\$1{,}736{,}487}$$

4.2. The fifth year's revenues are

$$(\$615{,}937)(1 + i_C)^4 + (\$262{,}800)(1 + i_F)^4$$
$$+ (\$219{,}000)(1 + i_G)^4 + (\$638{,}750)(1 + i_A)^4$$
$$= (\$615{,}937)(1 + 0.092)^4 + (\$262{,}800)(1 + 0.134)^4$$
$$+ (\$219{,}000)(1 + 0.071)^4 + (\$638{,}750)(1 + 0.176)^4$$
$$= \boxed{\$2{,}820{,}259}$$

4.3. The tenth year's revenues are

$$(\$615{,}937)(1 + i_C)^9 + (\$262{,}800)(1 + i_F)^9$$
$$+ (\$219{,}000)(1 + i_G)^9 + (\$638{,}750)(1 + i_A)^9$$
$$= (\$615{,}937)(1 + 0.092)^9 + (\$262{,}800)(1 + 0.134)^9$$
$$+ (\$219{,}000)(1 + 0.071)^9 + (\$638{,}750)(1 + 0.176)^9$$
$$= \boxed{\$5{,}328{,}872}$$

SOLUTION 5

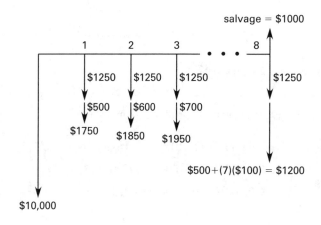

5.1. The total EUAC for eight years is

$$A_8 = (\$10{,}000)(A/P, 10\%, 8) + \$1250 + \$500$$
$$+ (\$100)(A/G, 10\%, 8)$$
$$- (\$1000)(A/F, 10\%, 8)$$
$$= (\$10{,}000)(0.1874) + \$1250 + \$500$$
$$+ (\$100)(3.0045) - (\$1000)(0.0874)$$
$$= \boxed{\$3837.05}$$

5.2. The total present worth for eight years is

$$P = A_8(P/A, i\%, n)$$
$$= -(\$3837.05)(P/A, 10\%, 8)$$
$$= -(\$3837.05)(5.3349)$$
$$= \boxed{-\$20{,}470.28}$$

5.3. The total EUAC for four years is

$$A_4 = (\$10,000)(A/P, 10\%, 4) + \$1750$$
$$+ (\$100)(A/G, 10\%, 4)$$
$$- (\$3000)(A/F, 10\%, 4)$$
$$= (\$10,000)(0.3155) + \$1750$$
$$+ (\$100)(1.3812) - (\$3000)(0.2155)$$
$$= \boxed{\$4396.62}$$

5.4. The total present worth for two years is

$$P = -\$10,000 - \frac{\$1750}{1.1} - \frac{\$1850}{(1.1)^2} + \frac{\$6500}{(1.1)^2}$$
$$= \boxed{-\$7747.93}$$

5.5. Calculate the EUAC for seven years.

$$A_7 = (\$10,000)(A/P, 10\%, 7) + (\$100)(A/G, 10\%, 7)$$
$$+ \$1750 - (\$1500)(A/F, 10\%, 7)$$
$$= (\$10,000)(0.2054) + (\$100)(2.6216)$$
$$+ \$1750 - (\$1500)(0.1054)$$
$$= \$3908.06$$

A_8 was calculated in Sol. 5.1 as \$3837.05. Because $A_8 < A_7$, the most economical life is $\boxed{\text{eight years.}}$

5.6.

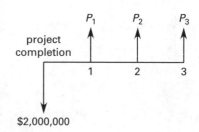

$$A_8 = (\$13,500)(A/P, 10\%, 8) + \$1250$$
$$- (\$1000)(A/F, 10\%, 8)$$
$$= (\$13,500)(0.1874) + \$1250 - (\$1000)(0.0874)$$
$$= \boxed{\$3692.50}$$

5.7. The total present worth for eight years is $P = A_8(P/A, i\%, n)$.

From Sol. 5.6,

$$A_8 = \$3692.50$$
$$P = -(\$3692.50)(P/A, 10\%, 8)$$
$$= -(\$3692.50)(5.3349)$$
$$= \boxed{-\$19,699.12}$$

5.8. Calculate A_7 and compare with A_8 as calculated in Sol. 5.6.

$$A_7 = (\$13,500)(A/P, 10\%, 7) + \$1250$$
$$- (\$1500)(A/F, 10\%, 7)$$
$$= (\$13,500)(0.2054) + \$1250 - (\$1500)(0.1054)$$
$$= \boxed{\$3864.80}$$

Since $A_7 < A_8$, keep the vehicle for $\boxed{\text{eight years.}}$

SOLUTION 6

6.1. Calculate the profit, tax, and after-tax profit.

$$\text{profit} = \$2,400,000 - \$2,000,000$$
$$= \$400,000$$
$$\text{tax} = (\$400,000)(0.45) = \$180,000$$
$$\text{after-tax profit} = \$400,000 - \$180,000$$
$$= \$220,000$$

The contractor's after-tax ROR is

$$\text{after-tax ROR} = \frac{\$220,000}{\$2,000,000}$$
$$= \boxed{11\%}$$

6.2.

The effective rate, i', considering interest, i, and inflation, e, is

$$i' = i + e + ie$$
$$= 0.10 + 0.05 + (0.10)(0.05)$$
$$= 0.155$$

The first payment is

$$P_1 = (0.50)(\$2,400,000)(1.155)$$
$$= \$1,386,000$$

The second payment is

$$P_2 = (0.25)(\$2,400,000)(1.155)^2$$
$$= \$800,415$$

The third payment is

$$P_3 = (0.25)(\$2,400,000)(1.155)^3$$
$$= \$924,479$$

The ROR is the rate that results in a zero present worth.

$$P = -\$2,000,000 + \frac{P_1}{1 + \text{ROR}} + \frac{P_2}{(1 + \text{ROR})^2}$$
$$+ \frac{P_3}{(1 + \text{ROR})^3}$$
$$= 0$$

Try an ROR of 25%.

$$P = -\$2,000,000 + \frac{\$1,386,000}{1.25} + \frac{\$800,415}{(1.25)^2}$$
$$+ \frac{\$924,479}{(1.25)^3}$$
$$= \$94,399$$

Try an ROR of 30%.

$$P = -\$2,000,000 + \frac{\$1,386,000}{1.30} + \frac{\$800,415}{(1.30)^2} + \frac{\$924,479}{(1.30)^3}$$
$$= -\$39,436$$

The before-tax ROR is between 25% and 30%; by further trial and error, the before-tax ROR is $\boxed{28.5\%.}$

6.3.

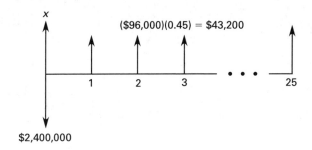

The depreciation over 25 years is

$$D = \frac{\$2,400,000}{25} = \$96,000$$

$$(x - \$2,400,000) + (\$96,000)(0.45)(P/A, 15\%, 25) = 0$$
$$(x - \$2,400,000) + (\$96,000)(0.45)(6.4641) = 0$$
$$x = \$2,120,750 \quad [\text{aftertax}]$$

The before-tax present worth of revenues is

$$\frac{\$2,120,750}{1 - 0.45} = \boxed{\$3,855,909}$$

BEARY GOOD BEARS!

Author	Christine Clementz Stack
Illustrator	Jenny Campbell
Editorial/Art Director	Karen Sevaly
Graphic Designers	Cory Jackson
	Jim Sevaly

Look for the entire series of Teacher's Friend
Early Childhood Theme Books at your local school supplier.

Copyright © 2001
Teacher's Friend Publications, Inc.
All rights reserved.
Printed in the United States of America
Published by Teacher's Friend Publications, Inc.
3240 Trade Center Dr., Riverside, CA 92507

ISBN 0-439-50006-0
TF4101

teacher's friend publications

Table of Contents

About the Author

Christine Clemente Stack is from a suburban community in upstate New York. She holds permanent certification as a Special Education Teacher. As an educator for over twelve years, she has had the opportunity to spend many hours dedicated to the encouragement and enhancement of the development of young children. She has been engaged in several professional opportunities, including teacher, supervisor, presenter/trainer, facilitator, collaborator, mentor, Early Intervention Specialist, and Child Development Specialist. Her most rewarding aspect of each position has been the time spent with children and their families. She encourages you to enjoy this book and hopes that it will help you to better serve each young child in your care.

Safety Warning!

Parents and professionals should use their best judgment to determine whether a particular idea or suggestion would be appropriate to use with an individual child or group of children.

It is important that children only use materials and products labeled child-safe and non-toxic. Small children should always be supervised by a competent adult and youngsters must never be allowed to put small objects or art materials in their mouths. Please consult the manufacturer's safety warnings on all materials and equipment used with young children.

When using food products, make sure that you take into consideration any children's allergies or food restrictions. Make sure that you take special note of the beliefs, values, and culture of the families you serve.

INTRODUCTION

This 48 page thematic book about bears contains a collection of skill-based activities and ideas for children ages 3 to 6 years. The ideas and suggestions are based on the philosophy of developing the whole child through play, exploration and varied experiences. These activities should enhance the child's ability to develop communication, motor/physical, cognitive, social/emotional, and adaptive/self-help skills. The teacher should approach and implement the activities in a way that takes into consideration each child's individual needs and abilities.

Many of the activities have been organized to meet the criteria of the interest areas (centers) found in most early childhood classrooms. They include:

• Art and Crafts	• Library/Writing	• Dramatic Play
• Housekeeping	• Math and Manipulatives	• Sand and Water
• Outdoor Play	• Music and Movement	• Computer Center
• Blocks and Building	• Table Toys	• Cooking and Nutrition

Opportunities to develop pre-literacy and literacy skills should be apparent in every center in the classroom. Using the activities in this book, children can explore the basic concepts of color, shape, number, quantity and position that naturally occur in a stimulating, early childhood environment. It is essential for early childhood educators to build each child's vocabulary every day. It is one of the basic components needed by children in order for them to become successful learners and literate adults. In almost every activity throughout the day, there will be an occasion to incorporate information related to basic language concepts. With this book, one will discover activities and games which will allow the children to explore ideas and vocabulary which will help them develop essential skills.

HOW TO USE THIS BOOK

This resource book includes an array of ideas, activities and reproducible pages that you and other early childhood educators can utilize to develop stimulating activities around a given theme. The graphics, clip art, game ideas and other reproducible pages throughout the book are meant to be copied for individual classroom use. You may adapt or modify the ideas or instructions to best meet the developmental level of the children you serve. As a time saver, protect your work by laminating the products or placing the reproducible pages in plastic page protectors. The following describes each section and some helpful tips and hints regarding how to implement the ideas into your curriculum:

Everything You Need to Know About...
This section includes a number of interesting facts or points about bears in general. It also includes a list of websites that may be used as a resource for additional information on the topic.

Dictation and/or Creative Writing Page　　Library and Writing
Use this page for children to write their own words or stories. Make several copies and place them in your writing center with appropriate writing tools.

Clip Art
There are many uses for these cute, simple thematic illustrations. Each graphic can be enlarged or reduced to meet your individual needs. You can use the clip art illustrations in the following ways:

In newsletters and notes	In rebus recipe charts	In thank you notes
On rebus stories or letters	On song charts	On homemade games
In experiential charts or books	In each center or area	With your calendar
As part of your routine chart	On charts and bulletin boards	On rewards/certificates

Awards and Certificates

Use these cute, thematic certificates to reward children or thank volunteers. You can use the awards to:

Recognize accomplishments   Praise good deeds   Identify strengths

Recognize good sportsmanship   Support teamwork   Thank volunteers

Acknowledge acts of kindness   Support positive behavior   Acknowledge good choices

Thematic Literature List   `Library and Writing`

This list of children's literature with the theme of bears, includes books to read to the class, picture books, and books for beginning readers. A list of vocabulary words pertinent to the theme of bears is also included. Here are a few tips for reading books aloud:

- Select good, well written and illustrated books
- Position yourself so that each child can see the book
- Review some of the words in the book before you read it
- Change the intonation of your voice to dramatize the story
- Read leaving words out to elicit responses from children
- Point out beginning consonant sounds (phoneme) and connect it with the letter (symbol)
- Comment on words that rhyme
- Point out the title, author and illustrator
- Point to words from left to right and top to bottom, etc.
- Keep the children on track - listening to the story
- Ask "wh" questions- "what, where, why, when"
- Expand on the concepts in the book in other centers
- Repeat readings of the same story

Songs, Poems and Fingerplays   `Music and Movement`

Songs and fingerplays are an essential piece of an early childhood curriculum. These simple thematic jingles are set to familiar children's songs. Short poems and fingerplays are also included in this section. They help children to learn about the rhythm and rhyme of our language. (Copies of the songs and poems can be shared with parents and used at home.) The songs can be written on large poster board so that children, families and volunteers can follow along. Clip art pictures can be added in place of words to create a rebus song chart. Develop a song basket filled with copies of each song. (Song cards can be made by copying each song to a large index card. Glue a clip art picture to the back of each card and laminate for durability. Place the song cards in a large basket for children to easily select a song they wish to sing.)

Activities and Games

These skill-based activities and games can be used with the entire class, small groups of children or by individual children. The activities have been organized as they relate to interest centers typically found in the early childhood classroom. Use the ideas and activities in each section to facilitate the following skill development and concepts:

`Dramatic Play`

exploring emotions   practicing skills   vocabulary development

dress-up & make believe   sharing & turn-taking   using imagination

using creativity   imitation   social skill development

acting out life experiences   resolving conflicts   self-help skills

pretend play & role play   receptive & expressive language development

Outdoor Play & Field Trips

exercise	exploration	gross-motor skills
social skills	following directions	teamwork
trying new things	fair play & sportsmanship	building concepts

Math and Manipulatives (Science)

using math & science tools	following directions	time & space
social skills	cognitive skill development	findings & predictions
language skills	math concepts & properties	living & non-living things
visual-perceptual skills	position, quality & quantity	gathering information
measurement relationships	recording information	turn-taking
experimentation & exploration	learning properties of substances	
learning about your surroundings	number, letter, shape recognition and matching	

Music and Movement

gross-motor development	singing & vocal expression	beat
following directions	rhyme & rhythm	math skills
movement of body parts	exploration of emotions	pre-literacy skills
language skills	individual expression	imitation

Library and Writing

literacy skills	use of writing tools	retelling stories
sound-symbol relationships	exploring books	following directions
creative writing	self-expression	attention span
concept & vocabulary development		
looking at, listening to and participating in stories		

Patterns and Crafts   Arts and Crafts

This section includes several patterns or craft ideas to utilize in your art center. One must always consider that, for the young artist and creator, the "process" is much more important than the product. The young artist develops skills in several development domains. Use the ideas and activities in this section to facilitate the following skill development and concepts:

creativity	pre-literacy skills
exploration & experimentation	unique products
learning properties of substances	using drawing, writing and cutting tools
building self-esteem	finding new ways to use materials
building fine motor skills	manipulation of various materials
following directions	

Bulletin Board Ideas

Several suggestions are included for creating interesting and informative bulletin boards based upon the theme of bears and can be used in a variety of ways in the classroom. Here are a few tips:

Keep them neat and professional looking	Display children's unique creations
Display photos of kids at play & work	Change the boards periodically
Display magazine pictures of "real" things	Use neat and concise letters & writing

Keep them current to theme or activities of the program
Display at the appropriate height for children's viewing
Display higher for family members and volunteers

My Book About...

Make copies of the pages in this section to create an interesting book about bears for your young learners. Simply duplicate the pages, cut and assemble them by number. Attach the pages together with staples or brads. Laminate the front and back for durability. Children will enjoy listening to the story and coloring the pages. Older children may have the skills to color, cut, assemble and read the story for themselves. Having children make their own book helps develop pre-emergent reading skills in early learners.

Goodies to Make and Eat!

Cooking and Nutrition

The recipes found in this section are devoted to cooking and nutrition. Clay or dough recipes, which you may use in your art or sensory area, are also included. The recipes can be copied onto poster board, large sheets of paper, or recipe-sized cards so that children, families and volunteers can follow along. You may want to add clip art pictures in place of words to create rebus recipe charts. Use the recipes in this section to facilitate these developmental skills:

following directions	performing activities in sequential order
cooperating with others	manners and taking turns
counting & measurement concepts	use of simple kitchen utensils
sensory exploration of foods	observation of food in different forms
pre-literacy and literacy skills	practicing different food preparations

At Home With...

This section is meant to provide families with ideas and activities to do at home with their children. These activities will help to encourage communication between family member and child, as well as support the skill development of the child. The activities should be optional for those families who wish to and have the time to participate. Educators may wish to incorporate the ideas into their newsletters or simply attach the "At Home With....." sheet to their notes.

Parent/Family Involvement

Throughout each theme book, educators will note the various opportunities to encourage parent/family involvement. Communication between parents and the program is one key to increasing parent involvement. One mechanism to increase communication with parents is through notes or newsletters.

Tips for Notes and Newsletters

Make them visually appealing - add clip art	Write blocks of information
Be specific when referring to a day or time	Typewritten or printed
Write about each center, activity, or type of news	Make it neat - professional
Add a response sheet for parents	Proof for typing errors

Request assistance - parent volunteers for at least one activity
Keep them easy to read - vocabulary/avoid lengthy sections
Avoid using last names on general newsletters without permission
Add questions for parents to ask their children to encourage conversation

---

Note:

All of the patterns and illustrations in this book can be enlarged to better suit the needs of young learners.

EVERYTHING YOU NEED TO KNOW ABOUT..."BEARS"

Bears, real or teddy, have long been a favorite of children and teachers alike... Here are some facts that may be of interest to you and your students.

The "Bear" Facts

 Bear features: large animal, strong legs, big head, small eyes and ears, sharp jaw and claws, and covered in heavy fur.

 Bears are black, brown, white or beige.

 Adult bears weigh from 100 to over 800 pounds.

 There are 8 different types of bears:

brown (grizzly) bears	American black bears
polar bears	giant panda bears
Asiatic black bears	sloth bears
spectacled bears	sun bears

*Koala bears are not real bears, they are marsupials (with a pouch to carry their young).

 Bears live where the winters are cold. Bears live all around the world, except Africa, Australia, and Antarctica.

 Bears eat one or more of the following types of food:

insects	nuts	honey
roots	bark	plants and grass
berries and fruit	bulbs	bamboo

salmon or trout (fish) and other animals
They are also known to scavenge through garbage and on dead animals.

Bears eat for 6 to 8 months to store the fat that they need to hibernate through the winter months.

 Most bears hibernate in their dens all winter.

 A mother bear can have 1, 2 or 3 cubs. Bear cubs are born while the mother is hibernating. A bear cub weighs about 1 1/2 pounds at birth.

 Hikers Beware! Bears will only attack if they are surprised. If they smell or hear a human, they will travel in the opposite direction. When hiking in areas where bears frequently roam, some have suggested talking among yourselves, clapping, humming, making noise, or wearing a "bear" bell on your pack or belt loop.

 Bear enemies: wolves, tigers, sometimes male bears, and people.

 More bears are dying each year than are being born. Some types of bears are considered endangered species.

Look for more interesting information on the following websites:

www.pbs.org
www.bear.org
www.nature-net.com/bears/
www.smokeybear.com
www.bearden.org
www.animaladventures.com
www.brevardzoo.org
www.colszoo.org
www.aza.org
www.buschgardens.org

Thematic Literature List

Bergman, Marg. **Bears, Bears Everywhere!**. Barron's Educational Series, Inc., 1999

Blackstone, Stella. **Bear In A Square**. New York: Scholastic, Inc., 1998

Fair, Jeff. **Bears for Kids**. Minocqua, Wisconsin: NorthWord Press, 1991

Helmer, Diana Star. **Black Bears**. New York: Rosen-Powerkids Press, 1997

Helmer, Diana Star. **Brown Bears**. New York: Rosen-Powerkids Press, 1997

Helmer, Diana Star. **Famous Bears**. New York: Rosen-Powerkids Press, 1997

Helmer, Diana Star. **Panda Bears**. New York: Rosen-Powerkids Press, 1997

Helmer, Diana Star. **Polar Bears**. New York: Rosen-Powerkids Press, 1997

Kulling, Monica. **Bears: Life in The Wild**. New York: Golden Books Publishing Co., 1998

Martin, Bill. **Brown Bear, Brown Bear, What do You See?**. New York: Henry Holt and Co., 1983

Martin, Bill. **Polar Bear, Polar Bear, What Do You Hear?**. New York: Henry Holt and Co., 1991

Milton, Joyce. **Bears are Curious**. New York: Random House, 1998

Nicholas, Christopher. **Bears!** Learning Horizons, Inc. 2000

Parker, Janice. **Grizzly Bears**. Raintree Sterk Vaughn Publishers, 2000

Phillips, Joan. **Lucky Bear**. New York: Random House, 1986

Rosen, Michael. **We're Going On a Bear Hunt**. New York: McElderry Books, 1989

Ward, Paul. **Bears of the World**. Blandford Press, 1999

Corduroy series by Don Freeman
Goldilocks and The Three Bears various versions available
Jesse Bear series by Nancy W. Carlstrom
Moon Bear series by Frank Asch
Winnie-The-Pooh series by A.A. Milne

Concepts & Vocabulary

den	cub	hibernation	cave	honey
endangered	extinct	growl	big	little
grizzly bear	claws	white	brown	black
giant panda bear	sun bear	polar bear	fur	woods
sloth bear	spectacled bear	brown bear		
American black bear		Asiatic black bear		

Dictation and/or Creative Writing Page

Name _____ Date _____

TF4101 Beary Good Bears Book

"BEARY" GOOD CLIP ART!

Name _____

has been a "beary" good student!

Date _____ Teacher _____

Name _____

did a great job today!

Date _____ Teacher _____

"Beary" Good Award!

awarded to

_____

for

_____

Date _____ Teacher _____

Songs, Poems and Fingerplays

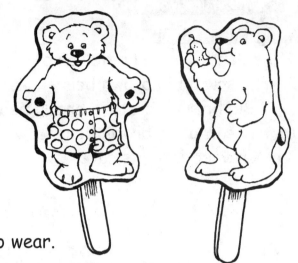

This Little Bear
(Tune: This Little Piggy)

This little bear had short hair,
This little bear had long hair,
This little bear went to the fair,
This little bear ate a pear,
And this little bear couldn't find anything to wear.

(Have the children touch each finger on their hand to represent the different bears. Copy the bear stick puppets below and tape the pictures to craft sticks or tongue depressors.)

A Big Brown Bear
(Tune: Mary Had a Little Lamb)
Johnny saw a big brown bear,
    Big brown bear, big brown bear.
Johnny saw a big brown bear,
Who crawled into his cave.

Substitute each child's name for Johnny's as you sing the song.

Black Bear
(Tune: One little, Two little,
Three little Indians)

One little, two little, three little black bears,
Four little, five little, six little black bears,
Seven little, eight little, nine little black bears,
Ten little black bears hiding in the den.

Copy this bear to make finger puppets for your class.

My Teddy Bear
My teddy bear,
My teddy bear,
I like to bring him everywhere,
I hug and kiss his furry hair,
I snuggle with him in my chair,
Yes, I just love my teddy bear!

(Show & Tell – Children can bring in their favorite teddy bear. Have the children tell about their teddy bears and write one sentence from their description on a large piece of paper. Then teach "My Teddy Bear" song to the them.)

ACTIVITIES AND GAMES

Fill the dramatic play area with the following items: different sized, colored and shaped stuffed bears; small, medium, and large bowls, cups, blankets, etc.; household items for pretend housekeeping play; or cover a large sturdy table or set of tables with blankets to create a cave; flashlight; bear storybooks

 Encourage the children to match the small bear with the small dress-up items or dishes and the large bear with the big items.

 Pretend to visit a bear cave. Bring a flashlight for more fun.

 Backpack bear – Place a bear and a notebook in a backpack. Each night, send the backpack home with a different child. Ask the family to write down what they did with the bear. Families may choose to add things to the backpack to depict the bear's travels and activities. Read the special passages to the group during circle time.

Outdoor Play & Field Trips

 Visit real bears at a local zoo or circus. Ask family members to come along.

 The group can take a walk around the neighborhood. Point out things that are the colors of bears or things that bears may eat. Discuss whether or not a bear could live in your neighborhood. Play "I see something brown (white or black)." The children will enjoy guessing what you see. Let them have a turn leading the group in the "I see something..." game.

 It may be fun just to take a walk in the woods and discover all the critters that live with bears. Explain to the children that real bears can be very scary and dangerous. A local wildlife expert or ranger may be able to visit with the class and explain how to be safe in the woods.

Sort small counting bears by color or use an egg carton or muffin tin to sort by number.

Note: Try substituting counting bears with blueberries, gummy bears, or small bear-shaped crackers.

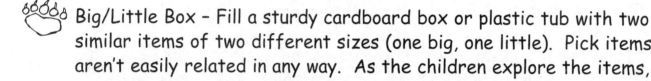 Big/Little Box – Fill a sturdy cardboard box or plastic tub with two similar items of two different sizes (one big, one little). Pick items that aren't easily related in any way. As the children explore the items, they may naturally recognize the size difference. Suggestions: one big stuffed bear, one little stuffed bear; one big button, one little button; one big square block, one little square block; one big feather, one little feather; one big crayon, one little crayon.

Collect the following items (mixture of real and magazine pictures) and share them with the children at circle time or in the science/sensory area. Talk about the many things that bears may eat. Allow the children to explore them. Write their discoveries on an experience chart. The children may match these graphic pictures with the real items or photos.

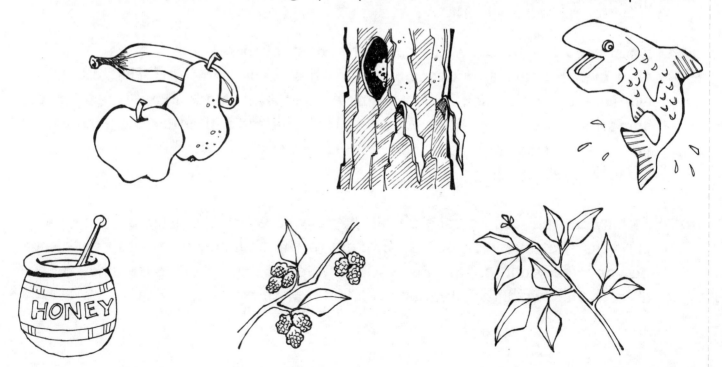

Matching Number Bears!

Copy the following "Bear Squares" onto heavy paper. Color, laminate and cut the pictures below into individual squares.

To play a matching game, have two or more players turn the pictures face down. Ask them to take turns turning the squares over two at a time to find matches. If the pictures match, then the player keeps the squares and goes again. If the pictures do not match, the next player takes a turn. Match number to number or number to word or number to dots.

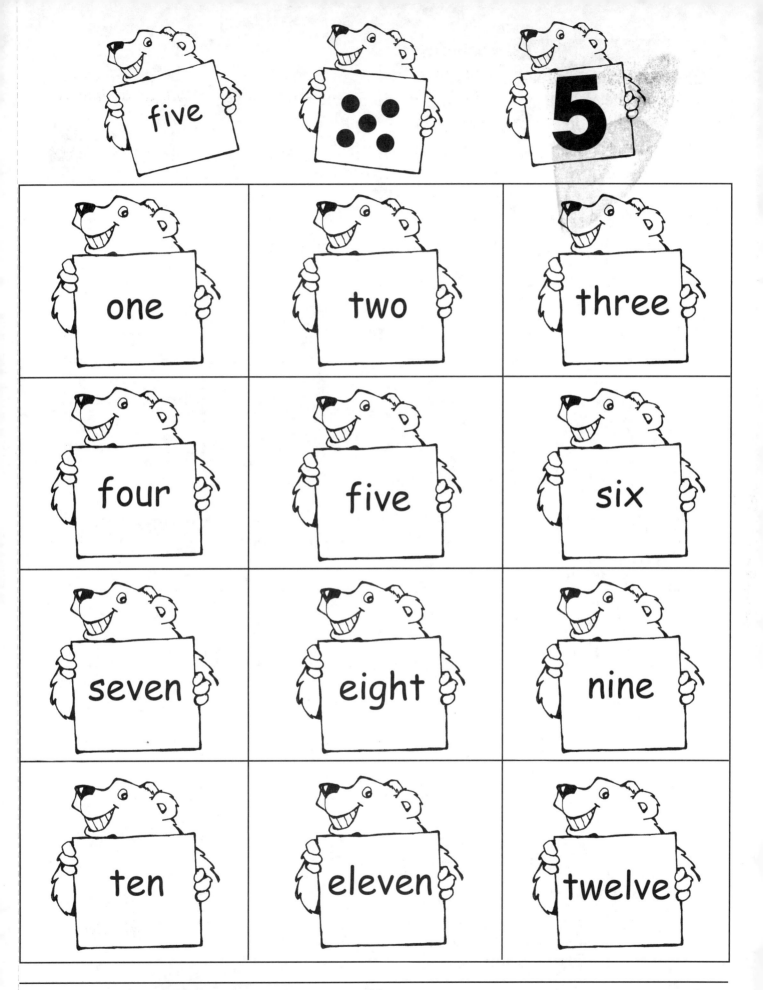

five

one

two

three

four

five

six

seven

eight

nine

ten

eleven

twelve

Where's The Bear?
For the youngest learners, provide a box and stuffed teddy bear. Give the child directions about where to place the bear, "under the box," "next to the box," etc. You may want to offer the child a box and b and observe the child playing with the items. Ask "Where is the bear?"

Where's The Bear?   Look at the pictures and words on this page. Draw a line to connect the matching pictures. Color the pictures.

Bear on top of the square.

Bear next to the square.

Bear inside the square.

Teddy Bear Hunt

Before the children arrive, cut out 20 or more bear footprints from brown construction paper. (Laminate them to protect them from being torn.) Tape the bear footprints in a path throughout the classroom or down the hall. The path can take the children under tables, through large refrigerator boxes or tunnels, up steps, etc. At various places in the path, ask the children if the teddy bear is "under the table," "behind the door," "at the bottom of the steps," etc. (Your comments will differ according to the path.) At the end of the path, hide a teddy bear under a clothes basket or box, or in a closet. The children will have great fun hunting for the bear and, at the same time, learn about many positional concepts (under, behind, etc.). The children can pretend to carry flashlights while following the bear's path.

Bear Footprints

The "Bear" Moves

Assemble the cube pattern on the next page. (Enlarge pattern to make a larger cube.) Cut the pattern from index paper or heavy tag board. Fold along dotted lines and glue or tape to form a cube. Children can glue the bear pictures below to the cube – one picture per side. (Laminate the box before assembling, if you wish.) Have the children take turns rolling the cube and imitating the "bear" movement.

Cube Pattern

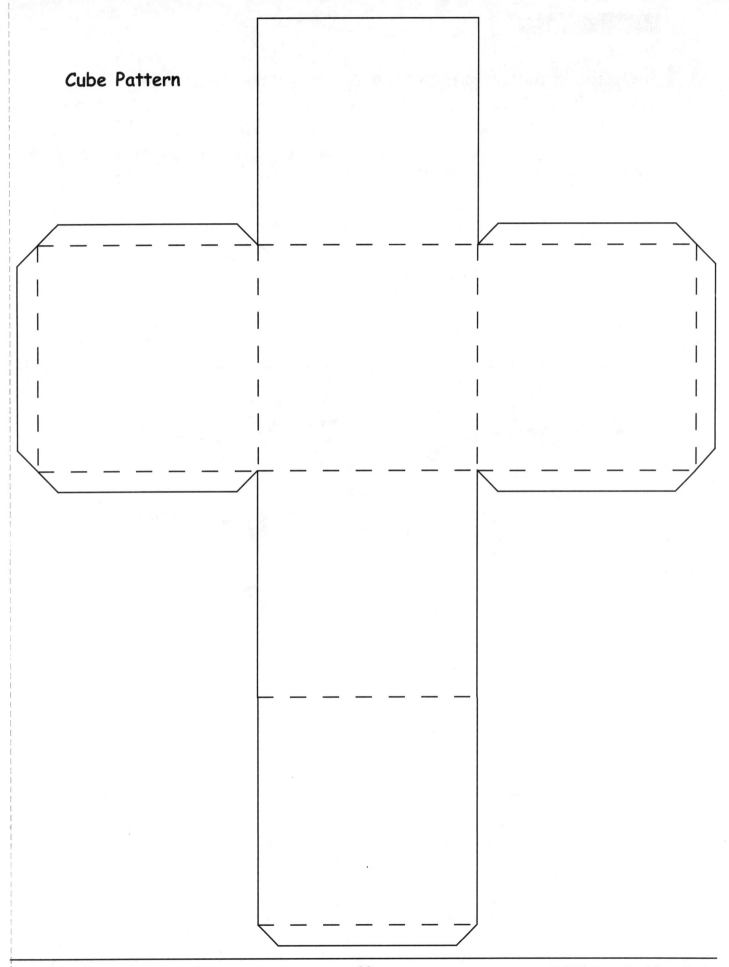

"B" Picture Chart

Paste the letters "B" and "b" to a large sheet of paper mounted on a class bulletin board. Ask the children what sound the letter "b" makes. Repeat the sound and ask the children to repeat the sound. Tell the children that you want to have them fill the paper with pictures of things that start with the letter "B" or sound "b," like in "bear." Offer scissors, gluesticks and several magazines. The children cut out pictures in magazines. Have them or you write the words to label the pictures underneath each picture. Review the picture chart – clearly pronouncing the "b" sound at the beginning of each word.

(Children can color the letters)

B-E-A-R Strips

Copy this page and have students practice writing the letters B-E-A-R and b-e-a-r. You may want to laminate the completed pages and have the children trace over the letters using wipe-off markers or crayons.

B   B   B

E   E   E

A   A   A

R   R   R

BEAR

Library and Writing

b-e-a-r Strip

b b b

e e e

a a a

r r r

bear

What Do Bears Eat? Look at the pictures below. Circle the things that bears eat.

HONEY

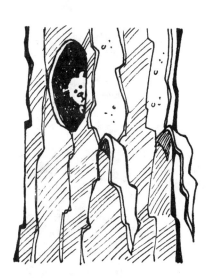

Color the pictures.

Beary Good Words

Copy, color and laminate several bear patterns. Write a word on each bear with a wipe-off marker. The words can be sight words, spelling words, vocabulary about bears, or new words that the children are learning. Spread the bears out on the floor. Each child takes a turn tossing a beanbag onto a bear. When the beanbag lands on the bear, the child says the word and/or uses the word in a sentence. (Older children can write the words and/or sentence on the class board.) Use the bears again and again as the vocabulary changes.

The Story of Goldilocks and The Three Bears!

Read or tell the story of Goldilocks and The Three Bears to your children. This is one version of the classic story.

Once upon a time, deep in a forest lived a bear family. There was a great big Papa bear, a medium sized Mama bear, and a wee little Baby bear. Now, one morning, Mama bear made porridge for breakfast. The porridge was very hot. Papa bear said, "Why don't we take a walk in the forest while the porridge is cooling down." So, they went for a walk.

Along came a little girl, whose name was Goldilocks, walking in the forest. She saw the bears' house and was very curious about who might live there. So, she knocked on the door, but nobody answered. She smelled the porridge and decided to go inside.

She sat down and tasted Papa bear's porridge, but it was too hot. So, she tried Mama bear's porridge, but it was too cold. Then, she tasted Baby bear's porridge and thought it was just right and ate it all up.

She explored some more and saw three chairs. She sat in Papa bear's chair and thought,"This chair is too hard." She tried Mama bear's chair and thought, "This chair is too soft." So, she tried Baby bear's chair and said, "Oooh, this chair is just right." But, oops! She was too heavy and broke the chair. She stretched her arms and yawned and thought," I'm feeling tired," so she looked for the bedroom. She laid on Papa bear's bed and thought," This bed is too hard." She laid on Mama bear's bed and thought," This bed is too soft." So, she tried Baby bear's bed and thought, "This bed is just right," and she fell asleep.

Meanwhile, the three bears decided to go back home to eat their warm porridge. When they came into the kitchen, Papa bear said, in a deep voice, "Somebody's been eating my porridge." Mama bear said, in a sweet voice, "Somebody's been eating my porridge." Then, Baby bear said, as he began to cry, "Somebody's been eating my porridge and ate it all up!"

So, they went into the living room and Papa bear said, in a deep voice, "Somebody's been sitting in my chair." Mama bear said, in a sweet voice, "Somebody's been sitting in my chair." Then, Baby bear said, as he began to cry, "Somebody's been sitting in my chair and broke it!"

So, they went into the bedroom and Papa bear said, in a deep voice, "Somebody's been sleeping in my bed." Mama bear said, in a sweet voice, "Somebody's been sleeping in my bed." Then, Baby bear said, as he began to cry, "Somebody's been sleeping in my bed and she's still there!"

Goldilocks heard the bears talking and woke up. She was so scared to see the three bears that she jumped out of Baby bear's bed, ran out of the bear house and all the way home. The bears never saw Goldilocks again.

Goldilocks and The Three Bears!

Copy, color and laminate these patterns. Add tape or felt to the back of each pattern and use them to retell the popular story of Goldilocks and The Three Bears.

Patterns and Crafts

 Bear Painting

Collect different size bear shaped cookie cutters. Using only one color, pour poster paint into a tray large enough to easily fit the biggest bear cookie cutter. Instruct the children to dip the cutters into the tray and then use it to print a bear pattern onto a sheet of construction paper. Talk to the children about the different sizes. It will be easier for them to distinguish the sizes because you are using only one color of paint.

 Nature Painting

Collect several items found in the woods: twigs, flowers, pinecones, leaves, acorns, nuts, cattails, etc. Pour poster paint into a tray large enough to easily fit the biggest nature item. Instruct the children to dip the nature objects into the tray and then use them to print nature patterns onto a sheet of construction paper. Watch the children create unique masterpieces.

Bear Pattern

Use this bear pattern to create an interesting shaped surface for children to paint, color, write or glue on. The pattern can be enlarged or reduced in size. You may want to copy the pattern onto different types of paper – construction, fingerpaint, sandpaper or wallpaper. Each day, have the children use a different medium – paint, crayons, tissue paper, dry pasta, styrofoam peanuts, pine needles, etc.

Bear Pattern Strips

Copy several "Bear Pattern Strips" onto heavy paper. Make 3 "Bear Pattern Strips" and 12 matching squares with two sheets of the "Bear Pattern Strips." First, color the bears in each vertical row on both sheets as follows - 1st row - red, red, green, red; 2nd row - blue, green, blue, green; and 3rd row - blue, blue, yellow, yellow. Then, laminate and cut the first sheet into three strips along the dotted lines. Laminate and cut the second sheet along both the solid and dotted lines to make the matching squares. Repeat these steps several times using different color patterns. Have the children match bear squares to the bear strips as shown.

Bear Pattern Strips

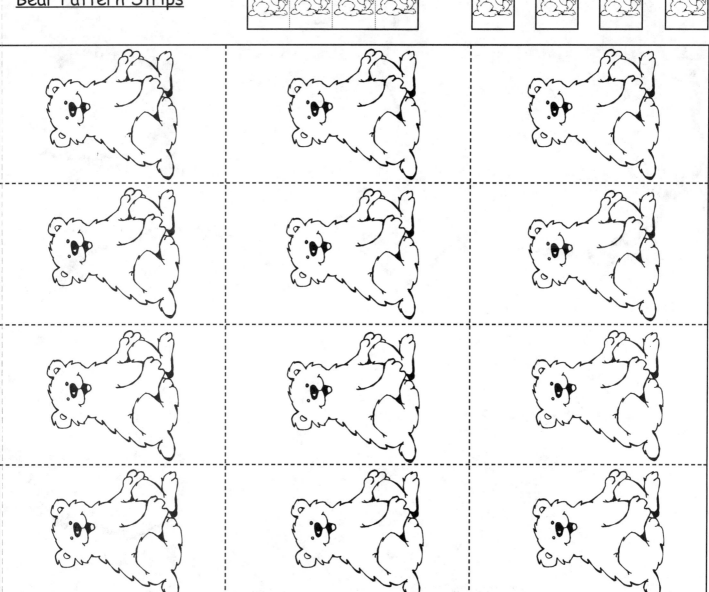

Bear Ears

Children will be delighted to wear these cute "Bear Ears." Trace the ears onto brown, white or black construction paper and cut them out. Cut the inside of each ear from pink construction paper and have the children glue them in place. Make a paper headband for each child and staple the ears to the sides as shown. Encourage the children to cut the ears and paper headband themselves.

Bear Mask

Trace the bear features on brown, white, or black construction paper. Cut eye holes in a paper plate and tape a tongue depressor to the back of the plate. Children can glue the patterns, as shown. Older children may be able to complete all the steps by themselves.

Encourage the children to act like real bears. Watch the fun as they growl and roar!

BULLETIN BOARD IDEAS

Different Types Of Bears

Cut out and laminate several magazine pictures of bears. Discuss the different species/types of bears. The children can help you sort and mount the pictures on a bulletin board. Ask families to help collect catalogs and/or magazines that may have bear pictures.

Display for Volunteers

Create a bulletin board displaying the various ways family and community members can volunteer to help in your classroom or school. Use it as a display or ask family or community members to sign up to volunteer.

"Beary" Great Volunteers

Copy the bear pattern on page 28 and write one way in which a person can volunteer on each bear. Display the bears on your class board attached to a sign-up sheet.

Here are some ideas of how families and friends can volunteer in the classroom:

Plan field trips	Supervise field trips
Plan special events	Plan fundraisers
Read to children during storytime	Classroom volunteers
Lunch or Snack Time Helpers	Help develop program policies
Help develop a program yearbook	Help develop curriculum
Office helpers	Share a talent, hobby, etc.
Become a member of the "P.T.A."	Participate in various programs
Help monitor children's tasks	Assist with food preparation

"Bearrific" Work

Display artwork, creative writing pages, or other handouts on a bulletin board with a "Bearrific Work" heading. Use some of the patterns in this book to decorate the board!

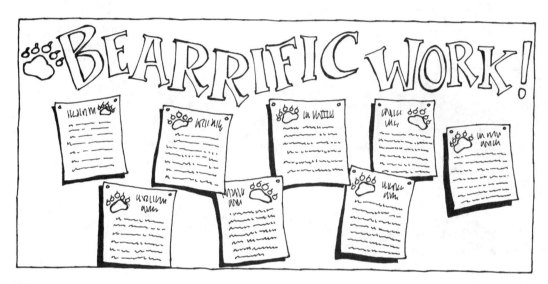

Bears On The Move

Take pictures of the children during movement or gym time. Display the photos on a bulletin board with a "Bears On The Move" heading. Decorate with some of the patterns from this book.

My Book About...BEARS!!!!

I know about bears.

Bears are black, brown or white.

2

Bears live in dens.

3

Bears eat ants and berries.

4

Baby bears are called cubs.

5

Bears sleep all winter.

6

Bears like to growl.

7

This is my picture of a bear.

8

My name is _____.

I completed my book about bears on:

_____.

Goodies to Make and Eat!

Bowl of Berries

(Makes 8 half cup servings)

Allergy Alert!

Mix in a large bowl.
1 cup strawberries, cut into pieces
1 cup blueberries
1 cup raspberries
1 cup blackberries

Bear Trail Mix

Use this treat when you adventure out
on a bear hunt! Here's what you'll need:
2 cups sunflower seeds
2 cups raisins
2 cups chocolate or carob chips
2 cups peanut butter chips
2 cups circle shaped cereal
2 cups mini marshmallows

Allergy Alert!

Let each child place a large spoonful of each ingredient in a sealable plastic
sandwich bag. Help them seal the bag closed. Tell the children to shake their
bags to mix the ingredients.

Blue"beary" Muffins

In a bowl, beat 1 egg with 2/3 cup milk, 1/2 cup vegetable oil and 1 teaspoon
vanilla. Stir in 2 cups flour, 1/4 cup white sugar, 1/4 cup brown sugar, 1 teaspoon cinnamon, 1 teaspoon baking powder and 1/2 teaspoon salt. Mixture will
be lumpy. Stir in blueberries. Scoop into a muffin tin (greased or lined with
cups). Mix a small amount of white sugar and cinnamon together and sprinkle
the mixture on top of each muffin. Bake at 400° for approximately 20 minutes.

"Bear" Treats

Complete "What Do Bears Eat?" on page 27. Encourage the children to taste foods that bears eat by having a tasting party. Here is all you will need:

a variety of berries
tuna fish
honey on toast or crackers

Let each child place a scoop or piece of each food onto a plate. Encourage the children to talk about how the different foods taste.

Breakfast for Goldilocks and The Three Bears

After reading or telling the story of Goldilocks and The Three Bears to your children, serve a "porridge" type breakfast. Follow the packaging directions to make Cream of Rice®, Cream of Wheat®, or warm oatmeal cereal. The children can top the special breakfast with fresh berries or fruit jam.

Cinnamon Clay Bears

This is a fun recipe to make with children, BUT please don't eat it!

Mix in a large bowl:
1 cup ground cinnamon
1 tablespoon nutmeg
1/2 cup applesauce
2 tablespoons white glue

Don't Eat!

Mixture will become clay-like. Roll out the dough with a rolling pin and have children use bear cookie cutters to cut out the bear. Use a drinking straw to cut a hole in the bear. Allow to dry at least 24 hours, turning occasionally. When bears are dry, insert a ribbon through the hole. The Cinnamon Clay Bear smells great and can be hung from doorknobs, on kitchen cabinets, or on a hook in a bathroom.

Blue"beary" Muffins

In a  , beat 1 with 2/3

1/2 vegetable oil and 1 teaspoon vanilla.

Stir in 2  flour, 1/4 white sugar,

1/4  brown sugar, 1 teaspoon cinnamon,

1 teaspoon baking powder and 1/2 teaspoon salt .

Mixture will be lumpy  . Stir in blueberries .

Scoop into  (greased or lined with cups).

Mix a small amount of white sugar and cinnamon together and

sprinkle the mixture on top of each

Bake at 400°  for approximately 20 minutes.

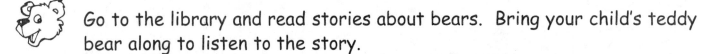

Ask your child what the first letter is in the word "bear". If your child doesn't know, tell him/her or say the "b" sound and ask him to guess. Repeat "b" sound for "bear." Ask him/her whether he/she wants to help you hunt for other things in your house that start with the "b" sound just like "bear." Look throughout your house and see how many things you can find.

Go to the library and read stories about bears. Bring your child's teddy bear along to listen to the story.

Have your child tell you a story about his/her teddy bear. Write down you child's words on paper and have your child draw a picture of him/her with his/her teddy bear.

Cut out the letters below. Name each letter or ask your child to name them (younger children may not know the names of the letters). Play a game with the letters. Look around your house for the same letters – "Bb" "Ee" "Aa" "Rr" "Ss." How many letter "Bb"s can you find? And where did you find them? Do the activity with each letter or until the child doesn't want to play anymore. The child can carry the letter around with him/her to match more easily. Make it fun, not work. Hint: places to find letters – on food/toy packaging, in books or magazines, on appliances or household equipment, on videos, tapes or CDs, or in brand names of products.

B	E	A	R	S